Engineering Materials

For further volumes:
http://www.springer.com/series/4288

Rafat Siddique · Mohammad Iqbal Khan

Supplementary Cementing Materials

Springer

Dr. Rafat Siddique
Department of Civil Engineering
Thapar University
Patiala 147004
India
e-mail: rsiddique@thapar.edu
siddique_66@yahoo.com

Dr. Mohammad Iqbal Khan
Structural Engineering
Center of Excellence for Concrete Research and Testing College of Engineering
King Saud University
Riyadh 11421
Saudi Arabia
e-mail: miqbal@ksu.edu.sa

ISSN 1612-1317
e-ISSN 1868-1212

ISBN 978-3-642-17865-8
e-ISBN 978-3-642-17866-5

DOI 10.1007/978-3-642-17866-5

Springer Heidelberg Dordrecht London New York

Cover design: deblik, Berlin

Printed on acid-free paper

Springer is part of Springer Science+Business Media (www.springer.com)

Dedicated to Our Parents and Families

Preface

Concrete is the most widely used construction material because of its versatility, economy, availability of raw materials, strength, and durability. Concrete can be designed to withstand the harshest environmental conditions while taking on the most inspirational and imaginable shapes and forms. Scientist/Engineers and academicians are continuously working for better concrete from strength and durability standpoint with the help of innovative chemical admixtures and supplementary cementing materials (SCMs). In addition, the use of SCMs conserves energy and has environmental benefits because of reduction in carbon dioxide emission as a result of reduction in manufacture of Portland cement. Strict air-pollution controls and regulations have produced an abundance of industrial byproducts that can be used as supplementary cementitious materials. Typical examples are fly ash, silica fume, ground granulated blastfurnace slag, metakaolin, rice husk ash and natural pozzolans which can be used incorporated in concrete addition or as partial cement replacement.

Supplementary cementing materials are often used in concrete mixes to reduce cement contents, improve workability, increase strength and enhance durability through hydraulic or pozzolanic activity. Utilization of these byproducts in cement/concrete not only prevents them from being land-filled but also enhances the properties of concrete in the fresh and hardened states.

This book is an attempt to consolidate the published research related to the use of SCMs in cement and concrete. This book is intended to cater to the needs of graduate students, researchers, concrete technologists and practicing engineers.

The book comprises of five chapters. Each chapter is devoted to a particular supplementing cementing material. It is based on the literature/research findings published in journals/conference proceeding, etc. Topics covered in the book are; coal fly ash, silica fume (SF), granulated blast furnace slag (GGBS), metakaolin (MK), and rice husk ash (RHA). Each chapter contains introduction, properties of the waste material/by-product, its potential usage, and its effect on the properties of fresh and hardened concrete and other cement based materials.

We would like to place on record our immense sense of gratitude to academicians, scientists, concrete technologists, and our colleagues and friends globally who have contributed significantly in the broader area of concrete technology, and our sincere appreciation and acknowledgement to the published work of the researchers on the subject, which has been referred in this book.

We are also extremely grateful to Springer for publishing the book in an excellent form in the shortest possible time.

We owe our sincere thanks and irrepayable gratitude to our families and friends whose consistent encouragement and love have been a tremendous impetus for the completion of this book.

Dr. Rafat Siddique
Dr. Mohammad Iqbal Khan

Contents

About the Authors

Dr. Rafat Siddique is Senior Professor of Civil Engineering & Dean of Faculty Affairs at Thapar University, Patiala, India. He earned Ph.D. degree from Birla Institute of Technology & Science, Pilani, India, and did 22-months post-doctoral work at the University of Wisconsin-Milwaukee, USA. He has been Visiting Professor to University of Cergy Pontoise, France; INSA Rennes, France; University of Wolverhampton, U.K.; Consolis Technology, Finland; and BAM Berlin, Germany. He is the author of a booktitled "Waste Materials & Byproducts in Concrete", by SPRINGER. He has published more than 125 research papers in journals and conference proceedings. He is reviewer of 20 leading International Journals.

Dr. Mohammad Iqbal Khan is Associate Professor in Structural Engineering, Department of Civil Engineering, King Saud University, Saudi Arabia. He is founding member and Managing Director of Center of Excellence for Concrete Research and Testing at King Saud University. He is formerly Lecturer of Structural Engineering, School of Civil Engineering, University of Nottingham, UK. He received his Ph.D. degree from the University of Sheffield, UK in 1999. He is actively involved in research since 1990 and has published more than 70 research papers in refereed international journals and conferences proceedings. He is reviewer of five leading International Journals. He has one granted United States Patent and two pending United States Patents.

Chapter 1
Fly Ash

1.1 Introduction

The *fly ash*, also known as pulverised fuel ash, is produced from burning pulverized coal in electric power generating plants. During combustion, mineral impurities in the coal (clay, feldspar, quartz, and shale) fuse in suspension and float out of the combustion chamber along with exhaust gases. As the fused material rises, it cools and solidifies into spherical glassy particles called fly ash. It is a fine-grained, powdery particulate material that is collected from the exhaust gases by electrostatic precipitators or bag filters. Depending upon the collection system, varying from mechanical to electrical precipitators or bag houses and fabric filters, approximately 85–99% of the ash from the flue gases in retrieved in the form of fly ash. Fly ash accounts for 75–85% of the total coal ash, and the remainder is collected as bottom ash or boiler slag.

Fly ash produced from thermal power plants is a variable material because of several factors. These factors are (1) type and mineralogical composition of the coal; (2) degree of coal pulverization; (3) type of furnace and oxidation conditions; and (4) the manner in which fly ash is collected, handled and stored before use. Since no two utilities or plants may have all these factors in common, fly ash from various power plants is likely to be different. Fly ash properties may also vary within the same plant because of load conditions over a 24-h-period.

Fly ash particles size is primarily depends upon the type of dust collection equipment. It is generally finer than Portland cement. Diameter of fly ash particles ranges from less than 1–150 μm. The chemical composition of fly ash is determined by the types and relative amounts of incombustible material in the coal used. Depending upon the source and makeup of the coal being burned, the components of fly ash vary considerably, but all fly ash includes substantial amounts of silicon dioxide (SiO_2) (both amorphous and crystalline), aluminium oxide (Al_2O_3) and calcium oxide (CaO), both being endemic ingredients in many coal bearing rock strata.

R. Siddique and M. Iqbal Khan, *Supplementary Cementing Materials*, Engineering Materials, DOI: 10.1007/978-3-642-17866-5_1,

1.1.1 Handling of Fly Ash

Generally, collected fly ash is typically conveyed pneumatically from the electro static precipitators ESP or filter fabric hoppers to storage silos where it is kept dry pending utilization or further processing, or to a system where the dry ash is mixed with water and conveyed (sluiced) to an on-site storage pond. The collected dry ash is normally stored and handled using equipment in a similar way the Portland cement is handled:

- Fly ash is stored in silos, domes and other bulk storage facilities
- It can be transferred using air slides, bucket conveyors and screw conveyors, or it can be pneumatically conveyed through pipelines under positive or negative pressure conditions
- It is transported to markets in bulk tanker trucks, rail cars and barges/ships
- It can be packaged in super sacks or smaller bags for specialty applications

Dry collected fly ash can also be moistened with water and wetting agents, whenever needed/applicable, using specialized equipment and hauled in covered dump trucks for special applications such as structural fills. Water conditioned fly ash can be stockpiled at jobsites. Exposed stockpiled material should be kept moist or covered with tarpaulins, plastic, or equivalent materials to prevent dust emission.

1.1.2 Environmental Benefits of Using Fly Ash

Utilization of fly ash in cement and concrete has significant environmental benefits such as

- Increasing the life of concrete roads and structures by improving concrete durability,
- Reduction in energy use and greenhouse gas and other adverse air emissions when fly ash is used to replace or displace manufactured cement.
- Reduction in amount of coal combustion products that must be disposed in landfills, and
- Conservation of other natural resources and materials.

1.2 Properties of Fly Ash

1.2.1 Size, Shape and Colour

Fly ash particle size is finer than ordinary Portland cement. Fly ash consists of silt-sized particles which are generally spherical in nature and their size typically

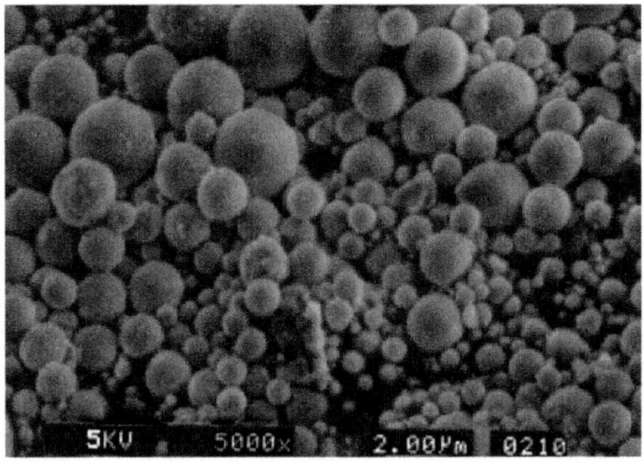

Fig. 1.1 Fly ash particles at ×5,000 magnification

ranges between 10 and 100 μm (Fig. 1.1). These small glass spheres improve the fluidity and workability of fresh concrete. Fineness is one of the important property contributing to the pozzolanic reactivity of fly ash.

Fly ash colour depends upon its chemical and mineral constituents. It can be tan to dark gray. Tan and light colours are generally associated with higher lime content, and brownish colour with the iron content. A dark gray to black color is attributed to elevated unburned carbon (LOI) content. Fly ash color is usually very consistent for each power plant and coal source.

1.2.2 Fineness

Fineness of fly ash is most closely related to the operating condition of the coal crushers and the grindability of the coal itself. Fineness of fly ash is related to its pozzolanic activity. For fly ash use in concrete applications, fineness is defined as the percent by weight of the material retained on the 5 μm (#325) sieve. ASTM C618 [5] limits the maximum amount of fly ash retained on the 45 μm (#325) mesh sieve on wet sieving as 34%. Generally, a large fraction of ash particle is smaller than 3 μm in size. In bituminous ashes, the particle sizes range from less than 1 to over 100 μm. Joshi [53] reported that average size lies between 7 and 12 μm. A coarser gradation can result in a less reactive ash and could contain higher carbon content.

1.2.3 Specific Gravity

The specific gravity of fly ash is related to shape, color and chemical composition of fly ash particle. In general, specific gravity of fly ash may vary from 1.3 to 4.8 [53].

Canadian fly ashes have specific gravity ranging between 1.94 and 2.94, whereas American ashes have specific gravity ranging between 2.14 and 2.69.

1.2.4 Pozzolanic Activity

The property of fly ashes, possessing little or no cementing value to react with calcium oxide in the presence of water, and produce highly cementitious water insoluble products, is called pozzolanic reactivity. The meta-stable silicates present in self-cementitious fly ash react with calcium ions in the presence of moisture to form water insoluble calcium-alumino-silicate hydrates.

The pozzolanic activity of a fly ash depends upon its (1) fineness; (2) calcium content; (3) structure; (4) specific surface; (5) particle size distribution; and (6) and LOI content [54]. Several investigators have reported that when fly ash is pulverized to increase fineness, its pozzolanic activity increases significantly. However, the effect of increase in specific surface area beyond 6,000 cm^2/g is reported to be insignificant.

1.2.5 Particle Morphology

Fly ash particles consist of clear glassy spheres and a spongy aggregate. Several morphological investigations have been carried out on particle shape and surface characteristics of various types of fly ashes using scanning electron microscope (SEM) and energy dispersive x-ray analysis (EDXA) [30, 53, 56, 80]. Scanning electron micrographs of different fly ashes show the typical spherical shape of fly ash particles, some of which are hollow. The hollow spherical particles are known as cenospheres or floaters as they are very light and tend to float on water surface. Cenospheres are unique free flowing powders composed of hard shelled, hollow, minute spheres. Cenospheres are made up of silica, iron and alumina. Cenospheres have a size range from 1 to 500 μm. Colors range from white to dark gray (Fig. 1.2).

Sometimes fly ashes may also contain many small spherical particles within a large glassy sphere, called pherospheres. The exterior surfaces of the solid and hollow spherical particles of low-calcium oxide fly ashes are generally smooth and

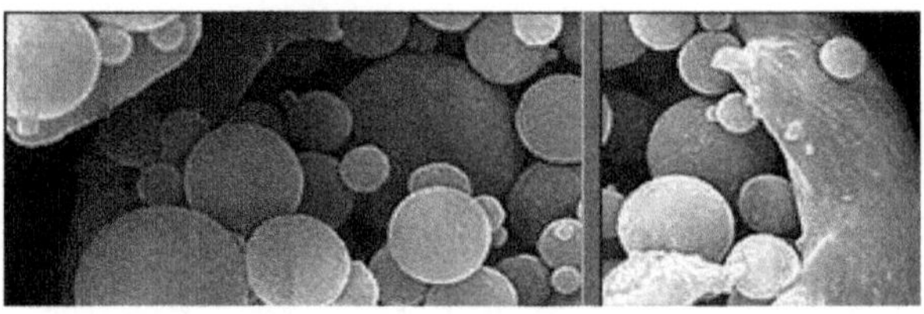

Fig. 1.2 Cenospheres from fly ash

better defined than those of high-calcium oxide fly ashes which may have surface coatings of material rich in calcium.

1.2.6 Moisture

Any amount of moisture in Class C fly ash will cause hardening from hydration of its cementitious compounds. Even surface spraying may cause caking. To prevent caking and packing of the fly ash during shipping and storage and to control uniformity of fly ash shipments, a 3.0% limit on moisture content is specified in ASTM C618. Therefore, it is important that such ashes have to be kept dry before being mixed with cement.

Physical properties of fly ashes reported by Electric Power Research Institute [35] are given in Table 1.1.

1.2.7 Chemical Composition

Chemical composition of fly ashes include silica (SiO_2), alumina (Al_2O_3), and oxides of calcium (CaO), iron (Fe_2O_3), magnesium (MgO), titanium (TiO_2), sulfur (SO_3), sodium (Na_2O), and potassium (K_2O), and unburned carbon (LOI). Amongst these SiO_2 and Al_2O_3 together make up about 45–80% of the total ash. The sub-bituminous and lignite coal ashes have relatively higher proportion of CaO and MgO and lesser proportions of SiO_2, Al_2O_3 and Fe_2O_3 as compared to the bituminous coal ashes. Table 1.2 presents the chemical analysis of different fly ashes [2].

1.2.8 Mineralogical Characteristics

X-ray diffraction study of the crystalline and glassy phases of a fly ash is known as mineralogical analysis. Mineralogical characterization determines the crystalline phases that contain the major constituents of fly ash. Generally, fly ashes have 15–45% crystalline matter. The high-calcium ashes (Class C) contain larger

Table 1.1 Physical properties of fly ashes [35]

Parameter	Range
Retained on #325 sieve (%)	3.55–36.90
Blaine fineness (cm^2/g)	1579–5550
Specific gravity	2.14–2.69
Moisture content (%)	0.0–0.38

Table 1.2 Composition of Class F and Class C fly ashes [2]

Parameter	Class F fly ash	Class C fly ash
Silicon dioxide (%)	45–64.4	23.1–50.5
Calcium oxide (%)	0.7–7.5	11.6–29.0
Aluminum oxide (%)	19.6–30.1	13.3–21.3
Iron oxide (%)	3.8–23.9	3.7–22.5
Sodium oxide (%)	0.3–2.8	0.5–7.3
Magnesium oxide (%)	0.7–1.7	1.5–7.5
Potassium oxide (%)	0.7–2.9	0.4–1.9
Loss on ignition (%)	0.4–7.2	0.3–1.9

amounts of crystalline matter ranging between 25 and 45%. Table 1.3 presents crystalline phases in fly ashes identified by XRD analysis [76].

Although high-calcium Class C ashes may have less glassy or amorphous material, they do contain certain crystalline phases such as anhydride ($CaSO_4$), tricalcium aluminate ($3CaOAl_2O_3$), calcium sulpho-aluminate ($CaSAl_2O_3$) and very small amount of free lime (CaO) that participate in producing cementitious compounds. Also, glassy phase in Class C ashes is usually more reactive. The glassy particles in Class C fly ashes contain large amount of calcium which possibly makes the surface of such particles highly strained, and probably, it is because of highly reactive nature of Class-C fly ashes.

Anhydrite ($CaSO_4$) is formed from the reaction of CaO, SO_2 and O_2 in the furnace or flue. Quantity of anhydrite increases with the increase in SO_3 and CaO contents. It plays a significant role in fly ash hydration behavior because it

Table 1.3 Crystalline phases in fly ashes from North America (McCarthy et al. 1988)

Class of fly ash and code	Name	Nominal composition
Low-calcium/Class F		
Hm	Hematite	Fe_2O_3
Mu	Mullite	$Al_6Si_2O_{13}$
Qz	Quartz	SiO_2
Sp	Ferrite spinel	$(Mg.Fe)(Fe,Al)_2O_4$
High-calcium/Class C		
Ah	Anhydrite	$CaSO_4$
AS	Alkali sulfate	$(Na,K)_2SO_4$
C_2S	Dicalcium silicate	Ca_2SiO_4
C_3A	Tricalcium aluminate	$Ca_3Al_2O_6$
Hm	Hematite	Fe_2O_3
Lm	Lime	CaO
Ml	Melilite	Ca_2 (Mg.Al) $(Al,Si)_2O_7$
Mu	Mullite	$Al_6Si_2O_{13}$
Mw	Merwinite	$Ca_3Mg(SiO_4)2$
Pc	Periclase	MgO
Qz	Quartz	SiO_2
So	Sodalite structure	$Ca_2(Ca,Na)_6(Al,Si)_{12}O_{24}(SO_4)_{1\text{-}2}$

Table 1.4 Concentration range of minerals in fly ashes [3]

Minerals	Concentration (%)
Mullite	6.5–9.0
Hematite	1.1–2.7
Magnetite	0.8–6.5
Quartz	2.2–8.5
Free CaO	Up to 3.5

participates along with tricalcium aluminate and other soluble aluminates to produce ettringite and calcium sulphoaluminate hydrate.

Tricalcium aluminate ($3CaOAl_2O_3$) is one of the most important crystalline phases to identify and quantify the fly ash because it contributes to ettringite formation, and also in self-hardening reactions as well as disruptive sulfate reactions in hardened concrete.

Periclase is the crystalline form of magnesium oxide (MgO). Presence of this form of MgO in fly ash affects the soundness of the resulting concrete through its expansive hydration to brucite, $Mg(OH_2)$. Crystalline iron oxide, ferrite spinel and/or hematite are generally found in all fly ashes. In most of the fly ashes, about 0.33–0.50% of iron is present as crystalline oxide. The reactivity of fly ash is, however, dependent on the glassy phases of Fe_2O_3.

The concentrations of important minerals found in fly ashes bituminous coal, as reported by Alonso and Wesche [3] are given in Table 1.4.

1.3 Classification of Fly Ash

ASTM C618 [5] categorizes natural pozzolans and fly ashes into the following three categories.

Class F Class F fly ashes are low in CaO. They are normally produced from burning anthracite or bituminous coal falls in this category. This class of fly ash exhibits pozzolanic property but rarely, if any, self hardening property. They are predominantly (>70%) noncrystalline silica which is the determining factor for pozzolanic activity. Their crystalline minerals are generally composed of quartz, hematite, mullite, magnetite [106].

Class C Class C fly ashes are generally produced from lignite or sub-bituminous coal. This class of fly ash has both pozzolanic and varying degree of self-cementitious properties. (Most Class C fly ashes contained more than 15% CaO. But some Class C fly ashes may contain as little as 10% CaO). Class C fly ashes contain predominantly calcium alumino-silca glass which is highly reactive. Crystalline phases in Class C ash includes quartz, lime, mullite, gehlenite, anhydrite, and cement materials such as C_3A, C_2S and C_4A_3S.

Table 1.5 Requirements for fly ash and natural pozzolans for use as mineral admixtures in Portland cement concrete (ASTM C618 [5])

Requirements	Fly ash classification		
	N	F	C
Chemical requirements $SiO_2 + Al_2O_3 + Fe_2O_3$, min (%)	70.0	70.0	50.0
SO_3, max (%)	4.0	5.0	5.0
Moisture content, max (%)	3.0	3.0	3.0
Loss on ignition, max (%)	10.0	6.0	6.0
Physical requirements Amount retained when wet sieved on 45-μm sieve, max (%)	34	34	34
Pozzolanic activity index, with Portland cement at 28 days, min (%) of control	75	75	75
Pozzolanic activity index with lime at 7 days, min (MPa)	5.5	5.5	–
Water requirement, max (%) of control	115	105	105
Autoclave expansion or contraction, max (%)	0.8	0.8	0.8
Specific gravity, max variation from average	5	5	5
Percentage retained on 45-μm sieve, max variation, percentage points from average	5	5	5

Class N Raw or calcined natural pozzolans such as some ditomaceous earths, opaline chert and shale, stuffs, volcanic ashes and pumice are included in this category. Calcined kaolin clay and laterite shale also fall in this category of pozzolans.

Table 1.5, presents chemical and physical requirements for fly ash and natural pozzolans for use as a mineral admixture in Portland cement concrete.

1.4 Reaction Mechanism

Setting or hardening of OPC concretes occurs due to the hydration reaction between water and cementitious compounds in cement which give rise to several types of hydrates of calcium silicate (CSH), calcium aluminate (CAH) besides calcium hydroxide (CH). These hydrates are generally called as "Tobermorite gel". The adhesive and cohesive properties of the gel bind the aggregate particles. Calcium hydroxide is a by-product of cement hydration.

When fly ash is incorporated in concrete, the calcium hydroxide liberated during hydration of OPC reacts slowly with the amorphous alumino-silicates, the pozzolanic compounds, present in the fly ash. The products of these reactions, termed as pozzolanic reaction products, are time dependent but are basically of the same type and characteristics as the products of the cement hydration.

Thus additional cementitious products become available which impart additional strength to concrete.

The following equations illustrate the pozzolanic reaction of fly ash with lime to produce additional calcium silicate hydrate (C–S–H) binder:

$$\text{Cement reaction:} \quad C_3S + H \xrightarrow{\text{(hydration)}} C - S - H + CaOH$$

$$\text{Pozzolanic reaction:} \quad CaOH + S \rightarrow C - S - H$$

silica from
ash constituents

1.5 Uses of Fly Ash

Coal fly ash as an engineering material can be used in following ways

- Portland cement
- Stabilized base course
- Flowable fill
- Structural fills/embankments
- Soil improvement
- Asphalt pavements
- Grouts for pavement subsealing

1.5.1 Uses of Fly Ash in Cement Concrete

Utilization of fly ash in cement or concrete can be categorized based on the volume of its usage.

1.5.1.1 Medium Volume Uses

This includes the use of fly ash

- as raw material in cement production
- as an admixture in blended cements
- as partial replacement of cement or as a mineral admixture in concrete
- in addition coal ash including fly ash may be used as partial replacement of fine aggregate in concrete
- for production of light weight aggregates for concrete and many other applications

1.5.1.2 High Volume Uses

High volume utilization of fly ash includes

- as structural fills in embankments, dams, dikes and levees, and
- as sub-base and base courses in road way construction

1.5.1.3 Low Volume Uses

This includes the coal ash utilization

- in high value added applications such as metal extractions. High value metal recovery of Aluminum (Al), Gold (Au), Silver (Ag), Vanadium (Va) and Strontium (Sr) fall in this category.
- Fly ash has potential uses for producing light weight refractory material and exotic high temperature resistant tiles.
- Cenospheres or floaters in fly ash are used as special refractory material and also as additives in forging to produce high strength alloys.

1.5.1.4 Miscellaneous Uses

Based upon its physical properties, coal ash is used

- as land fill for land reclamations for residential, commercial and recreational development projects.
- as filler in asphalt, plastics, paints and rubber products.
- in water treatment and as absorbent for oil and chemical spills.

1.6 Objectives of Using Fly Ash in Cement/Concrete

The objective of using fly ash in concrete is to achieve one or more of the following benefits:

- Reducing the cement content to reduce costs
- Improving workability
- Obtaining reduced heat of hydration, especially in mass concreting
- Attaining required levels of strength in concrete at ages beyond 56 days

1.7 Benefits of Using Fly Ash in Cement/Concrete

Inclusion of fly in cement or concrete has several benefits. Benefits to concrete vary depending on the type of fly ash, proportion used, other mix ingredients, mixing procedure, field conditions and placement. Some of the benefits of fly ash in concrete are:

1.7.1 Reduced Bleeding and Segregation

Bleeding and segregation are considerably reduced with the use of fly ash as a mineral admixture in concrete and thus improving the pumpability of concrete. This is due to (1) the lubricating effect of the glassy spherical fly ash particles; and (2) increased ratio of solids to liquid make the concrete less prone to segregation and increase concrete pumpability.

1.7.2 Improved Workability

The spherical shape and glassy surface of fly ash particles permit greater workability for equal w/c ratio. In other words, w/c ratio may be reduced for equal workability.

1.7.3 Reduced Heat of Hydration

Hydration of cement paste is accompanied by liberation of heat that raises the temperature of concrete. Because of the slower pozzolanic reactions, partial replacement of cement by fly ash results in release of heat over a longer period of time, and the concrete temperature remains lower slowly. This is of immense importance in mass concrete where cooling, following a large temperature rise, can lead to cracking. Low-calcium Class F fly ashes generally tend to reduce the rate of temperature rise more as compared to high-calcium Class C fly ashes

1.7.4 Higher Ultimate Strength

The additional binder produced by the fly ash reaction with available lime allows fly ash concrete to continue to gain strength over time. Mixtures designed to produce equivalent strength at early ages (less than 90 days) will ultimately exceed the strength of straight cement concrete mixes.

1.7.5 Reduced Permeability

The decrease in water content combined with the production of additional cementitious compounds reduces the pore interconnectivity due to refinement of pore structure of concrete resulting in reduced permeability. The reduced permeability results in improved long-term durability and resistance to various forms of deterioration.

1.7.6 Increased Resistance to Sulfate Attack

Fly ash in concrete increases the sulphate resistance and potentially corrosive salts that penetrate into concrete and cause steel corrosion with accompanying cracking and spalling of concrete. Fly ash induces three phenomena that improve sulfate resistance (1) consumes the free lime making it unavailable to react with sulfate; (2) reduced permeability prevents sulfate penetration into the concrete; and (3) replacement of cement reduces the amount of reactive aluminates available

1.7.7 Improved Resistance to Corrosion

Fly ash addition to concrete improves the long term corrosion resistance of concrete. The reaction of fly ash with $Ca(OH)_2$ produces a denser concrete and thus inhibits the ingress of chloride ions takes place at a slower rate.

1.7.8 Increased Resistance to Alkali-Silica Reactivity (ASR)

Fly ash reacts with available alkali in the concrete, which makes them less available to react with certain silica minerals contained in the aggregates.

1.8 Effect of Fly Ash on the Fresh Properties of Cement Concrete

Fresh concrete is a concentrated suspension of particulate materials of different densities, particle sizes and chemical composition in a solution of lime and other compounds. Fresh concrete properties include workability, air-entrainment, bleeding and segregation, pumpability, compactability, and finishability. In fresh concrete, fly ash plays an important role in the fluidity of concrete.

1.8.1 Workability

Workability is defined as the ease with which a freshly mixed concrete can be properly compacted, transported, placed, and finished. Workability is one of the governing factors of concrete mix design. Workability is determined by the rheological behavior of fresh concrete. Water content of concrete plays a dominant role in controlling workability. Workability depends on (1) water content; (2) aggregate shape and size; (3) cementitious content; and (4) age (level of hydration), and can be modified by adding mineral/chemical admixtures. The spherical shape and glassy surface of most fly ash particles, usually finer than cement, permit greater workability or slump for equal water–cement ratios.

Lane and Best (1980) concluded that use of fly ash as partial replacement of cement usually reduces the water content for a given consistency. With proper proportioning of the concrete, cohesion and plasticity are adequate, and bleeding is reduced. Helmuth [45] has reported that if the water reduction were due to the spherical shape of the fly ash particles or the lack of chemical reactivity, the water reduction would progressively increase with increase in fly ash content.

Brown [15] conducted several studies with fly ash replacing cement and fine aggregate at levels of 10–40% by volume. He concluded that for each 10% of ash substituted for cement, the compacting factor or workability changed to the same order as it would by increasing the water content of the mix by 3–4%. When fly ash was substituted for sand or total aggregate, workability increased to reach a maximum value at about 8% ash by volume of aggregate. Further substitution caused rapid decrease in workability.

Yuan et al. [130] showed that the water demand decreased when the quantity of fly ash was between 15 and 20%, and increased when fly ash content was more than 20%. This was attributed to increase in water demand to the introduction of additional specific surface and porous grains, and decrease to deflocculation by adsorption of fine grains of fly ash on cement clusters.

Owens [96] reported that with the use of fly ash containing large fraction of particles coarser than 45 μm or a fly ash with high amount of unburned carbon, exhibiting loss on ignition more than 1%, higher water demand was observed. Water demand was noticeably increased to maintain the desired level of fluidity.

Fig. 1.3 Influence of coarse particulate of fly ash on the water required for equal workability in concrete [96]

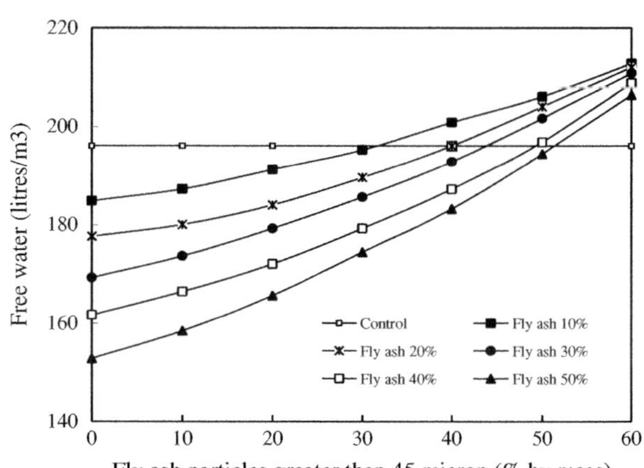

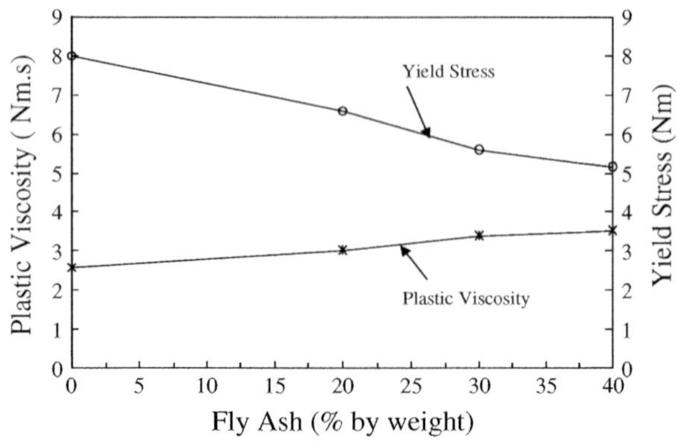

Fig. 1.4 Influence of fly ash content on plastic viscosity and yield stress of concrete [61]

Figure 1.3 shows the effects of coarse fly ash particles on the water demand of concrete mixtures.

In order to develop a theoretical concept and understanding of the rheology of fresh fly ash concrete, Tattersall and Benfill [119] suggested mathematical expressions relating the yield value (τ) and the plastic viscosity (μ) varying with volumetric parameters of concrete, in terms of Bingham model. Yield stress and plastic viscosity varied with volumetric parameters of concrete, water-to-cement ratio, replacement levels of cement/aggregates with fly ash, and fineness of fly ash. An increase in the volume of the paste at a constant ratio of ash to total cementitious material resulted in an increase in plastic viscosity. Apparently because of fine particle size and smooth glassy texture as well as spherical shape, fly ash acts to plasticize concrete at given water content when used as partial replacement of cement or fine aggregate. Khan [61] reported that the incorporation of 20, 30 and 40% fly ash increased the plastic viscosity of concrete whilst the yield stress of concrete reduced with increase in fly ash content as shown in Fig. 1.4

1.8.2 Bleeding and Segregation

Incorporation of fly ash in mortar or concrete significantly reduces the bleeding and segregation. This is due to the lubricative effect of the glassy spherical fly ash particles and the increased ratio of solids to liquid make the concrete less prone to segregation and increase concrete pumpability. Figure 1.5 shows the bleeding rate of fly ash concrete compared to that of control concrete.

Joshi and Lohtia [57] used Alberta fly ashes in making high-volume fly ash concrete mixes, and concluded that fly ash concrete mixes were more cohesive than control mixes. During the slump test, the fly ash concrete mixes sub-sided more slowly and gradually than the control mixes which exhibited abrupt fall or subsidence.

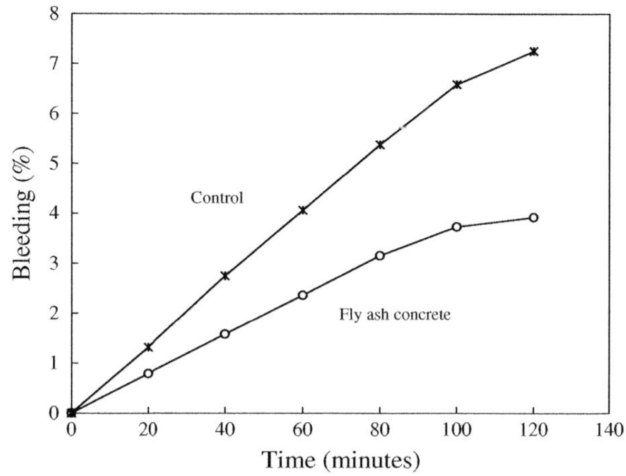

Fig. 1.5 Relative bleeding of control and fly ash concretes [23]

1.8.3 Air Entrainment

Fly ash addition affects both air-content and loss of air-content with time in fresh concrete depending upon type of fly ash concrete. For entraining a specified amount of air content, usually around 4–6% more air-entraining agent (AEA) is required in fly ash concrete than for a similar concrete containing no fly ash concrete. This is (1) because of the greater surface area of fly ash in concrete. Fly ash is generally finer than cement and volume of fly ash added is normally more than the volume of cement replaced. Because of this, surface area of the binder within the concrete mix is increased. Thus greater volume of air entraining agent is needed to provide the same concentrations of the air voids in the mortar or concrete containing fly ash; and (2) secondly, the main reason leading to the increased demand of AEA is related to the carbon content, expressed in terms of loss on ignition (LOI) of fly ash. The carbon absorbs a portion of the air-entraining agent, which limits its availability for producing the needed air bubbles. The amount of absorption varies with the amount of carbon content.

Gebler and Klieger [40] investigated the requirements of AEA for Class C and Class F fly ashes. They reported that (1) concretes made with Class C fly ash generally require less AEA than those made with Class F fly ashes; (2) for 6% air content in concrete, the AEA varied from 126 to 173% for fly ashes having more than 10% CaO, whereas it was in the range of 177 to 553% for fly ashes containing less than 10% CaO; and (3) increase in both total alkalies and SO_3 contents in fly ash affect the air entrainment favorably. A concrete containing a Class F fly ash that has relative high CaO content and less organic matter or carbon tends to be less vulnerable to loss of air.

Joshi et al. [58] reported that for some Class F fly ashes replacing about 50% cement, the average requirement of AEA was found to be more than double of the equivalent plain concrete. A wide range of AEA demand was reported by Carette and Malhotra [19] from the study of concretes made with Canadian fly ashes to entrain about 6% air content.

Burns et al. [16] made attempts to neutralize the absorption characteristics of activated carbon in fly ash by using chlorine gas, calcium hypochlorite and some other surface active agents to resolve the problem of increased AEA demand. However, these studies need to be substantiated for practical use. The deactivating agents used for carbon in fly ash should not interfere with air entrainment and concrete durability on their own. Economy is another important factor to be considered in developing suitable remedial additives for solving the problem of AEA demand in concrete.

1.8.4 Temperature Rise

Hydration of cement paste is accompanied by liberation of heat that raises the temperature of concrete. Because of the slower pozzolanic reactions, partial replacement of cement by fly ash results in release of heat over a longer period of time and the concrete temperature remains lower slowly. This is of immense importance in mass concrete where cooling, following a large temperature rise, can lead to cracking. Low-calcium Class F fly ashes generally tend to reduce the rate of temperature rise more as compared to high-calcium Class C fly ashes.

Compton and Macinnis [24] reported temperature–time curves (Fig. 1.6) for control and fly ash concretes. In fly ash concrete, cement was replaced with 30% fly ash. This particular phenomenon is very useful in mass concreting, where cooling, following a significant temperature rise due to the generation of the heat of hydration occurs, stresses can develop and cause cracking. Temperature rise, in fact, depends upon more factors than the rate of heat generation associated with

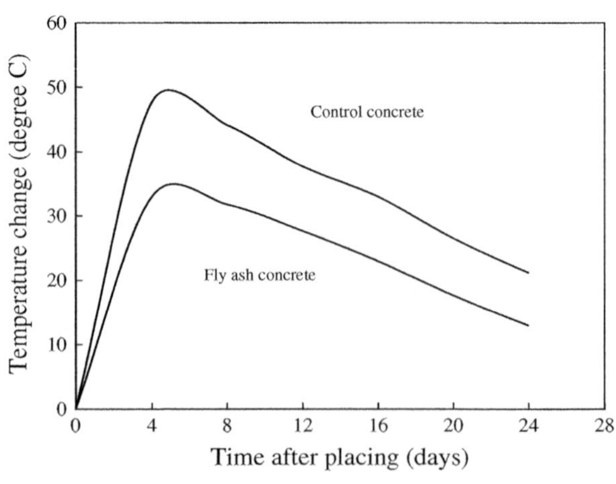

Fig. 1.6 Temperature rise curve for fly ash and plain concrete test sections [24]

hydration and pozzolanic reactions, including the rate of heat loss and the thermal properties of the concrete and the surrounding medium.

Fly ash retards the hydration of C_3S in the early stages but accelerates it at later stages [51, 94]. Jawed and Skalny [51] found that two chemically similar Class F fly ashes with different surface areas (314 and 205 m^2/kg) had similar retarding effects on C_3S hydration, but the effect changed dramatically in 0.5 M NaOH solutions. They attributed the retardation to two factors (1) chemisorptions of calcium ions on the fly ash particles, resulting in a reduction in its concentration in the liquid phase and a delay in $Ca(OH)_2$ nucleation and (2) poisoning of the nucleation and growth of $Ca(OH)_2$ and C–S–H by soluble silicates and aluminates.

Takemoto and Uchikawa [118] concluded that hydration of C_3A in the presence of calcium hydroxide, gypsum, and fly ash was accelerated in the presence of pozzolan due to the adsorption of calcium ions from the solution and generation of sites for ettringite or other hydrates to precipitate.

Bamforth [12] investigated the temperature rise in large size foundation made with concrete containing fly ash and slag. Three types of concretes were used (1) control concrete with a Portland cement content of 400 kg/m^3; (2) mix with 75% of the Portland cement replaced by ground granulated blast furnace slag (GGBS); and (3) mix with 30% of the Portland cement replaced by a bituminous fly ash. It was observed that with an increase in the quantity of cement replaced by fly ash and slag, the rate of heat release was slowed down and as a result the maximum temperature reached at any point in the concrete mass was lower than the control concrete.

Sivasundram et al. [114] and Langley et al. [68] reported favorable effect of fly ash incorporation in concrete on temperature rise of not only massive concrete dams, but also in concrete mat foundations and massive columns in lower storey of the tall buildings. Low-calcium Class F fly ashes generally tend to reduce the rate of temperature rise more as compared to high-calcium Class C fly ashes [26]. Some high-calcium Class C fly ashes with self cementitious properties may react very rapidly with water, thus releasing excessive heat just like normal Portland cement hydration.

ACI Committee 211.1.81 [1] estimated that on the basis of equivalent mass, fly ash contributes to early age heat liberation in the range of 15 to 30% compared to normal Portland cement. Low-calcium bituminous fly ash was successfully used to control the rise of temperature at early age in the construction of concrete structures, particularly dams. In general, incorporation of fly ash as replacement of normal Portland cement exhibits less temperature rise than concrete without fly ash. Where early age strength is not the main design consideration, large quantities of cement replaced by fly ash are anticipated to reduce the rate and also amount of heat hydration significantly.

Atis [6] studied the heat evolution of high-volume fly ash (HVFA) concrete. Heat evolution of concrete was studied by measuring the temperature increase in concrete under adiabatic curing condition. They concluded that (1) characteristic of heat evolution of fly ash concrete was found to be strongly dependent on the replacement level of fly ash and dosage of superplasticizer; (2) use of fly ash as cement replacement resulted in reduction in maximum temperature rise. Increasing the replacement level of fly ash caused lower temperature rise in concrete; (3) superplasticizer caused a delay in peak temperature rise time. This is taken as an indicator that high-dosage superplasticizer used in concrete caused retardation in hydration of cement. Concretes having similar ingredients showed similar peak temperature rise whether they are superplasticized or not.

1.8.5 Setting Time

With the addition of water to concrete, hydration reaction starts and the cement paste begins to stiffen accompanied by heat release. The rate of stiffening of cement paste is expressed in terms of setting time. Generally, the effect of fly ash on the setting time depends upon the characteristics and amount of fly ash used. The interacting effects of fly ash with other chemical and mineral admixtures may also influence the setting of concrete.

Investigations have revealed that the addition of low-calcium Class F fly ashes generally show some degree of retarding effect on cement setting. High-calcium fly ashes, generally low in carbon and high in reactive and/or cementitious components sometimes exhibit opposite behavior of reduced setting time. Not all Class C fly ashes cause rapid setting.

Ramakrishan et al. [102] reported an increase in setting time with the use of high-calcium fly ash in concrete. Lane and Best [66] concluded that the influence of fly ash on setting time is less than the influence due to cement fineness, water content, and ambient temperature.

Carette and Malhotra [19] studied the effect of Canadian fly ashes on the fresh concrete properties. Fly ashes were collected from 11 different sources. Cement was replaced with 20% fly ash in all the mixes. Chemical properties of fly ashes are given in Table 1.6. Fresh concrete properties are given in Table 1.7.

Rodway and Fedirko (1989) studied the setting times of concretes made with varying percentages (0, 56, 68 and 76%) of fly ash of the total cementitious material. High fly ash concrete mixes exhibited increasingly greater initial setting times of 22–42.5 h with increasing fly ash content from 56 to 76% compared to 7.6 h for the control mix without fly ash. They observed that delays appeared to be related to the problem of compatibility between cementitious materials and superplasticizer to maintain workability.

Sivasundram et al. [115] investigated the setting time of high-volume fly ash (HVFA) concrete mixes, and concluded that the initial setting time of 7.50 h was comparable to that of the control concrete, whereas the final setting time was extended by about 3 h as compared to that of the control concrete.

Joshi et al. [58] examined tree different sub-bituminous Alberta coal ashes at replacement levels of 40–60% by cement weight to produce superplasticized and air entrained concretes. They observed that fly ash concrete achieved an initial setting time of 5–11 h as compared to about 5 h for non-fly ash concrete.

Table 1.6 Properties of some Canadian fly ashes [19]

Fly ash source	Type of coal	Major chemical compounds (% by weight)					
		SiO_2	Al_2O_3	Fe_2O_3	CaO	MgO	LOI
1	Bituminous	47.1	23.0	20.4	1.21	1.17	2.88
2	Bituminous	44.1	21.4	26.8	1.95	0.99	0.70
3	Bituminous	35.5	12.5	44.7	1.89	0.63	0.75
4	Bituminous	38.3	12.8	39.7	4.49	0.43	0.88
5	Bituminous	45.1	22.2	15.7	3.77	0.91	9.72
6	Bituminous	48.0	21.5	10.6	6.72	0.96	6.89
7	Sub-bituminous	55.7	20.4	4.61	10.7	1.53	0.44
8	Sub-bituminous	55.6	23.1	3.48	12.3	1.21	0.29
9	Sub-bituminous	62.1	21.4	2.99	11.0	1.76	0.70
10	Lignite	46.3	22.1	3.10	13.3	3.11	0.65
11	Lignite	44.5	21.1	3.38	12.9	3.10	0.82

Table 1.7 Properties of concrete incorporating Canadian fly ashes [19]

Mix no.	Cement (kg/m^3)	Slump (mm)	Air (%)	Bleeding (%)	Setting time (h:min)	
					Initial	Final
Control	295	70	6.4	2.9	4:10	6:00
F1	236	100	6.2	3.1	4:50	8:00
F2	237	105	6.2	4.6	7:15	10:15
F3	237	100	6.2	5.1	5:20	8:10
F4	238	110	6.3	4.3	6:20	8:25
F5	237	65	6.4	2.7	5:15	8:55
F6	238	75	6.5	2.6	4:30	6:50
F7	239	100	6.1	2.9	4:15	6:20
F8	236	115	6.2	5.6	5:10	7:30
F9	236	100	6.4	4.4	5:25	9:00
F10	237	130	6.5	2.5	4:45	7:00
F11	237	140	6.6	0.6	4:00	6:05

The final setting time varied from 10 to 13 h as against 7 h for control mixes without fly ash.

1.9 Effect of Fly Ash on Properties of Cement Concrete in Hardened State

1.9.1 Compressive Strength

The pozzolanic reaction has several characteristics that affect the strength. Mehta [79] has identified as

- the reaction is slow, so that the rates of both heat liberation and strength development are correspondingly slow
- the reaction consumes lime rather than producing it
- the reaction products are efficient in filing up space and subdividing pores

Rate of strength development depends upon following factors:

- fly ash characteristics such as its chemical and mineralogical composition fineness, pozzolanic reactivity
- type of cement
- replacement level of cement with fly ash
- mixture proportions
- ambient temperature
- curing environment

The low-calcium fly ashes do not exhibit significant pozzolanic activity to affect strength until about 2 weeks after hydration, but highly pozzolanic fly ashes start their contribution to strength development almost from the onset of Portland cement hydration. Some high-calcium fly ashes, with calcium oxide content more than 15%, may start contributing to compressive strength development as early as 3 days after mixing because of their self hardening and pozzolanic properties.

Because of its fineness as well pozzolanic reactivity, fly ash in cement concrete significantly improves the quality of cement paste and the micro-structure of the transition zone between the binder matrix and the aggregate. As a result of the continual process of pore refinement, due to the inclusion of fly ash hydration products in concrete, a gain in strength development with curing age is achieved.

When high-calcium Class C fly ashes are used, the strength development with time is likely to be different from the one using Class F fly ash. The self-hardening reactions in the Class C fly ashes are likely to occur within the same time frame as the normal Portland cement hydration reactions, giving equal or sometimes greater strengths at early ages. The pozzolanic activity of such cementitious fly ashes further enhances strength at later ages.

Lane and Best [66] observed that proportioning fly ash concrete on strength basis requires a replacement ratio greater than one-to-one by mass so that the fly ash in effect replaces some of the fine aggregate. When fly ash replaces cement on a one-to-one basis, the rates of hardening and strength gain at early ages are reduced. When replacement is on two or three-to-one basis and the fine aggregate content are reduced accordingly, 3-day strength is slightly reduced compared to the control, 28-day is strength is comparable, and later-age strength is higher.

Cook [25] conducted investigations using a high-calcium fly ash with CaO content of 30.3% at 25% replacement level to develop concrete mixes with 28-day strength in the range of 55–75 MPa. He reported that as cement factor was increased, sand content and the water-to-cementitious material ratio was reduced. To maintain slump around 100 mm, water reducing admixture was

Table 1.8 Compressive Strength of hardened concrete [19]

Mixture no.	Compressive strength (MPa)			
	7 days	28 days	91 days	365 days
Control	23.4	30.6	34.9	39.2
F1	18.4	25.7	31.4	38.3
F2	16.9	25.2	34.8	37.0
F3	14.4	21.0	27.6	34.4
F4	17.8	23.3	32.3	36.9
F5	20.1	28.0	33.9	44.3
F6	18.4	24.8	31.8	39.2
F7	16.7	24.1	29.1	35.7
F8	17.9	27.7	29.0	40.4
F9	16.7	24.9	31.1	35.6
F10	19.2	28.5	33.7	39.7
F11	21.1	29.4	35.3	40.1

Table 1.9 Compressive of high-calcium fly ash concrete [101]

Cement (kg/m^3)	Fly ash (kg/m^3)	Compressive strength (kg/m^2)		
		Percentage of control		Percentage gained from 28 to 56 days
		28 days	56 days	
91	50	318	354	15.2
	59	352	393	16.0
	68	401	435	13.6
	77	471	494	10.5
136	50	210	228	10.2
	59	231	256	11.8
	68	245	269	11.1
	77	253	274	10.3
182	50	149	155	8.2
	59	153	167	12.4
	68	163	177	12.0
	77	187	190	6.2

used in all the mixes. The 180-day strength of the mixes ranged between 68 and 86 MPa.

Carette and Malhotra [19] studied the effect of Canadian fly ashes on the compressive strength of concrete mixes. Cement was replaced with 20% fly ash in all the mixes. Compressive strength was measured up to the age of 365 days, and results are given in Table 1.8. It can be seen from this table that compressive strength continued to increase with age, indicating pozzolanic action of fly ashes.

Raba et al. [101] determined the compressive strength of concrete made with bituminous fly ash (CaO 20%). In the mixes, fine aggregate was replaced with fly ash by volume, and mass of cement and coarse aggregate was kept constant for each series (Table 1.9).

Swamy and Mahmud [117] reported that concrete containing 50% low-calcium bituminous fly ash as partial replacement of cement developed 20–30 MPa compressive strength at 3 days, 60 MPa at the age of 28 days.

Joshi et al. [58] tested a large number of fly concrete mixes made by using three different Alberta fly ashes containing about 10% calcium oxide. The replacement level varied between 40 and 60% by weight of cement. The mixes were superpalasticized and air-entrained to obtain 100 to 120 mm slump and $6 \pm 1\%$ air content. The cementitious material content varied from 380 to 466 kg/m^3, water-to-cementitious material ratio from 0.27 to 0.37, coarse aggregate ranged from 1,012 to 1,194 kg/m^3, and fine aggregate or sand varied from 712 to 643 kg/m^3. They reported that (1) at 7 days, the fly ash concretes obtained strength between 27.9 and 41.0 MPa compared to 44.1 MPa of control concrete. However at the age of 28 days, the fly ash concretes developed strength varying from 37.6 to 50.7 MPa against 58.7 MPa for control concrete. At 120 days, strength of fly ash concrete ranged from 54.8 to 74.6 MPa whereas it was 74.6 MPa of control concrete.

Mehta [82] have reported that no significant contribution to strength development was noticed up to 7 days with the use of low-calcium fly ash in concrete. At 28 days and beyond, most fly ashes at the replacement levels of up to 30% by cement weight exhibited strength gain in concrete and the strength generally equaled that of control concrete.

Haque et al. [44] concluded that for concrete mixes with 40–75% bituminous fly ash replacing cement, the increase in flexural strength was slightly less than the increase in compressive strength between 28 and 91 days of curing.

Klieger and Perenchio [63] found that concrete made with Type-I fly ash cement had lower strength than the control at all ages through 3 years. Lower casting and initial-curing temperatures resulted in higher strengths at later ages for both types of concretes. Korac and Ukraincik [65] found that the early-age strengths of 50% fly ash concrete were lower than the controls; after 90 days strengths were comparable.

Erdoğdu and Türker [37] studied the effect of particle size of high and low-calcium fly ash on the compressive strength of mortar. Major chemical compounds in high-calcium ash were Al_2O_3 (19.43%), SiO_2 (46.45%), CaO (12.69%) and Fe_2O_3 (9.32%), whereas low-calcium fly ash had Al_2O_3 (27.72%), SiO_2 (57.30%), CaO (2.24%) and Fe_2O_3 (5.72%). Fly ashes were sieved from the 125, 90, 63, and 45-μm sieve by a sieve-shaker. Six different size groups, including the all-in ash, were obtained for both the ashes and the materials retained on each sieve. They were; material retained on 125 μm sieve, material having 125–90, 90–63, and 63–45, and material passing 45-μm sieve. For strength tests, mixtures composed of 25% fly ash and 75% Portland cement, by weight, were prepared. Tests were conducted up to the age of 90 days, and results are given in Table 1.10. It is evident from these results that (1) high-calcium ash incorporated mortars exhibited higher strength than mortars with low-calcium ash at all ages; (2) for both of the ashes, the finer the size of a fraction, higher was the compressive strength.

Table 1.10 Compressive strength results of the mortars [37]

		Compressive strength (N/mm^2)			
		2 days	7 days	28 days	90 days
25% High + 75% PC	All-in ash	27.6	36.7	46.7	55.7
	Above 125 μm	19.2	26.6	31.2	33.2
	125–90 μm	22.6	27.6	35.0	38.4
	90–63 μm	24.8	31.3	35.4	42.6
	63–45 μm	26.6	35.1	41.4	47.5
	Under 45 μm	32.0	43.0	55.8	64.6
25% Low + 75% PC	All-in ash	23.5	33.2	39.6	47.4
	Above 125 μm	17.9	24.4	26.7	29.9
	125–90 μm	21.6	29.2	32.0	37.1
	90–63 μm	22.6	29.6	34.9	40.6
	63–45 μm	23.8	33.5	37.4	46.8
	Under 45 μm	26.6	37.3	43.4	57.2
100% PC 32.5		38.0	44.4	51.6	58.2

Saraswathy et al. [107] investigated the influence of activated fly ash on the compressive strength of concrete. Various activation techniques, such as physical, thermal and chemical were adopted. Concrete specimens were prepared with 10, 20, 30 and 40% of activated fly ash replacement levels with cement. Compressive strength was determined at 7, 14, 28 and 90 days. They concluded that (1) activation of fly ash improved the strength of concrete. However, the compressive strength of fly ash concrete was less than that of ordinary portland cement (OPC) even after 90 days of curing; and (2) among the activation systems, chemically activated coal fly ash (CFA) improved the compressive strength to a certain extent, only with 10 and 20% replacements. Since the CFA surface layer is etched by a strong alkali to facilitate more cement particles to join together and also the addition of CaO which is further promoting the growth of CSH gel and $Ca(OH)_2$ which is more advantageous to enhance the strength development.

Siddique [112] studied the effect of partial replacement of fine aggregate (sand) with varying percentages of Class F fly ash on the compressive strength of concrete up to the age of 365 days. Fine aggregate (sand) was replaced with five levels of percentages (10, 20, 30, 40, and 50%) of Class F fly ash by weight. Control mix (without fly ash) was proportioned to have a 28-day cube compressive strength of 26.4 MPa. Compressive strength results are shown in Fig. 1.7. Based on the results, it was concluded that (1) compressive strength of fine aggregate (sand) replaced fly ash concrete specimens was higher than the plain concrete (control mix) specimens at all the ages. The strength differential between the fly ash concrete specimens and plain concrete specimens became more distinct after 28-days; (2) compressive strength continued to increase with age for all fly ash replacement levels; (3) The maximum compressive strength occurs with 50% fly ash content at all ages. It was 40.0 MPa at 28-day, 51.4 MPa at 91-day, and 54.8 MPa at 365-day; and (4) results of this investigation suggests that Class-F fly ash could be very conveniently used in structural concrete.

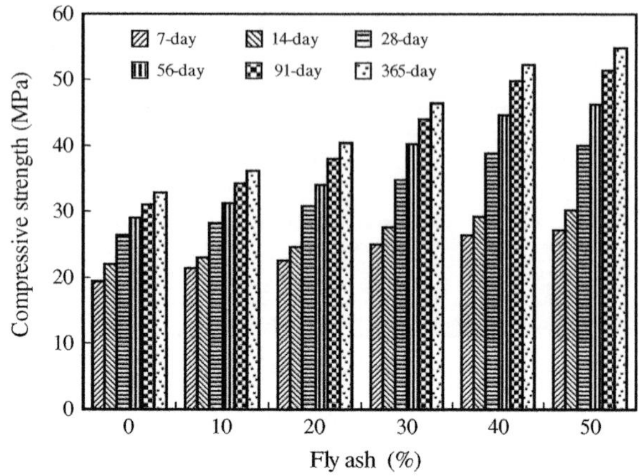

Fig. 1.7 Compressive strength versus fly ash percentage [112]

Demirboğa et al. [28] investigated the role of high-volumes of Class C fly ash on the compressive strength of concrete. Cement was replaced with 0, 50, 60, and 70% fly ash. Compressive strength of concrete mixtures was determined at 3, 7, 28 and 120 days. Based on the investigation, they reported that (1) fly ash FA reduced compressive strength of concrete at all levels of replacement at 3, 7, 28 and 120 days; (2) reductions were very high at early ages, but with the increase in curing period, the reduction percent decreased. Reductions at 3-day curing period were 69, 84 and 91% for 50, 60 and 70% fly ash replacement of Portland cement, respectively. At 28-days curing period, these values reduced to 52, 68 and 78% for 50, 60 and 70% fly ash replacement of Portland cement, respectively. At 120-days curing periods, reductions were 36, 43 and 50% for 50, 60 and 70% fly ash replacement of Portland cement, respectively; and (3) cement paste containing fly ash showed a steady reduction in strength at 3, 7, 28 and 120 days as a function of replacement percentage. This can be directly related to the properties of fly ash that decrease the heat of hydration of cement and required long curing period. Results of numerous studies have indicated that fly ash slows the rate of hardening and reduces the early age compressive strength of concrete.

Chindaprasirt et al. [22] studied the effect of fly ash fineness on the compressive strength of concrete. Three fly ash finenesses: coarse, medium and fine were used. The coarse fly ash was 100% original fly ash (100FA). The medium fly ash was the 45% fine portion of the original fly ash (45FA). The fine fly ash was the 10% fine portion of the original fly ash (10FA). Three concrete mix series viz. a low, normal and a high strength concrete mix series were made. For the low- and normal-strength concrete, the water-to-cement ratios of 0.54 and 0.48, respectively, were used for Portland cement mixes. For the high strength concrete, the water-to-cement ratio was 0.25 with the use of superplasticizer. The fly ash dosage of 30% by weight of binder was used for all fly ash concrete mixes. Compressive strength of concrete mixes was determined up to the age of 90 days (Table 1.11). It can be seen from this table that the strength of fly ash concrete were higher than those of the portland cement concrete in the same group. For the normal-strength concrete,

Table 1.11 Compressive strength of concrete [22]

Mix	Compressive strength (MPa)		
	7 days	28 days	90 days
PC1	24.5	33.5	35.2
100FA1	24.5	41.0	47.0
45FA1	32.5	40.5	49.0
10FA1	31.0	46.5	51.5
PC2	33.0	48.5	53.0
100FA2	31.0	46.0	55.0
45FA2	32.5	44.0	57.5
10FA2	35.5	52.0	61.5
PC3	62.5	79.5	85.5
100FA3	51.5	84.0	88.5
45FA3	50.5	72.5	83.5
10FA3	59.5	71.0	82.5

the 28-day strength of the PC2 concrete was 48.5 MPa, whereas those of fly ash concretes were between 44.0 and 52.0 MPa. For the high strength concrete, the strength of the PC3 concrete was 79.5 MPa, whereas those of the fly ash concretes were between 71.0 and 84.0 MPa. For the low strength concrete series, the strength of PC1 concrete was lower than the fly ash concrete owing to a very large reduction in the water content of the fly ash concrete. The reduction in the water content, the good dispersing and the filling effect of the fly ash contribute to the relatively good strength development of the fly ash concrete in this series. With the use of finer fly ash, the water content was further reduced and the strength of concrete enhanced further.

1.9.2 Effect of Curing Temperature at Early Age on Strength

The rate of pozzolanic reaction of fly ash in cement concrete is significantly influenced by curing temperatures at early ages as is the case of cement hydration reactions. Pozzolanic reactions are highly temperature dependent. Higher the curing temperature, higher is the rate of the pozzolanic reactions. When concrete made with Portland cement is cured at temperatures in excess of 30°C, an increase is seen in strength at early ages but a marked decrease in strength is observed in mature concrete [92].

Ravina [103] observed that fly ash concrete subjected to high temperature at an early age of curing, exhibited increased rate of strength gain possibly due to the higher rate of pozzolanic reactions. It was suggested that when concrete was cured at elevated temperatures, large quantities of fly ash may be incorporated with a significant improvement in strength compared to the rather limited contribution under normal curing conditions up to 28 days.

William and Owens [125] reported that concrete containing fly ash behaved significantly different than concrete made with Portland cement (Fig. 1.8). In contrast

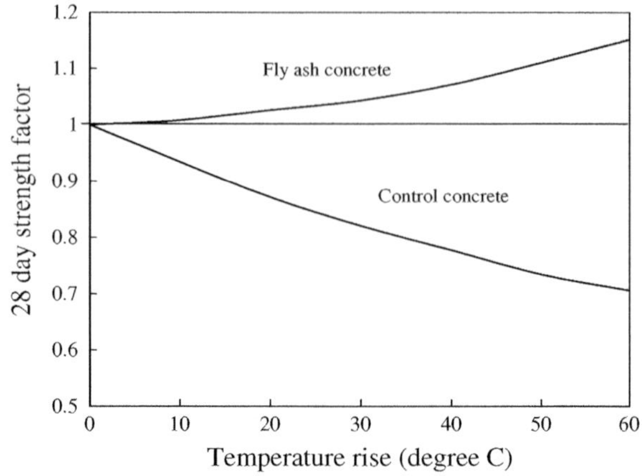

Fig. 1.8 Effect of temperature rise during curing on the compressive strength development of concretes [125]

to the loss of strength that occurred with ordinary Portland cement concrete, fly ash concretes exhibited strength gains as a consequence of heating. The favorable effects of fly ash in concrete cured at moderately elevated temperatures can be advantageously used in the construction of mass concrete or concrete construction at elevated temperatures.

1.9.3 Effects of Curing Conditions on Compressive Strength

Ozer and Ozkul [95] reported the influence of initial water-curing on the strength development of ordinary Portland cement and pozzolanic cement concretes. They concluded that poor curing conditions adversely affect the strength of concrete made from pozzolanic cement than that of ordinary Portland cement. Atis [9] worked on strength properties of high-volume fly ash roller compacted and workable concrete, and the influence of curing condition. He reported that fly ash–cement concrete was more sensitive to dry curing conditions than conventional concrete.

Termkhajornkit et al. [121] studied the effect of water-curing condition on compressive strength of fly ash–cement paste. Replacement ratios of fly ash were 0, 25 and 50% of total powders. The water-to-binder ratio was 0.80 and 1.00 by volume. Mix proportion and curing condition details are given in Table 1.12. Compressive strength results of fly ash–cement paste are given in Table 1.13. They reported that (1) for the samples prepared with water curing condition, the compressive strength of 0% fly ash increased until 91 days but suddenly dropped at 182 days. The compressive strength of 25% fly ash and 50% fly ash continuously increased until 182 days. At 182 days, the compressive strength of 25% fly ash was higher than that of the cement paste. For the samples prepared under 7 days initial curing, the compressive strength of 0% fly ash was almost constant. The compressive strength of 50% fly ash increased until 28 days, and after that became

Table 1.12 Mix proportion and curing condition [121]

Water/binder (W/B)	Fly ash (%)	Curing condition	Code
0.8	0	In water	0.8-0-W
0.8	25	In water	0.8-25-W
0.8	50	In water	0.8-50-W
1.0	0	In water	1.0-0-W
1.0	25	In water	1.0-25-W
1.0	50	In water	1.0-50-W
1.0	0	In water 7 days	1.0-0-W7
1.0	25	In water 7 days	1.0-25-W7
1.0	50	In water 7 days	1.0-50-W7
1.0	0	In water 3 days	1.0-0-W3
1.0	25	In water 3 days	1.0-25-W3
1.0	50	In water 3 days	1.0-50-W3

Table 1.13 Compressive strength of fly ash–cement paste [121]

Mix/curing condition code	Compressive strength (MPa)				
	7 days	28 days	56 days	91 days	182 days
0.8-0-W	117.03	150.91	145.99	129.91	150.11
0.8-25-W	80.89	113.36	115.06	116.7	163.29
0.8-50-W	41.45	65.42	69.41	83.72	104.81
1.0-0-W	91.84	114.71	127.24	128.63	90.19
1.0-25-W	55.39	78.58	88.72	94.2	94.7
1.0-50-W	29.93	44.7	46.51	55.69	69.24
1.0-0-W7	91.84	95.89	94.97	96.68	97.94
1.0-25-W7	55.39	71.54	69.37	84.75	89.51
1.0-50-W7	29.93	42.84	48.45	48.29	46.77
1.0-0-W3	85.02	95.47	94.61	97.04	94.01
1.0-25-W3	65.17	71.82	87.68	77.18	98.51
1.0-50-W3	30.78	40.19	46.9	45.04	48.55

nearly constant. In contrast, the compressive strength of 25% fly ash constantly increased until 56 days. At this age, its compressive strength became close to that of cement paste; and (2) regarding the effect of curing condition on the compressive strength, for 0% fly ash, the compressive strength of samples cured in water was higher than others except at 182 days. In contrast, there was no significant difference among compressive strength of samples cured in water 7 and 3 days. As for 25% fly ash and 50% fly ash, the effect of water curing condition is small, especially for 25% fly ash.

Yazici et al. [126] investigated the effect of curing condition on the compressive strength of high-volume fly ash concrete mixtures. Cement was replaced with up to 70% fly ash, and concrete mixtures with 360 kg/m^3 cementitious content and a constant water/binder ratio of 0.40 were made. The curing conditions were standard curing in water and steam curing. Compressive strength of concrete mixtures was determined up to the age of 90 days. Based on the test results, they concluded that (1) under standard curing conditions, the 1-day compressive strength decreased sharply with increasing fly ash content. However, at later ages (beyond 3 days), this difference diminished. The strength of 30% fly ash mixture was approximately equal to control mix strength at 3 days. At 7 days, 40% fly ash mixture exceeded the control mix strength, while the 50 and 60% fly ash mixtures reached the control mix strength at 28 and 56 days, respectively. Strength was approximately 10 MPa at 1-day and 20 MPa at 3 days with 50 and 60% fly ash content, which is quite satisfactory for most cast-in place structural elements; and (2) under steam curing, with increasing amount of fly ash content, compressive strength at early ages decreased. However, application of steam curing increased the 1-day compressive strength of 50% fly ash concrete to 20 MPa, which is adequate for formwork removal, and therefore, beneficial for fabrication of precast products. Application of steam curing did not improve the later-age compressive strength of high volume fly ash concrete as much as standard curing. For example, steam-cured high-volume fly ash concrete mixture containing 50% fly ash showed only 40 MPa strength at 90 days compared to 60 MPa for the standard-cured mixture at the same age. Figure 1.9 shows the effectiveness of steam-curing versus standard curing for different ages. It is evident from this figure that the high-volume fly ash systems containing 40% or more fly ash, steam-curing may be of interest only when 1-day strength is the sole consideration.

Swamy and Mahmud [117] examined the effect of curing regime on strength development of high volume fly ash concrete mixes (Table 1.14). They reported 50–100% increase in strength over the 28-day strength of fly ash concrete after 1 year under-continuous moist or fog curing compared to only 18–25% increase for the control or plain concrete under similar curing conditions. Under the other two curing regimes, one with 7-day moist curing followed by air drying and the other with continuous dry curing, the corresponding increase in strength of fly ash

Fig. 1.9 The effectiveness of steam curing [126]

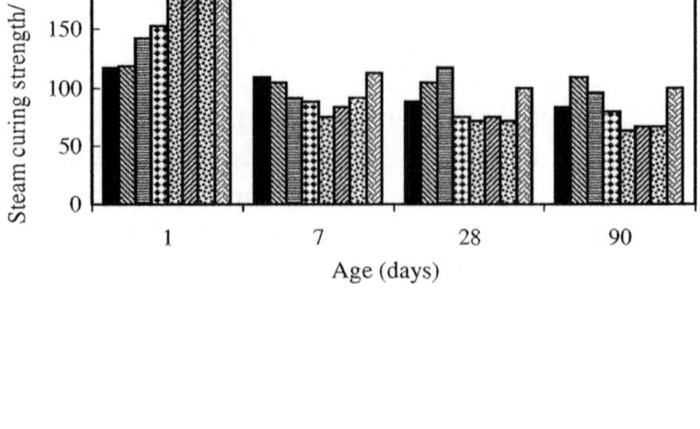

Table 1.14 Rate of strength development of 28-day strength [117]

Age (days)	Strength (MPa)								
	20			40			60		
	Fog	Dry	7F + D	Fog	Dry	7F + D	Fog	Dry	7F + D
1	19	–	–	25	–	–	34	–	–
7	55	71	–	67	80	–	68	79	–
28	100	100	100	100	100	100	100	100	100
150	176	125	104	140	119	105	135	123	117
270	197	114	110	150	125	116	140	113	121
365	209	118	106	162	122	107	146	116	118

7F + D = 7 days fog followed by air dry curing

concrete after 1 year varied between 6 and 22% of the 28-day strength of the reference concrete.

Haque et al. [44] reported that mixes with Alberta fly ashes replacing up to 50% cement showed smaller reduction in strength at lower ash contents when curing was done at 50% relative humidity at room temperature of about 23°C.

Gifford et al. [42] studied the effect of dry curing on unprotected concrete specimens at 0 and 5°C made with 40% cement replacement with fly ash. They found that within 50 h after casting, about 30 and 60% of the mixing water was lost through evaporation from unprotected surfaces on curing at 50 and 10% relative humidity, respectively. In addition to strength loss due to evaporation of water, it was reported that the combination of low curing temperature and the cooling effect of concomitant evaporation of water at low humidity would significantly retard the rate of hydration and thereby strength development. The importance of curing concrete in an enclosed environment particularly at low temperatures is therefore stressed in order to mitigate the effect of water evaporation during the initial hours after the placing of concrete.

Langley et al. [68] reported that minimum duration of moist curing for fly ash concrete was 3 days after which normal curing practices as for ordinary plain concrete might be employed without any significant adverse effects. They also pointed out that long-term strength development in mass fly ash concrete was less influenced by dry curing than much smaller test specimens used in the laboratory.

Gopalan [43] studied the effect of curing conditions on the compressive strength of concretes made with fly ash. Compressive strength was determined for three grades (designated as M1, M2 and M3) of cement concrete having water cement ratios of 0.53, 0.62 and 0.88. Each grade of concrete was then incorporated with two levels (20 and 40%) of fly ash as cement replacement by weight. Cylinders of size 200 × 100 mm were cast. Samples were subjected to standard fog curing at 23°C ± 2°C and relative humidity of 95 ± 3% for 7 days. Samples were also cured (dry curing) at 23°C ± 2°C and relative humidity of 50 ± 3%. Compressive strength results are given in Table 1.15. It is evident from these results that fly ash replacements up to 20% did not have a significant impact on the long term strength of fog cured mixes. The results from the fog cured samples showed that the rate of

Table 1.15 Compressive strength of fly ash concretes [43]

Mix	Compressive strength (MPa)					
	28 days		91 days		180 days	
	Fog	Dry	Fog	Dry	Fog	Dry
M1-00	49.9	45.1	58.8	61.2	51.9	
M2-00	33.1	31.1	40.2	37.0	42.6	34.3
M3-00	18.9	15.4	21.0	18.6	20.8	16.9
M1-20	36.2	32.1	44.0	40.2	53.9	41.1
M2-20	25.2	22.4	33.1	26.6	42.4	27.4
M3-20	13.6	11.3	17.2	14.0	22.0	13.5
M1-40	20.4	16.7	28.3	21.5	35.8	20.9
M2-40	19.2	16.6	27.5	20.8	35.5	19.6
M3-40	10.3	7.5	13.1	9.5	18.1	9.2

Table 1.16 Flexural Strength of hardened concrete [19]

Mixture no.	Flexural strength (MPa)		
	14-day	28-day	91-day
Control	4.9	5.4	5.9
F1	4.4	4.4	5.4
F2	3.9	4.8	5.5
F3	4.0	5.0	5.3
F4	4.1	4.4	5.2
F5	3.5	4.4	5.3
F6	3.5	4.6	5.6
F7	3.9	4.5	5.4
F8	4.6	5.0	6.1
F9	4.3	4.2	5.7
F10	4.1	5.1	5.8
F11	4.8	5.3	6.6

strength-loss was higher for richer mixes. Under drying conditions, there was no noticeable increase in strength after 91 days and strength development pattern did not depend upon the grade of the mix.

1.9.4 Tensile Strength Properties

Carette and Malhotra [19] studied the effect of Canadian fly ashes on the flexural strength of concrete mixes up to the age of 91 days (Table 1.16). It is evident that flexural strength continued to increase with age, indicating pozzolanic action of fly ash.

Siddique [112] investigated the effect of partial replacement of fine aggregate (sand) with varying percentages of Class F fly ash on the splitting tensile strength

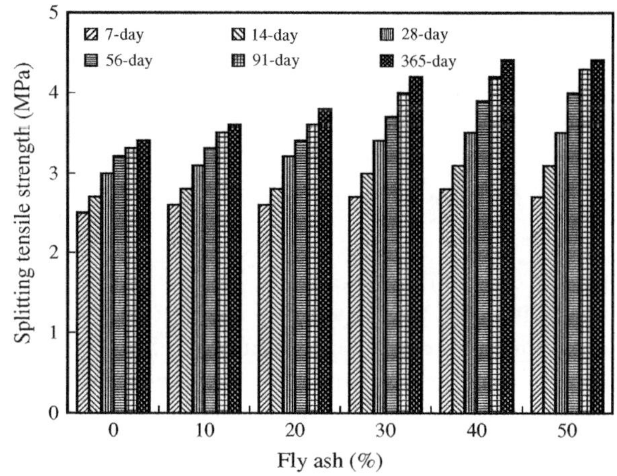

Fig. 1.10 Splitting tensile strength versus fly ash percentage [112]

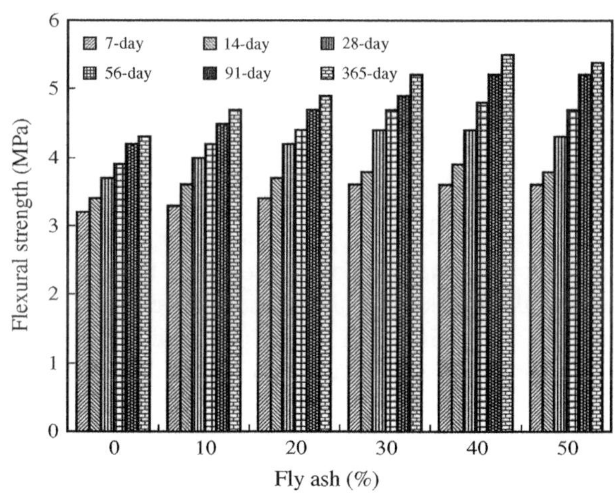

Fig. 1.11 Flexural strength versus fly ash percentage [112]

and flexural strength of concrete. Fine aggregate (sand) was replaced with five levels of percentages (10, 20, 30, 40, and 50%) of Class F fly ash by weight. A control mix without fly ash was proportioned to have a 28-day cube compressive strength of 26.4 MPa. Tests were performed up to the age of 365 days. Splitting tensile and flexural strength results are shown in Figs. 1.10 and 1.11, respectively. Based on the results, it was concluded that (1) splitting tensile strength, and flexural strength of fine aggregate (sand) replaced fly ash concrete specimens was higher than the plain concrete (control mix) specimens at all the ages. The strength differential between the fly ash concrete specimens and plain concrete specimens became more distinct after 28-days; (2) both splitting and flexural strengths continued to increase with age for all fly ash percentages; (3) at all the ages, the maximum splitting tensile strength was observed with 50% fly ash content. It was

3.5 MPa at 28-day, 4.3 MPa at 91-day, and 4.4 MPa at 365-day; (5) maximum flexural strength was found to occur with 50% fly ash content at all ages. It was 4.3 MPa at 28-day, 5.2 MPa at 91-day, and 5.4 MPa at 365-day.

1.9.5 Elastic Properties

Lohtia et al. [72], Ghosh and Timusk [41], Lane and Best (1981), Nasser and Marzouk [90], Langley et al. [68] have reported that the effect of fly ash as replacement of cement on modulus of elasticity of concrete is almost the similar as that of compressive strength. The modulus of elasticity of fly ash concrete is generally lower at an early ages and is slightly higher at late ages.

Lohtia et al. [72] reported that as compared to compressive strength gain, the increase in modulus of elasticity was less with the incorporation of 15–25% Class F fly ash in concrete at the age of 90 days. In general, fly ash increased the modulus of elasticity of concrete when concretes of the same strength with and without fly ash are compared. Ghosh and Timusk [41] reported that for all strength levels, the modulus of elasticity of fly ash concrete was generally equivalent to that of the corresponding reference concrete. They also found that the observed modulus exceeded the value given by the ACI formula, $Ec = 0.043\ W(fc)^{3/4}$ MPa, where W is unit weight of concrete in kg/m^3 and fc is compressive strength in MPa.

Normally fly ash properties affecting the compressive strength of concrete also influence the modulus of elasticity but to a lower extent. Crow and Dunstan [26] reported that like Portland cement concrete, fly ash concrete had increased modulus of elasticity with age. The modulus of elasticity ranged from a low of 18.8 GPa at 28-day to a high of 39.6 GPa at 365-day. The majority of the fly ash concrete had a 28-day Poisson's ratio ranging from 0.14 to 0.25. At elevated temperatures, the modulus of elasticity of fly ash concrete using Saskatchewan lignite fly ash decreased in a similar way as that of plain concrete. Nasser and Marzouk [90] reported that when fly ash concrete was heated from 21 to 232°C in the sealed containers to prevent loss of moisture, the modulus of elasticity was reduced up to 40%.

Langley et al. [68] found that at the age of 28 days, the modulus of elasticity of concretes made with 50% fly ash constituting the cementitious material varied between 27.9 and 36.1 GPa compared to 31.5–36.8 GPa for control concrete mixes. However, at 365 days, fly ash concrete mixes exhibited significant increase in modulus of elasticity compared to control concrete mixes.

Siddique [112] studied the effect of partial replacement of fine aggregate (sand) with varying percentages of Class F fly ash on the modulus of elasticity of concrete. Fine aggregate (sand) was replaced with five levels of percentages (10, 20, 30, 40, and 50%) of Class F fly ash by weight. A control mix without fly ash was proportioned to have a 28-day cube compressive strength of 26.4 MPa. 150 × 300 mm cylinders were cast for modulus of elasticity. Tests were

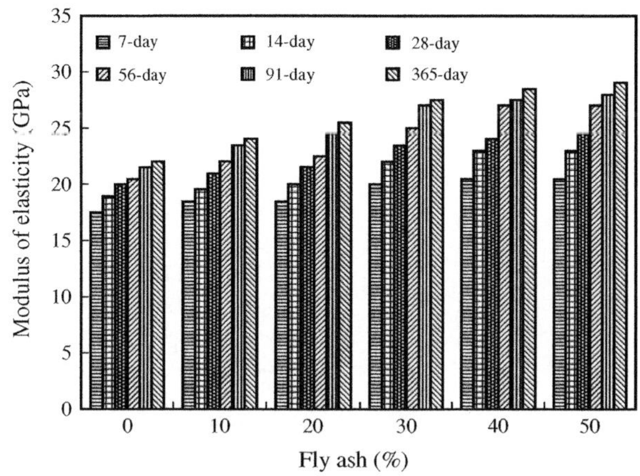

Fig. 1.12 Modulus of elasticity versus fly ash percentage [112]

performed up to the age of 365 days, and results are shown in Fig. 1.12. He concluded that (1) modulus of elasticity of fine aggregate (sand) replaced fly ash concrete specimens was higher than the plain concrete (control mix) specimens at all the ages. The differential between the fly ash concrete specimens and plain concrete specimens became more distinct after 28-days; (2) modulus of elasticity of fine aggregate (sand) replaced fly ash concrete continued to increase with age for all fly ash percentages; and (3) at all ages, the maximum value of modulus of elasticity occurs with 50% fly ash content. It is 24.5 GPa at 28-day, 28.0 GPa at 91-day, and 29.0 GPa at 365-day.

1.9.6 Sorptivity and Porosity

Gopalan [43] studied the sorptivity of concretes made with fly ash. Sorptivity measurements based on capillary movement of water were made on three grades (designated as M1, M2 and M3) of cement concrete having water cement ratios of 0.53, 0.62 and 0.88. Each grade of concrete was then incorporated with two levels (20 and 40%) of fly replacement with cement by weight. 100 × 25 mm discs were made for absorption tests. Samples were subjected to standard fog curing at 23°C ± 2°C and relative humidity of 95 ± 3% for 7 days. Samples were also cured (dry curing) at 23°C ± 2°C and relative humidity of 50 ± 3%. Measured sorptivity results are given in Table 1.17. Based on the test results, he concluded that (1) addition of fly ash influenced the sorptivity of the hardened concrete which strongly depended on the curing conditions. When fog cured concretes of identical strengths were considered, sorptivity of fly ash concretes was found to be lower than that of cement concrete. Under drying-curing conditions, the fly ash concrete had higher sorptivity than cement concrete; (2) addition of fly ash up to 20% did not change the sorptivity characteristics substantially, but for high fly ash concretes, sorptivity was lower than cement concrete of identical strength; under

Table 1.17 Sorptivity ($mm/h^{1/2}$) of fly ash concretes [43]

Mix	Sorptivity ($mm/h^{1/2}$)					
	28 days		91 days		180 days	
	Fog	Dry	Fog	Dry	Fog	Dry
M1-00	10.4	12.1	10.1	10.6	10.5	10.9
M2-00	10.5	12.8	9.7	11.3	11.4	12.1
M3-00	12.0	14.1	10.4	13.2	12.2	15.2
M1-20	11.0	15.3	9.0	11.7	8.0	12.4
M2-20	11.6	15.6	10.0	14.0	10.2	17.1
M3-20	13.4	15.9	9.7	16.9	9.6	18.2
M1-40	8.5	13.3	5.9	16.7	8.9	18.0
M2-40	9.6	18.9	8.0	17.0	9.3	22.0
M3-40	11.4	17.2	7.9	22.2	9.4	21.5

drying conditions, sorptivity increased by an average of 22%; (3) for fly ash replacement of 20%, there was no noticeable difference between fog cured fly ash and cement concretes; (4) for samples under drying conditions, the increase in sorptivity was 19%; and (5) thus, inadequate curing for a 20% fly ash concrete resulted in an increase of 20% in sorptivity. The corresponding value for 40% fly ash concrete was 60%.

Shafiq and Cabrera [111] investigated the influence of curing conditions on the porosity and degree of saturation of normal concrete (100% OPC) and blended cement concrete (OPC/FA). Three different concrete mixes were prepared using 0, 40, and 50% fly ash content as partial substitution for cement. Mix proportion of control concrete was 1:2.33:3.5. Concrete slabs of dimensions 400-mm long, 250-mm wide, and 40-mm thick, were cast in a wooden mould. After 24 h, the moulds were stripped and slabs were cured for 28 days under two different curing conditions, wet cured (in the fog room) and dry cured (at 65% RH at 20°C). At the end of this initial curing, 50-mm-diameter cylindrical discs were cored out from the slabs. After initial curing, samples were exposed to different climatic conditions (75, 65, 40, and 12% relative humidity at a constant temperature of 20°C) until the equilibrium moisture condition was achieved. Total porosity and degree of saturation was determined using vacuum saturation technique [104]. Porosity of concrete samples is shown in Fig. 1.13. Degree of saturation results are given in Table 1.18. They concluded that (1) initial curing conditions affect the total porosity of concrete mixes. For OPC concrete, total porosity of dry cured samples was 5–10% higher than that of their corresponding wet cured samples. For 40 FA concrete, the total porosity of dry cured samples was increased by 9–20% as compared to that of wet cured samples, whereas a significantly higher porosity value of dry cured 50 FA concrete samples was obtained with respect to the corresponding wet cured samples; it ranged from 23 to 40% higher porosity value; (2) average value of the degree of saturation of wet cured samples equilibrated at 75% RH was obtained as 68%, where as for the corresponding dry cured samples it was determined as 56%. Similarly, an average value of 15 and 13% was determined respectively for wet cured and dry cured samples, those were equilibrated at

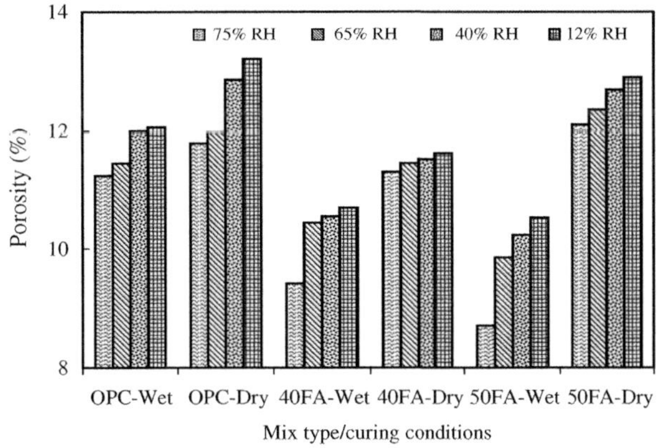

Fig. 1.13 Total porosity of different concrete mixes [111]

Table 1.18 Degree of saturation of OPC and FA blended cement concrete cured in wet and dry conditions [111]

RH (%)	Measured degree of saturation (%)					
	OPC		40 FA		50 FA	
	Wet	Dry	Wet	Dry	Wet	Dry
75	68.87	59.07	70.24	53.23	66.04	51.45
65	59.34	53.01	61.54	49.04	53.03	47.15
40	34.98	29.97	29.59	28.29	27.97	24.54
12	14.71	13.20	11.69	11.61	15.52	12.89

12% RH; and (3) initial curing condition is one significant factor that controls the porosity and pore network formation of different types of concrete.

1.9.7 Shrinkage

Volume changes in concrete due to hydration reactions, drying, wetting and drying, and thermal variation occur even without application of any loads. Under normal field conditions, drying shrinkage is an important phenomenon which induces time dependent deformations and is sometimes called creep under zero loads.

Klieger and Perenchio [63] found insignificant differences between the shrinkage of concretes made with or with out Class F fly ash blended cements.

Ghosh and Timusk [41] reported that shrinkage of concrete containing fly ash was lower than that of concrete without fly ash for the same maximum size of aggregate and for all strength levels. Munday et al. [83] concluded that incorporation of fly ash did not significantly affect the shrinkage and expansions due to drying, wetting/drying and thermal changes in concrete.

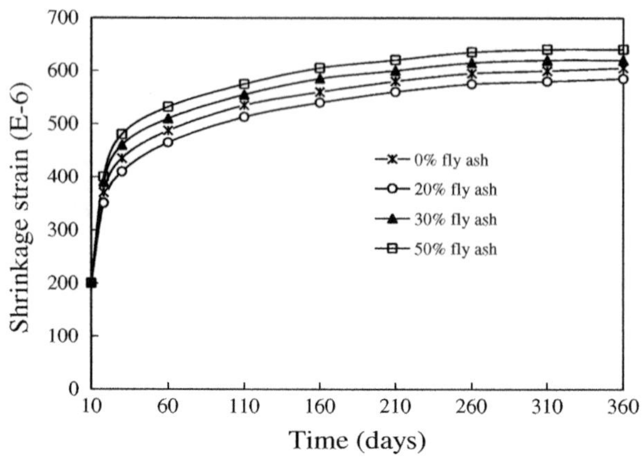

Fig. 1.14 Drying shrinkage of concretes incorporating high-calcium fly ash [129]

Yuan and Cook [129] reported that fly ash concrete containing 30 and 50% fly ash exhibited more shrinkage than either the control concrete or concrete containing 20% fly ash, as shown in Fig. 1.14.

Nasser and Al-Manasser [91] reported that up to 20% replacement of cement with fly ash did not have significant effect on the drying shrinkage of concrete. Increase in drying shrinkage with fly ash addition may occur from increase in the paste volume if water content is the same. However, when water content was reduced, shrinkage was observed minimal.

Haque et al. [44] investigated the shrinkage of concrete containing 40–75% cement replacement with a bituminous fly ash (CaO 10%). They concluded that drying shrinkage of concrete decreased with increase in fly ash content.

Atis et al. [8] assessed the drying shrinkage of mortar mixtures containing high-calcium nonstandard fly ash up to the age of 5 months. Five mortar mixtures including control Portland cement and fly ash mortar mixtures were prepared. Fly ash replaced cement on mass basis at the replacement ratios of 10, 20, 30 and 40%. Water–cementitious materials ratio was 0.4. Mixtures were cured at 65% relative humidity and 20 ± 2°C. They reported that shrinkage of Portland cement mortar at 5 months was 0.1228%. Shrinkage of fly ash mortar decreased with the increase in fly ash content. Shrinkages of mortar containing 10, 20 and 30% fly ash were 25, 37 and 43%, lower than the shrinkage of Portland cement mortar at the end of 5 months. The reduction in shrinkage with the use of fly ash in mortar could be explained by the dilution effect of fly ash. The expansive property of fly ash most probably contributed to the reduction in drying shrinkage.

1.9.8 Creep

Creep is the time dependent strain due to sustained loading. The effect of fly ash on creep of concrete are limited primarily to the extent to which fly ash influences the ultimate strength and rate of strength gain. For the same strength concrete made with and without fly ash, concrete without fly ash would produce less creep strain at all subsequent ages.

Lohtia et al. [71] studied the creep and creep recovery of plain and fly ash concretes at stress-strength ratios of 20 and 35%. Fly ash content was varied between 0 and 25%. They concluded that (1) replacement of 15% of cement with fly ash was optimum with respect to strength, elasticity, shrinkage and creep of fly ash concrete; (2) creep–time curves for plain and fly ash concretes were similar, and creep linearly related to the logarithm of time; (3) with fly ash content up to 15%, increase in creep was negligible. However, slightly higher creep occurred with fly ash content more than 15%; (4) creep coefficients were similar for the materials with fly ash content in the range of 0–25%; and (5) creep recovery was found to vary from 22 to 43% of the corresponding 150-day creep. For replacement beyond 15%, the creep recovery was smaller. No definite trend of creep recovery as a function of stress-strength ratio was observed.

Ghosh and Timusk [41] concluded that fly ash concrete proportioned for equivalent 28-day strength of plain concretes ranging from 20 to 55 MPa exhibited lesser creep than that of the plain concrete. This was possibly because of the favorable effect of pozzolanic reactions due to fly ash which caused higher rate of strength gain after the time of loading. Yuan and Cook [129] showed that high strength concretes containing 20 to 50% high-calcium fly ash exhibited opposite trend in that fly ash concrete containing 30 to 50% ash had more creep than control concrete. However, the 20% fly ash concrete exhibited about the same creep as control concrete.

Nasser and Marzouk [90] observed that creep of concrete with 20% fly ash under sealed conditions showed a continuing reduction with an increase in temperature except at 177°C, while that of unsealed specimens increased with temperature up to 71°C and decreased thereafter within the range of temperature between 21.4 and 232°C.

Nasser and Al-Manasser [91] concluded that concrete containing 20% fly ash in unsealed conditions had on average 72% greater creep than the equivalent sealed concrete for stress-strength rates varying from 10 to 60%. However concrete with 50% fly ash had 13% less creep for unsealed specimens and 39% less for sealed ones. Similar behavior was reported by Carette and Malhotra [20] with various types of Canadian fly ashes including Saskatchewan fly ash. In their study, concrete with 20% fly ash produced consistently lower creep compared to that of control concrete.

1.9.9 Thermal Conductivity

Thermal conductivity (TC) and other thermal transport properties of construction materials are essential in predicting the temperature profile and heat flow through the material. Mineralogical character of the aggregate greatly influences the TC of concrete [92]. Thermal conductivity of concrete depends upon (1) aggregate type;

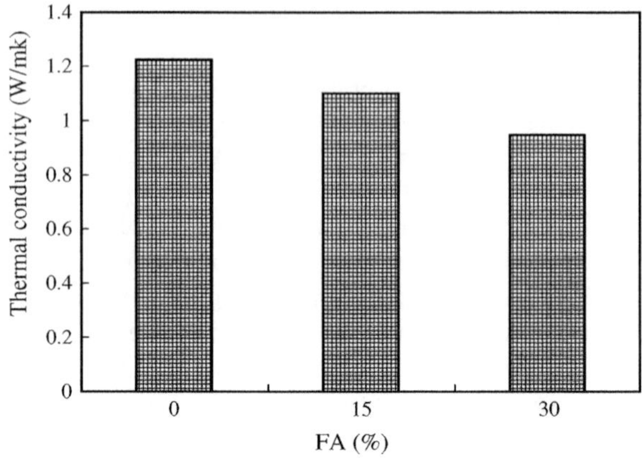

Fig. 1.15 Thermal conductivity of concrete with fly ash [27]

(2) moisture content; and (3) porosity. Aggregate with crystalline structure shows higher heat conduction than amorphous and vitreous aggregate of the same composition. Porosity and moisture content are other important factors that influence the TC of concrete.

Demirboğa [27] studied the influence of partial replacement of cement with fly ash on the TCof concrete. Cement was replaced with 0, 15, and 30% fly ash. Variation of thermal conductivity with fly ash is shown in Fig. 1.15. It can be seen from this figure that TC decreased with the increase in fly ash content. For 15 and 30% FA replacement, the reductions were 12 and 23%, respectively, compared to the corresponding control specimens.

Demirboğa et al. [28] investigated the TCof HVFA concrete at the age of 28 days. Cement was replaced with 0, 50, 60, and 70% of Class C fly ash. They concluded that TC of concrete decreased to 32, 33, and 39% for 50, 60 and 70% fly ash replacement, respectively.

1.10 Durability Properties of Concrete made with Fly Ash

1.10.1 Permeability

Permeability is the key to the durability of concrete exposed to harsh environments. Permeability of concrete is very important in determining the rates of mass-transport relevant to destructive chemical action. Mehta [79] identified water as either the agent of destruction or a necessary participant in many different types of deterioration in concrete such as frost damage, leaching of CH, acid attack, sulfate attack, corrosion of reinforcement, and alkali–aggregate reaction.

The permeability of concrete primarily depends on the size, distribution and continuity of the pores of the hydrated paste of the concrete. The important factors which control the pore structure of the paste are degree of hydration and water cement ratio.

In ordinary Portland cement concrete, calcium hydroxide formed during hydration of Portland cement can be leached out over a period of time. This creates channels available for the ingress of water and deleterious salt solutions. However, when fly ash is incorporated, it reacts with the calcium hydroxide in the water filled capillary channels to produce calcium silicate and aluminate hydrates of the same or similar type that are formed in the normal hydration of cement. Thus the calcium hydroxide is consumed in the pozzolanic reactions and converted to water insoluble hydration products. The reactions reduce the risk of leaching calcium hydroxide. The reaction products also tend to fill capillaries, thereby reducing permeability to aggressive fluids such as chloride or sulfate solutions.

The addition of fly ash results in considerable pore refinement. It transforms bigger pores into smaller ones due to the formation of pozzolanic reaction products concomitant with the progress of cement hydration. Since strength and impermeability are inversely related to the volume of pores larger than 100 Å in the hydrated paste, the phenomenon of pore refinement in fly ash concrete leads to the improvement in these characteristics.

Permeability of concrete containing fly ash can be determined by following methods

- Water permeability
- Rapid chloride permeability
- Air/gas permeability

1.10.1.1 Water Permeability

Rodway and Fedriko [105] investigated the water permeability of concrete incorporating Class C fly ash for 68% cement replacement. They reported permeability of fly ash concrete as 3.65×10^{-12} m/s. Ellis et al. [36] observed decrease in permeability of concrete with increase in both either Class C or Class F fly ash contents for a fixed amount of cement. They concluded that (1) concrete containing Class F was more effective than concrete with Class C fly ash in reducing permeability; and (2) permeability values of Class F fly ash concrete were either comparable to or superior to those achieved by using ether silica fume or GGBS.

Shafiq and Cabrera [111] investigated the effect of curing conditions on the water permeability of normal concrete and blended cement concrete (OPC/FA). Concrete mixes were made with 0, 40, and 50% fly ash content as partial substitution for cement. Mix proportion of control concrete was 1:2.33:3.5. Concrete slabs of dimensions 400-mm long, 250-mm wide, and 40-mm thick, were cast, and were cured for 28 days under two different curing conditions, wet cured (in the fog room) and dry cured (at 65% RH at 20°C). cylindrical discs of 50-mm diameter were cored out from the slabs after the initial curing. Samples were then exposed

Table 1.19 Coefficient of water permeability, Kop ($\times 10^{-19}$ m^2) and permeability ratio of OPC and fly ash blended cement concrete cured in wet and dry conditions [111]

RH (%)	Concrete type/curing condition								
	OPC			40 FA			50 FA		
	Wet	Dry	Dry/wet	Wet	Dry	Dry/wet	Wet	Dry	Dry/wet
75	11.0	210.0	19.1	2.8	442.0	157.9	10.0	3080.0	309.5
65	35.0	305.0	8.7	11.4	468.0	41.1	20.9	5140.0	245.9
40	166.0	658.0	3.9	22.8	678.0	29.7	62.0	12300.0	198.4
12	764.0	3870.0	5.1	425.0	6830.0	16.1	231.0	13300.0	57.6

to different climatic conditions (75, 65, 40, and 12% relative humidity at a constant temperature of 20°C) until the equilibrium moisture condition was achieved. Water permeability of concrete was determined by a penetration method. Coefficients of water permeability are given in Table 1.19. Based on the test results, they concluded that (1) water permeability ratio as calculated for OPC and fly ash blended concrete followed a similar trend but on different scales. For initially dry cured OPC concrete, the coefficients of water permeability were 2–19 times higher than the coefficients obtained for the initially wet cured concrete samples. In contrast, the coefficients of water permeability of dry cured fly ash blended cement concrete were 16–210 times greater than the coefficients of the corresponding wet-cured concrete samples; (2) a large difference in the coefficient of fluid permeability and/or diffusion between the dry cured and wet cured concrete was observed when concrete samples were equilibrated at 75% RH. In contrast a small difference between the fluid transport coefficients of dry cured and wet cured concrete samples was observed when they were equilibrated at 40 and 12% relative humidity (RH); and (3) there was approximately 12% difference between the average degree of saturation of dry cured concrete samples equilibrated at 75% RH and the average degree of saturation of the corresponding wet cured concrete samples, which resulted in the much higher values of the fluid permeability and diffusion coefficients of dry cured samples relative to wet-cured samples. This is probably due to the fact that the wet cured concrete samples when equilibrated at higher RH, have a tight pore network with small pore diameters and exhibit very high degree of saturation.

1.10.1.2 Chloride Permeability

Bilinear and Amphora (1992) measured water and chloride permeability of concretes having 55–60% cement replacement with various sources of fly ash. They reported coefficient of water permeability of fly ash concretes in the range of 1.6×10^{-14} to 5.7×10^{-13} m/s. The values of chloride were less than 650 C at 91 days.

Naik et al. [86] evaluated the influence of addition of large amounts (50 and 70% cement replacement) of Class C fly ash on the chloride permeability of

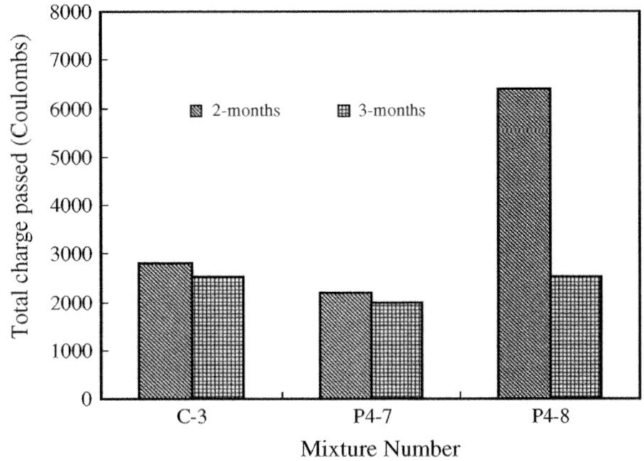

Fig. 1.16 Effect of fly ash addition on chloride permeability of concrete [86]

concrete. Concrete mixtures were designated as C-3 (0% fly ash), P4-7 (50% fly ash) and P4-8 (70% fly ash). Chloride permeability was determined in accordance with ASTM C1202, and chloride permeability results are shown in Fig. 1.16 Chloride permeability decreased with age. At the age of 2 months, all concrete mixtures except the 70% fly ash mixture exhibited moderate (2,000–4,000 C) permeability in accordance with ASTM C1202 specifications. The 50% fly ash concrete mixture showed lower permeability relative to the no-fly ash concrete at all ages. The 70% fly ash mixture also performed better than that of the no-fly ash concrete after 3 months.

Chindaprasirt et al. [22] examined the effect of fly ash fineness on the chloride permeability of concrete. Three fly ash finenesses: coarse, medium and fine were used. The coarse fly ash was 100% original fly ash (100 FA). The medium fly ash was the 45% fine portion of the original fly ash (45FA). The fine fly ash was the 10% fine portion of the original fly ash (10 FA). Three concrete mix series viz. a low, normal and a high strength concrete mix series were made. For the low- and normal-strength concrete, the water-to-cement ratios of 0.54 and 0.48, respectively, were used for PC mixes. For the high strength concrete, the water-to-cement ratio was 0.25 with the use of superplasticizer. Rapid chloride permeability (Coulomb charge) test was conducted at the age of 28 days as per ASTM C1202, and test results are shown in Fig. 1.17. Based on the test results they concluded that (1) coulomb charge for the PC concrete was higher than those of the fly ash concrete in the same group; (2) for low-strength concrete, the coulomb charge of the PC1 concrete was 5,800 indicating a rather poor chloride penetration characteristic. Incorporation of fly ashes resulted in drastic reductions in the coulomb charges. The Coulomb charge of the original fly ash concrete (100FA1) was reduced to 2,700, indicating moderate chloride penetration resistance. The charge was further reduced with the increase in the fly ash fineness. The charges of the 45FA1 and 10FA1 concrete are reduced to 2,300 and 700, respectively; (2) for the normal strength concrete, the coulomb charge of the PC2 concrete was 4,200 and the charges of the fly ash concrete were 1,250, 800 and 500 for the 100FA2, 45FA2

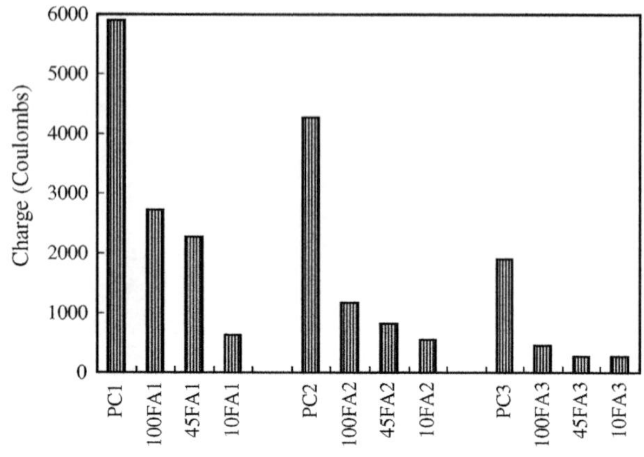

Fig. 1.17 Coulomb charge of concrete at the age of 28 days [22]

and 10FA2 mixes, respectively; and (3) for the high strength concrete group, the coulomb charge of the PC3 concrete was slightly higher than 1,000 and the charges of the fly ash concrete were only 500, 250 and 240 for the 100FA3, 45FA3 and 10FA3 mixes, respectively. Even at this high strength level, the reduction in the coulomb charge owing to the incorporation of the fly ash especially with the fine fly ashes was quite evident.

1.10.1.3 Gas permeability

Kasai et al. [60] studied the air/gas permeability of mortar made with blended cements containing fly ash and blast furnace slag. The study was important from durability aspects of concrete with respect to carbonation. Up to the age of 7 days, blended cement mortars exhibited more permeability than plain cement mortars. However, as curing age increased, the permeability of blended cement mortars decreased. In general terms, the permeability was found to be directly related to the compressive strength development of the mortars.

Shafiq and Cabrera [111] investigated the effect of curing conditions on the oxygen permeability of normal concrete (100% OPC) and blended cement concrete (OPC/FA). Mix proportion of control concrete was 1:2.33:3.5. Fly ash content was 0, 40, and 50% as partial substitution for cement. The gas permeameter developed by Cabrera and Lynsdale [18] was used to measure the oxygen permeability of cylindrical concrete samples of 50-mm diameter and 40-mm thick. Water permeability of concrete was determined by a penetration method. Coefficients of oxygen permeability are given in Table 1.20. Based on the test results, they concluded that (1) oxygen permeability ratios as calculated for OPC and fly ash blended concrete followed a similar trend but on different scales. For initially dry cured OPC concrete, the coefficients of oxygen permeability were 2–19 times higher than the coefficients obtained for the initially wet cured concrete samples. In contrast, the coefficients of oxygen permeability of dry cured fly ash blended

Table 1.20 Coefficient of oxygen permeability, Ko × 10^{-19} m^2 and permeability ratio of OPC and fly ash blended cement concrete cured in wet and dry conditions [111]

RH (%)	Concrete type/curing condition								
	OPC			40 FA			50 FA		
	Wet	Dry	Dry/wet	Wet	Dry	Dry/wet	Wet	Dry	ry/wet
75	13.3	230.0	17.3	2.5	518.0	208.9	4.1	906.0	220.4
65	62.1	308.0	5.0	18.2	665.0	36.5	31.8	1070.0	33.6
40	192.0	486.0	2.5	104.0	1720.0	16.5	91.8	2270.0	24.7
12	337.0	772.0	2.3	136.0	2210.0	16.3	117.0	2690.0	23.0

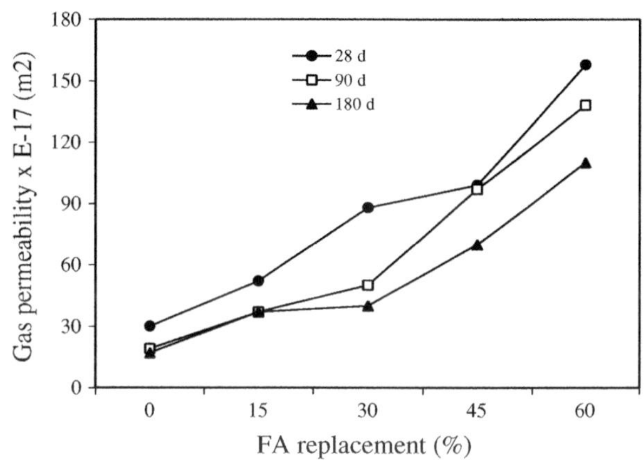

Fig. 1.18 Influence of FA on nitrogen gas permeability of HPC [48]

cement concrete were 16–210 times greater than the coefficients of the corresponding wet cured concrete samples; (2) a large difference in the coefficient of fluid permeability and/or diffusion between the dry cured and wet cured concrete was observed when concrete samples were equilibrated at 75% RH.

Hui-sheng et al. [48] examined the influence of fly ash on the nitrogen gas permeability of high-performance concrete (HPC) up to the age of 180 days. Control mixture contained cement (550 kg/m^3), fine aggregate (687 kg/m^3), and coarse aggregates (1,030 kg/m^3). Fly ash replacement levels were 0, 15, 30, 45 and 60% by weight of cement, with water-binder ratio of 0.30. Influence of FA on gas permeability of HPC is shown in Fig. 1.18. It can be observed that gas permeability coefficients significantly increased with the increase in FA replacement. According to Bamforth [13] and McCarthy and Dhir [77], fly ash can have significant reduction in permeability of normal concrete, however, their effects on HPC were not obvious. For instance, Thomas (1992) reported that fly ash had a significant reduction in gas permeability of normal concrete. However, according to Khan [62], incorporation of fly ash in HPC only had marginal reduction in oxygen permeability. The underlying mechanism may be attributed to the fact that HPC is much denser and hydrates slower than normal concrete, thus the filler effect and pozzolanic effect in HPC may not work as well as that in normal concrete.

Thomas and Matthews [122] investigated oxygen permeability of fly ash concrete. Concrete mixtures were made with 15, 30 and 50% fly ash as cement replacements, exhibited reduction in the permeability values by 50, 60 and 86%, respectively, compared to concrete with out fly ash.

1.10.2 Carbonation

Carbonation is a chemical reaction that takes place between portlandite and CO_2. Portlandite is present in the hydrated cement. The gas CO_2 is present in the atmosphere. When CO_2 penetrates into the hardened concrete, it reacts with portlandite in the presence of moisture forming $CaCO_3$. This is expressed as

$$Ca(OH)_2 + CO_2 \rightarrow CaCO_3 + H_2O$$

Carbonation requires the presence of water because CO_2 must dissolve in water and form $H_2CO_3^-$. According to Taylor [120], OH^- and Ca^{2+} ions required by these reactions are obtained by the dissolution of CH and decomposition of the hydrated silicate and aluminate phases. Decalcification of the C–S–H is evidenced initially by a reduction in the Ca/Si ratio, and ultimately by conversion into a highly porous form of silica. Phenolphthalein tests show that the pH falls to 8.5 or below, accelerating the rate of corrosion.

The rate at which concrete carbonates depends upon (1) its permeability; (2) degree of saturation with water; and (3) mass of calcium hydroxide available for reaction; (4) relative humidity; and (5) temperature of the environment where concrete is placed.

Gebauer [39] reported that an increase in water–cement ratio of concrete mix resulted in an increase in the depth of carbonation. Kasai et al. [60] studied the carbonation of mortar specimens made with different types of cement and fly ash after 7 days of moist curing. It was observed that (1) carbonation was observed to progress rapidly up to 3 months and after that it slowed down; and (2) greater the coefficient of permeability of the specimen, the greater was its susceptibility of CO_2 attack manifested in terms of increased depth of carbonation. The specimens made with fly ash cement exhibited greater carbonation effect than ordinary Portland cement specimens.

Nagataki and Ohga [85] found that rate of carbonation increases with fly ash content of mortars. However, longer curing periods reduce the rate of carbonation, and mortars with greater fly ash contents are more sensitive to the length of the curing period. These results are consistent with the slower hydration and development of a discontinuos pore system in fly ash cement pastes.

Schubert [110] believed that the consumption of $Ca(OH)_2$ in the pozzolanic reaction acts to increase the rate of carbonation, while the blocking of capillary pores to acts to decrease it. Kokubu and Nagataki [64] reported the 20-year data on outdoor exposure of concretes at various locations show that the carbonation depth increases with water-to-cementitious ratio, slump, and level of fly ash replacement. Carbonation depth also increases with reductions in the total content of cementitious materials.

Ho and Lewis [46] investigated the carbonation rates of three types of concrete mixes (1) plain concrete; (2) the second containing a water reducing admixture; and (3) third in which fly ash was used to replace part of the cement. Accelerated carbonation was induced by storing specimens in an enriched CO_2 atmosphere (4%) at 20°C and 50% RH for 8 weeks. One week under these conditions was approximately equivalent to 1 year in a normal atmosphere (0.003% CO_2). They concluded that (1) concretes having the same strength and water-to-cement ratio do not necessarily carbonate at the same rate; (2) concrete containing fly ash showed significant improvement in quality when curing was extended from 7 to 90 days. This improvement was much greater than that achieved for the plain concrete; and (3) depth of carbonation is a function of the cement content for concretes moist-cured for 7 days. However, with further curing to 90 days, concrete containing fly ash showed a slower rate of carbonation as compared to plain and water-reduced concretes.

Nagataki et al. [84] reported a direct relationship between 28-day compressive strength and depth of carbonation irrespective of fly ash replacement in concrete, and also mentioned that the extent of carbonation decreased with an increase in compressive strength.

Buttler et al. [17] studied the carbonation behavior of cement sand mortars made with 25% pulverized fuel ash or fly ash replacing cement and at water cement ratios of 0.35 to 0.55. They concluded that more carbonation occurred at lower water contents when the fly ash concrete was desiccated with calcium chloride ($CaCl_2$). However, in the case of non-desiccated specimens, the effect of fly ash on carbonation at lower water content was insignificant. Of course in non-desiccated specimens, more carbonation was observed at high water cement ratios.

Joshi et al. [59] found that up to about 7 days, the extent of carbonation measured by the affected depth from the outer surface in concrete, after subjecting the specimen to 4% CO_2 at 20°C and 50% RH, was more in concrete containing fly ash than the control concrete without fly ash. However, after 90 days curing, the trend reversed in that the fly ash concrete exhibited less carbonation than the control concrete.

Atis [7] carried out an accelerated carbonation test (using a controlled environment) to assess the carbonation of fly ash (FA) concrete. The concrete mixtures were made with 0, 50 and 70% replacement of normal Portland cement (NPC) with fly ash. Water-cementitious material ratios ranged from 0.28 to 0.55. The proportions of the control NPC concrete mixture (M0) were: 1:1.5:3 NPC, sand and gravel, respectively. The quantity of NPC was 400 kg/m^3, w/c was 0.55. M0 mixture was conventional Portland cement concrete. M1, M2, M3 and M4 concrete mixtures were made with fly ash. M1 and M3 mixtures also contained superplasticizer. Test results are given in Table 1.21. It is evident from these results that fly ash concrete made with 70% replacement ratio was carbonated

Table 1.21 Accelerated carbon depth of concrete cured at 100% RH with 20°C [7]

Mix name	Accelerated carbon depth (mm)			
	3 days	7 days	28 days	3 months
M0 (0% FA, w/c 0.55)	9.10	7.40	4.50	3.30
M1(70% FA, w/c 0.28)	13.30	10.90	6.50	4.60
M2(70% FA, w/c 0.29	13.80	11.70	7.30	5.00
M3(50% FA, w/c 0.33	8.70	8.40	3.20	1.80
M4(50% FA, w/c 0.30	9.60	7.50	2.10	1.60

Table 1.22 Experimental results of carbonation of concrete [52]

Mix no	Depth of carbonation (mm)							
	Concentration of CO_2 (20%)				Concentration of CO_2 (20%)			
	3d	7d	14d	28d	3d	7d	14d	28d
C1	7.7	10.4	12.7	14.8	7.8	11.2	12.3	13.8
C2	12.2	15.7	19.4	22.4	11.3	14.1	15.8	17.7
C3	10.3	12.4	14.1	15.9	8.1	11.4	12.4	13.7

more than that of 50% fly ash replacement concrete and normal Portland cement (NPC) concrete. In contrast, 50% fly ash replacement concrete showed lower or similar carbonation to NPC concrete. Before exposing the concrete to the accelerated carbonation testing, the longer initial curing period resulted in lower carbonation depth. The effect is more marked with moist curing.

Jiang et al. [52] studied the carbonation of concrete incorporating large volumes of low-quality fly ash (LVLQFA). Three mixtures were prepared. First mixture (C1) was control, second mixture (C2) and third mixture (C3) contained 40% fly ash of the total cementitious materials. Third mixture (C3) also contained activator (11.6%). Tests were conducted up to the age of 28 days with two concentrations of $CO_{2.}$ Carbonation results are given in Table 1.22. It is evident that at the 20% CO_2 concentration, carbonation depth of the LVLQFA concrete was greater than of control concrete, especially at the early-age carbonation. The activator can improve the carbonation resistance of LVLQFA concrete. At 28 days, the carbonation depth of the LVLQFA concrete was close to the control concrete. At the concentration of 3% CO_2, the depth of carbonation of LVLQFA concrete without an activator was greater than others. The carbonation depth of LVLQFA concrete with an activator was close to the control concrete. It is seen, therefore, that the concentration of CO_2 used in the experiment has considerable effect on estimating the carbonation resistance of LVLQFA concrete.

Hui-sheng et al. [48] studied the influence of fly ash on the carbonation of high-performance concrete up to the age of 56 days. Cement, fine aggregate, and coarse aggregates contents were 550, 687, and 1,030 kg/m^3, respectively. Fly ash replacement levels were 0, 15, 30, 45 and 60% by weight of cement, with water–binder ratio of 0.30. Influence of fly ash on carbonation resistance of HPC is shown in Fig. 1.19. Carbonation phenomenon of the control HPC was not observed at the

Fig. 1.19 Influence of FA on carbonation depth of HPC at w/b ratio of 0.30 [48]

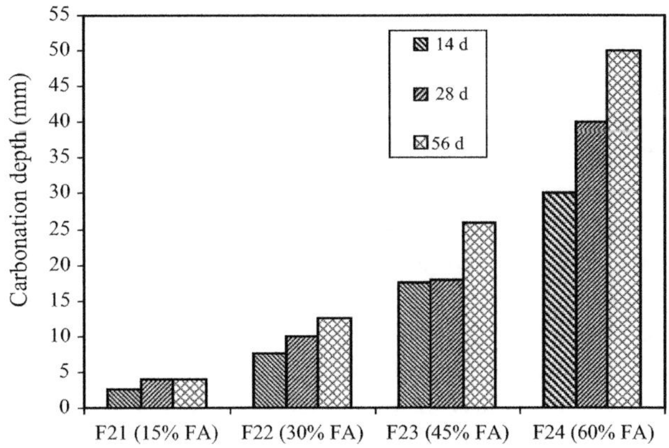

end of each carbonation period. It can be seen that the incorporation of fly ash decreased the carbonation resistance of HPC, but at significantly different levels. For HPC with fly ash at w/b of 0.30, carbonation depth significantly increased with the increase of fly ash replacement.

1.10.3 Corrosion Resistance

The corrosion of reinforcement is one of the most widespread causes of deterioration in concrete [120]. Nielsen (1979) has given the anodic equation for the corrosion of ions as:

$$Fe \rightarrow Fe^{2+} + 2e^{-}$$

And the cathodic equation as

$$O_2 + 2H_2O + 4e^{-} \rightarrow 4OH^{-}$$

For the usual case when oxygen is available for reaction. Corrosion in the absence of oxygen is much slower. Thus, the availability of both oxygen and water is critical to the corrosion process. The net reaction is

$$Fe^{2+} + 2OH^{-} \rightarrow Fe\,(OH)_2$$

Corrosion of reinforcing steel embedded in fly ash concrete has great cause of concern for applications of concrete in the construction of reinforced and prestressed concrete structures, particularly, those subject to chloride induced corrosion of steel caused by the use of deicing salts or sea water exposure of structures in a marine environment. The bond between the surrounding concrete and steel reinforcement and the high alkalinity of the concrete provide protection from corrosion.

In a hydrated Portland cement paste, about 20% $Ca(OH)_2$ by weight of the hydration products is present to provide the reserve basicity for steel protection. Diamond [29] reported that the high alkalinity of the pore solution in cement paste primarily results from the presence of sodium and potassium ions rather than from the presence of calcium hydroxide alone. In the two fly ash systems examined, it was observed that pH of pore solution was reduced from 13.75 in a control system to about 13.55 in the presence of fly ash. Under highly alkaline conditions i.e. pH lager than 11.5 of pore solution in concrete, a protective iron oxide film forms on the surface of reinforcing steel that makes it passive against further corrosion. This passive iron oxide layer is susceptible to the attack by chloride ions in concrete or to carbonation when the pH of the surrounding concrete is reduced below 11.0. Once this passive layer is destroyed, a galvanic cell can form between different areas on reinforcing bars causing reduction at anodic area. Furthermore, the rate and extent of corrosion of embedded steel depends on the electrical conductivity of the surrounding concrete and the permeation of moisture and air through the concrete. In practice, adequate cover of high quality and impervious concrete over the reinforcing steel has been found to provide adequate protection against corrosion.

Saraswathy et al. [107] investigated the influence of activated fly ash on the corrosion resistance of mortar by using anodic polarization technique. Mortar specimens with 0, 10, 20, 30, 40 and 50% fly ash as replacement of cement w/c ratio of 0.45, were subjected to anodic polarization. Mild steel rods were embedded in cylindrical mortar (1:3) specimens of size 58 mm diameters and 60 mm height. Anodic polarization studies have been carried out in 3% NaCl solution. The current flowing at +300 and +600 mV were recorded for mild steel embedded in OPC and OPC replaced by various fly ash systems at 10, 20, 30, 40 and 50% replacement levels and the corresponding magnitude of current for a fixed duration of 12 h are given in Table 1.23. From the table it can be seen that for OPC, the current measurement was 0.43 and 1.04 mA. For as-received fly ash (AFA), the current measurement at 30% replacement was 0.34 and 0.99 mA. On the other hand thermally activated fl ash (TFA) and chemically activated fl ash (CFA) systems showed superior properties even up to 50% replacement level. For example in the case of CFA system, the current measured was found to be 0.40 and 0.56 mA at 50% replacement level. These data clearly illustrated that activated fly ashes improved the corrosion–resistance properties even up to 50% replacement level.

Andrade [4] tested concrete mixes with and without fly ash for corrosion using polarization resistance techniques. The addition of fly ash promoted the corrosion of steel in mortars but had no effect on concrete specimens. The decrease in the alkalinity due to introduction of fly ash was reported to have a major effect in promoting corrosion in fly ash mortar mixes.

Jiang et al. [52] studied the carbonation of concrete incorporating large volumes of low-quality fly ash (LVLQFA). They concluded that (1) corrosion resistance of LVLQFA concrete in 5% Na_2SO_4 and 5% HCl solution was better than that of the control concrete; and (2) an activator can improve the corrosion resistance of steel

Table 1.23 Anodic polarization test parameters for OPC and various activated fly ash blended cement concrete in 3% NaCl solution [107]

System	As-received fly ash (AFA)		Physically activated fly ash (PFA)		Thermally activated fl ash (TFA)		Chemically activated fly ash (CFA)	
	+300 mV shift current (mA)	+600 mV shift current (mA)	+300 mV shift current (mA)	+300 mV shift current (mA)	+300 mV shift current (mA)	+300 mV shift current (mA)	+300 mV shift current (mA)	+300 mV shift current (mA)
OPC (100%	0.43	1.04	0.43	1.04	0.43	1.04	0.43	1.04
OPC + 10% FA	0.24	0.55	0.16	0.40	0.06	0.08	0.05	0.07
OPC + 20% FA	0.27	0.75	0.21	0.62	0.09	0.14	0.07	0.10
OPC + 30% FA	0.34	0.99	0.30	0.83	0.18	0.26	0.11	0.15
OPC + 40% FA	0.46	1.08	0.38	0.99	0.24	0.40	0.19	0.20
OPC + 50% FA	1.26	2.50	1.00	1.50	0.40	0.66	0.40	0.55

reinforcement in LVLQFA concrete caused by carbonation and seawater. The corrosion resistance of steel reinforcement in LVLQFA concrete with an activator was close to the control concrete.

Chalee et al. [21] studied the effect of W/C ratio on covering depth required against the corrosion of embedded steel of fly ash concrete in marine environment up to 4-year exposure. Fly ash was used to partially replace Portland cement type I at 0, 15, 25, 35, and 50% by weight of cementitious material. Water-to-cementitious material ratios (w/c) of fly ash concretes were varied at 0.45, 0.55, and 0.65. Tests were conducted for corrosion of embedded steel bar after being exposed to tidal zone for 2, 3, and 4 years. Based on the tests, they concluded that (1) covering depth required for the initial corrosion of embedded steel bar in concrete could be reduced with fly ash; (2) decrease in W/C ratio resulted in reducing the covering depth required for initial corrosion, and generally affected the cement concrete rather than the fly ash concrete; (3) fly ash concretes with 35 and 50% replacements and W/C ratio of 0.65, provided the result of corrosion resistance at 4-year exposure as good as cement concrete with W/C ratio of 0.45; and (4) concrete with compressive strength of 30 MPa could reduce the covering depth from 50 to 30 mm by using fly ash to replace Portland cement of 50%.

1.10.4 Freezing and Thawing Resistance

Freezing–thawing resistance of concrete depends on both cement paste and aggregate. The actual behaviour in a particular situation depends upon the location of escape boundaries, pore structure of the system, degree of saturation, the soundness of aggregate, degree of hydration, strength of binding paste, and tensile strength of the paste. Escape boundaries are provided by entraining numerous air bubbles in the paste. The pore structure is controlled by the water–cement ratio and curing.

Schiepl and Hardtle [108] concluded that main effect of pozzolanic reactions due to fly ash addition is to change the pore size distributions, the total porosity remaining mostly uncharged. In particular, the capillary water along with calcium hydroxide liberated during cement hydration is largely consumed in pozzolanic reactions. The phenomenon of pore refinement as a result of pozzolanic reactions leads to the breaking of continuity of the capillary pore structure. Malhotra et al. [74] reported that the pozzolanic reactions alter the pore structure of the cement paste and thus densify the transition zone between the paste and aggregates.

Air entrainment has the greater role on freeze–thaw durability of concrete mixtures. Perencho and Klieger [98] found that resistance to freezing and thawing in water was comparable and excellent for air-entrained concretes made with either Type 1 or Type II cements containing fly ash. Majko and Pistilli [73] found that properly air-entrained concretes made with Class C fly ash showed excellent freeze/thaw durability even at high fly ash contents.

Gebler and Klieger [40] investigated the influence of air entrainment on the air-void parameters of hardened concretes made with both Class C and Class F fly ashes. The concretes were cast after 30, 60 and 90 min of initial mixing. They found that air void spacing factors were almost constant for the majority of concretes containing fly ash which were cast after 90 min. At early periods of casting concretes with Class F fly ash exhibited greater variability in air void parameters than concretes with Class C fly ash.

Yuan and Cook [129] found that up to 400 freeze–thaw cycles, there was no significant difference in mass loss or dynamic modulus for air-entrained concrete with or with out Class C fly ash. Concrete with 20% fly ash exhibited better frost resistance than the control concrete. After 400 cycles, concrete made with 50% fly ash showed significant scaling damage. Virtanen [124] evaluated the freezing and thawing resistance concrete made with fly ash. They concluded that (1) air content has the greatest influence on the freeze–thaw resistance of concrete; (2) addition of fly ash had no major influence on the freeze–thaw resistance of concrete if the strength and air content are kept constant.

Carette and Malhotra [19] conducted freeze–thaw tests on concrete mixes with number of Canadian fly ashes. Chemical composition of fly ashes has already been given in Table 1.7. Durability factor for fly ash concrete mixes was determined at 300 cycles and results are given in Table 1.24. It can be seen that with constant air content around 6.4%, all the fly ash concretes had about the same durability factor as control concretes.

Joshi [55] investigated the freezing and thawing resistance of concrete made with 50% fly ash as partial replacement of cement. He concluded that (1) no significant differences were observed in freeze–thaw performance of air-entrained high volume fly ash concrete, and the concrete containing both air-entraining and water-reducing agents; (2) it exhibited relative dynamic modulus of elasticity values in excess of 60% after 300 cycles; and (3) fly ash concrete mixes showed

Table 1.24 Freeze–thaw durability factors for fly ash concretes [19]

Mix no.	Cement (kg/m^3)	Air (%)	Durability factor
Control	295	6.4	98.1
F1	236	6.2	96.4
F2	237	6.2	98.8
F3	237	6.2	96.8
F4	238	6.3	98.8
F5	237	6.4	97.2
F6	238	6.5	96.8
F7	239	6.1	97.6
F8	236	6.2	96.9
F9	236	6.4	97.6
F10	237	6.5	97.2
F11	237	6.6	95.8

some scaling after 150 to 200 freeze thaw cycles and exhibited about 2% weight loss at the end of the 300 cycles.

Malhotra et al. [75] examined the durability of high volume Class F fly ash concretes and concluded that superplasticized and air-entrained HFCC showed satisfactory durability against freeze–thaw attack. Schmidt [109] reported that mass loss of concrete cubes subjected to 100 cycles of freezing and thawing was higher for the concretes made with inter-ground fly ash cement than for the Portland cement controls.

Joshi et al. [58] investigated freezing and thawing resistance of air-entrained and superplasticized concretes containing with 40 to 60% fly ash as partial replacement of cement. The specimens were moist cured for 14 days prior to freeze–thaw tests. The air content of the mixes was maintained at $6 \pm 1\%$. The use of fly ash in concrete increased the air entraining agent demand to maintain constant air content. They concluded that most of the mixes had relative dynamic modulus of elasticity values in excess of 60% after 300 freeze–thaw cycles. In general, all the concrete mixes containing Alberta fly ash and with air content more than 5% exhibited acceptable freeze–thaw durability performance.

1.10.5 Alkali–Silica Reaction

Alkali–silica reaction (ASR) is caused by a reaction between the hydroxyl ions in the alkaline cement pore solution in the concrete and reactive forms of silica in aggregate such as chert, quartzite, opal, strained quartz crystals. Alkali ions (Na^+ and K^+) from cement, mixing water increase the concentration of OH^- ions in the pore solution. The OH^- dissolves amorphous silica (SiO^{-2}) in the aggregate, forming a gel of variable chemical composition. The gel imbibes water from the environment, expanding and generating hydraulic stresses in the cement paste which may cause cracking.

According to Mehta [79], the solubility of alkali silicate gel in water accounts for its mobility from the interior of the aggregate to micro-cracked regions within the aggregate or the cement paste. Continued availability of water to the gel causes further expansion, resulting in further crack growth.

The ASR reaction is the same as the Pozzolanic reaction which is a simple acid–base reaction between calcium hydroxide, also known as Portlandite, or ($Ca(OH)_2$), and silicic acid [H_4SiO_4, or $Si(OH)_4$]. For the sake of simplicity, this reaction can be schematically represented as following:

$$Ca(OH)_2 + H_4SiO_4 \rightarrow Ca^{2+} + H_2SiO_4^{2-} + 2\,H_2O \rightarrow CaH_2SiO_4 \cdot 2\,H_2O$$

This reaction causes the expansion of the altered aggregate by the formation of a swelling gel of calcium silicate hydrate (CSH). This gel increases in volume with water and exerts an expansive pressure inside the material, causing spalling and loss of strength of the concrete, finally leading to its failure.

Alkali–silica reaction is also known as alkali–aggregate reaction (AAR). The conditions required for ASR to occur are:

- Sufficiently high alkali content of the cement
- A reactive aggregate, such as chert
- Water—ASR will not occur if there is no available water in the concrete, since alkali–silica gel formation requires water

The mechanism of ASR causing the deterioration of concrete can be described as:

- Alkaline solution attacks the siliceous aggregate to convert it to viscous alkali silicate gel.
- Consumption of alkali by the reaction induces the dissolution of Ca^{2+} ions into the cement pore water. Calcium ions then react with the gel to convert it to hard calcium silicate hydrate.
- Penetrated alkaline solution converts the remaining siliceous minerals into bulky alkali silicate gel. The resultant expansive pressure is stored in the aggregate.
- Accumulated pressure cracks the aggregate and the surrounding cement paste when the pressure exceeds the tolerance of the aggregate.

However, ASR can be mitigated in concrete by following approaches:

- Limiting the alkali metal content of the cement.
- Limiting the reactive silica content of the aggregate: certain volcanic rocks are particularly susceptible to ASR because they contain volcanic glass and should not be used as aggregate.
- Addition of very fine siliceous materials: to neutralize the excessive alkalinity of cement with silicic acid by voluntary provoking a controlled pozzolanic reaction at the early stage of the cement setting. Convenient pozzolanic materials to add to the mix may be, e.g., pozzolan, silica fume, fly ashes, or metakaolin.

The use of pozzolans such as fly ash in the concrete mix as a partial replacement of cement can reduce the likelihood of ASR occurring as they reduce the alkalinity of the pore fluid. Numerous early studies suggested the effectiveness of fly ash in inhibiting or reducing expansion resulting from ASR [97, 116].

Stanton [116] recognized the deterioration of concrete because of reaction between the alkaline hydroxyl ions in the pore water of concrete and certain forms of silica occasionally present in the aggregate. Investigation was conducted using aggregates containing opaline material and cement with acid soluble alkali content of more than 0.6%. Deleterious expansion due to the alkali–silica reaction could be reduced or eliminated by the addition of finely divided mineral admixtures including fly ash containing siliceous material.

Dunstan [34] identified aggregates and their mineralogical constituents that can react with alkalis in concrete. They are (1) silica materials—opaline or chalcecondic cherts, tridymite, cristobalite, siliceous limestone; (2) glassy to crypocrystalline rhyolites, dacites, and their tuffs; (3) zeolite and neulandite; and

(4) certain phylites. An effect of cement alkalies and fly ash on alkali–aggregate reaction as reported by Dunstan [34] is shown in Fig. 1.20.

Hobbs [47] studied the effect of different Class F fly ashes on the alkali silica reactivity of mortars made with opaline aggregates and high alkali cement. Based on the test results, he concluded that (1) partial replacement of a high-alkali cement with fly ash reduced the long-term expansion due to alkali-silica reactivity but, even when 30 or 40% of the cement was replaced, most of the blended cement mortars cracked at earlier or similar ages as compared to the Portland cement mortars; (2) where part of the cement was replaced by fly ash, the lowest mortar alkali content, expressed as equivalent Na_2O, at which cracking was observed was 2.85 kg/m^3. This related only to acid soluble alkalis contributed by the Portland cement and compared with 3.5 kg/m^3 for a Portland cement mortar; (3) if it is assumed that fly ash acts effectively like a cement with an alkali content of 0.2% by weight, the lowest alkali content at which cracking was observed was 3.4 kg/m^3; and (4) fly ash acts as alkali diluters, slag being more effective in reducing damage due to alkali–silica reactivity than fly ash.

Oberholster and Westra [93] observed that fly ash addition effectively suppressed expansion at cement replacement levels of 20% or more on an equal volume basis. Mehta [78] concluded that alkali–silica reaction progresses slowly and the continuous formation of swelling type gel causes internal disruption and cracking of concrete. The damage to concrete due to this reaction is appreciably aggravated when other causes of deterioration such as weathering effects of freeze–thaw, sulfate attack, and other aggressive physical and chemical processes are concurrently active.

Perry et al. [99] investigated the influence of 12 different Canadian fly ashes on the alkali–aggregate reaction of mortar mixes made with reactive opaline aggregate. Replacement levels of cement with fly ash were 20–40%. They concluded that reduction in expansion after 1 year ranged from 5 to 81 at 20% replacement

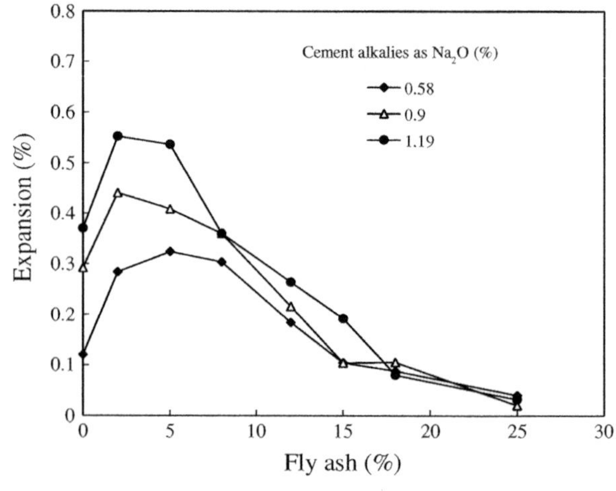

Fig. 1.20 Effects of cement alkalies and fly ash on alkali–aggregate reaction [34]

level, 34 to 89 at 30% replacement level, and 47 to 92% at 40% replacement level, respectively, compared to that of control mortar mix (no fly ash).

Idorn [49] reported that in field cases of deleterious silica reaction, the cement paste is chemically unaffected while the reacting aggregate particles are internally fractured and/or partially dissolved. High alkaline Portland cements generally having more than 0.6% sodium equivalent alkali content is highly vulnerable to attack by aggregates containing reactive amorphous as well as crystalline silica such as chalcedony, crypo-crystalline fibrous, and tridymite. Alkalies in most cases are derived from Portland cement itself, but they can also be augmented by their presence in the mixing water, admixtures, salt contaminated aggregate and deicing salts used on concrete.

1.10.6 Resistance to Aggressive Chemicals

Loss in durability of concrete by chemical attack can occur either due to the decomposition of cement paste or due to the disruptive internal expansion caused by chemical reactions in the paste or by combination of both the actions. Deleterious chemicals can react with $Ca(OH)_2$ to form water soluble salts that can be leached out of the concrete over a period of time, thereby increasing the permeability of concrete and aggravating the damage by increased and faster ingress of harmful chemicals. Sulfates react with $Ca(OH)_2$ and calcium aluminate compounds in concrete to form gypsum and calcium sulpho-aluminate, ettringite that can cause internal disruption of the concrete by concomitant volume increase of the paste.

1.10.7 Sulfate Resistance

Portland cement mortar and concrete are attacked by solutions containing sulfates (sodium or magnesium sulfate). Sulfate attack can lead to expansion, cracking, strength loss, and disintegration. The constituents of hydrated cement paste taking part in the expansive reactions are monosulfoaluminate hydrate, calcium aluminate hydrate, and calcium hydroxide.

Use of fly ash in concrete increases its resistance to sulfate attack and potentially corrosive salts that penetrate into the concrete and cause steel corrosion with accompanying cracking and spalling of concrete. It is well established that the reaction of fly ash with calcium hydroxide released during cement hydration results in the formation of additional calcium alumino-silicate hydrates and accompanying reduction in permeability of the concrete.

Dunstan [32, 33] investigated the sulfate resistance of concrete mixes made with lignite and sub-bituminous fly ashes. He concluded that lignite and sub-bituminous Class C fly ash generally reduced sulfate resistance when used in normal proportion. Dustan reported that as the calcium oxide in the fly ash increased above a lower limit of 5% and the ferric oxide (Fe_2O_3) decreased, sulfate resistance was reduced. He proposed the use of an indicator 'R' defined as $R = \%$ CaO–5/% Fe_2O_3.

For the fly ashes used by Dunstan [33], those having '*R*' values of 1.5 or less generally improved sulfate resistance while those with higher values did not. The general applicability of the 'R' factor to predict sulfate resistance of all fly ashes is further required to be investigated. With the use of fly ash at 25% cement replacement level, the sulfate resistance of concrete made with ASTM Type II cement at 0.45 water cement ratio has been related to the '*R*' factor as follows:

R limits	Sulfate resistance
<0.75	Greatly improved
0.75 to 1.5	Moderately improved
1.5 to 3.0	No significant change
>3.0	Reduced

Mehta [79] observed that if a fly ash is high in reactive aluminate phases it will not improve the sulfate resistance of concrete. He found that sulfate resistance depends on the type of aluminate phases at the time of sulfate exposure. Upon immersion in sulfate solution, pastes containing monosulfoaluminate or calcium aluminate hydrate suffered from strength loss due to ettringite formation. In contrast blended cement pastes containing fly ash that promoted the formation of ettringite prior to immersion into sulfate solution showed superior resistance.

Joshi [55] reported that with the incorporation of Alberta fly ash at 15% replacement level by weight of cement, the sulfate resistance of cement–sand mortar was improved significantly when exposed to sodium sulfate and magnesium sulfate solutions of concentration below 10%. For higher sulfate concentrations, the mortars made with Type-V cement and 15% Alberta fly ash developed adequate sulfate resistance.

Larsen [69] reported that the change in the pore structure of cement paste as a result of fly ash addition can not be the sole reason for the observed favorable performance of fly ash concrete subject to chemical attack, particularly sulfate attack. He observed the favorable effect of the fly ash on sulfate resistance of concrete mixes even at ages when the pozzolanic reaction of fly ash was still not particularly high. Accordingly, it is suggested that additional chemical–mineral interaction including the reduction in tricalcium aluminate (C_3A) and calcium hydroxide content contribute to sulfate resistance. The fly ash may combine with some alumina phases such as C_3A in the cement during the first few days of cement hydration to form primary ettringite, thus reducing the potential for expansive sulfate-alumina reactions responsible for sulfate attack. With the increased conversion of C_3A in the initial stages, less of the aluminate containing phase is available for reaction during subsequent sulfate attack.

Fay and Pierce [38] conducted a study on air-entrained fly ash concrete. They used three fly ashes from the United States, representing a range of CaO contents of 11 to 28.8% and used 10 to 100% fly ash by weight of the total cementitious material. The specimens were cured for 14 days in a 100% humidity room and for another 14 days in a 50% humidity room before they were immersed in 10% Na_2SO_4 solution or subjected to cyclic soaking and drying phases in 2.1% Na_2SO_4 solution to generate data by accelerated test. The test results showed that (1) both low-calcium Class C and Class F fly ashes can be effective cement replacements in controlling sulfate expansion; (2) Class F fly ash was the most effective in reducing sulfate expansion at the lower cementitious content (251.5 kg/m^3). Likewise at the higher cementitious content (387 kg/m^3), Class F fly ash at 30% replacement level was found to be the most effective; (3) high-calcium Class C fly ash concrete mixes, with fly ash replacement levels blow 50%, generally exhibited more expansion than the control mixes, opposite to the behavior observed with low-calcium class F concrete; (4) for low-calcium Class C fly ash replacement levels were suggested to be greater than 30% and for high-calcium Class C fly ashes the corresponding suggested levels were greater than 75% to achieve the most improved sulfate durability; (5) the replacement level for a particular type of fly ash was found to depend on cementitious material content of the mix. With lower cementitious material content, 50% or more, while for richer mixes 50% or less Class C was suggested for improving the sulfate resistance of concrete; and (6) Class F fly ash at replacement level of 30% significantly improved the sulfate durability for the cementitious levels and appeared to be optimum for both test conditions employed in the investigation. The high-calcium Class C fly ash concretes generally performed much worse than the control mix under sulfate attack indicating the least effectiveness of these types of ash in reducing expansion.

Mehta [81] concluded that fly ashes are amongst the group of pozzolans that significantly increase the life expectancy of concrete exposed to sulfate attack. In general, Class F type fly ash meeting the specification requirements will improve the sulfate resistance of any concrete/mortar mix in which it is included, although the degree of improvement may vary with either the cement used or the fly ash. The situation with Class C fly ash is different. A few studies indicated that some Class C fly ashes may rather reduce sulfate resistance when used in normal proportions.

Prunsinski and Carrasquillo [100] found that sulfate resistance of concretes made with class C fly ash was strongly influenced by the gypsum content of the mix. In their studies, fly ash was inter-ground and blended in varying quantities (25–75%) with Type I and II cements. In some blends, additional gypsum was inter-ground to supply the fresh concrete with sufficient sulfate ions to stabilize the ettringite. Increasing the gypsum content decreased the expansion of the specimens, specifically, mixes having gypsum contents of twice the sulfation point had lower expansions than the controls made with Type II cement alone, while specimens made with Class C fly ash and Type II cement with out additional gypsum failed.

1.10.7.1 Resistance to Sulfuric Acid

Sulfuric acid is a very aggressive acid that reacts with the free lime $Ca(OH)_2$, in cement paste forming gypsum ($CaSO_4 \cdot 2H_2O$). This reaction is associated with an increase in volume of the concrete by a factor of 2.2 [10]. Another destructive action is the reaction between calcium aluminate present in cement paste and gypsum crystals. These two products form the less soluble reaction product, ettringite ($3CaO \cdot Al_2O_3 \cdot 3CaSO_4 \cdot 32H_2O$). These very expansive compounds cause internal pressure in the concrete, which leads to the formation of cracks. The reacted surface becomes soft and white. Because of these, concrete structure loses its mechanical strength. Another important phenomenon where sulfuric acid is responsible for concrete corrosion is biogenic sulfuric acid corrosion, which occurs often in sewer systems. Due to different chemical and microbiological reactions, hydrogen sulfide releases into the atmosphere of sewer structures above the water level. This gas reacts with oxygen to form elemental sulfur, which is deposited on the walls of the sewer structures. In the slime layer coating these walls, aerobic sulfur-oxidizing bacteria (*Thiobacillus* spp.) metabolize the sulfur to sulfuric acid. Repair and sometimes complete replacement of the damaged structures becomes necessary after this acid attack. Pozzolans combine and stabilize the calcium hydroxide liberated during the hydration of cement in concrete, to form additional cementitious compound, mainly in the form of calcium silicate hydrate (CSH). The resultant binder matrix is more chemically resistant, by virtue of its denser microscopic pore structure.

Aydin et al. [11] investigated the effect of Class C fly ash on the sulfuric acid resistance of concrete. Cement was replaced with fly ash up to 70%. Cylinders (100 × 200 mm) and cubes (71 mm) were cast. After 28 days of water curing, specimens were immersed in a 5% sulfuric acid (H_2SO_4) solution for 60 days in a plexiglass container. After 60 days of exposure, acid-attacked surfaces of the specimens were cleaned with distilled water. Chemical resistance was evaluated by determining the weight loss (WL) and compressive strength loss (SL) of the specimens. Based on the results, it was concluded that (1) loss in compressive strength was related with the FA content. The FA replacement seemed to improve the acid resistance of the steam-cured samples as loss percentages dropped to 21% (FA70) from 58% (FA0); (2) negligible differences in weight-loss were observed for standard cured concrete with increasing FA content. Similar trends have been observed in terms of weight loss for both curing cases. Weight loss dropped from 5% (FA0) to 3.3% (FA70) for standard curing and 8.3% (FA0) to 1.1% (FA70) for SC conditions; and (3) strength loss and weight loss of Portland cement concrete cured in water were lower than that of steam-cured ones due to somewhat higher permeability of steam-cured samples. However, over 30 and 40% FA replacement levels, weight and strength losses of steam-cured concrete were lower than water-cured ones, respectively. In other words, sulfuric acid resistance of steam-cured concrete could be improved significantly by incorporation of FA. The positive effect of FA on acid resistance may arise from pozzolanic reaction between FA and calcium hydroxide liberated during the hydration of cement, which forms additional cementitious compound, mainly CSH.

1.10.8 Abrasion Resistance

Abrasion occurs due to rubbing, scraping, skidding or sliding of objects on its surface. Abrasion resistance of concrete is influenced by number of factors such as water-cement ratio, types of aggregates, air-entrainment, compressive strength, surfacing finish, types of hardeners, and curing.

In general, hardened paste has low abrasion resistance. The use of hard aggregates and low water to cement ratio has been found to be quite effective in increasing abrasion/erosion resistance of both types of concrete, i.e., with and without fly ash. With the use of fly ash in concrete the quality of cement paste is improved, and the leachability of calcium hydroxide is impeded with age. Because of the dense structure of cement paste and good bonding characteristics of fly ash concrete, it is believed that the tendency of coarse aggregate to be plucked out of the binding matrix by abrasive action is reduced. At equal compressive strengths, properly cured and finished concrete with and without fly ash will exhibit essentially equal resistance to abrasive-erosive forces. Use of fly ash thus affects this aspect of durability only to the extent that it usually improves compressive strength of concrete due to its pozzolanic activity with time.

Liu [70] compared the abrasion resistance of non-fly ash concrete with a fly ash concrete made with 25% cement replacement. He concluded that abrasion of concrete with or without fly ash was similar up to 36 h of abrasion testing, but after 72 h of testing, the fly ash concrete lost about 25% more weight than the concrete without fly ash. Tikalsky et al. [123] concluded that concrete containing Class C fly ash possessed superior abrasion resistance compared to either ordinary Portland cement concrete or concrete containing Class F fly ash.

Nanni [89] investigated the abrasion resistance of roller-compacted concrete using laboratory and field specimens made by replacing cement with 50% Class C fly ash. He concluded that (1) testing under air-dry conditions produced 30–50% less wear than under wet conditions; addition of steel or synthetic fibers did not cause any appreciable change in the abrasion resistance of concrete; and (3) improper moist-curing conditions produced more negative effects on the surface quality than the compressive strength of concrete.

Langan et al. [67] investigated the influence of compressive strength on the durability of concrete containing 50% fly ash as replacement of cement, and concluded that the presence of fly ash at high levels of cement replacement increased the weight loss due to abrasion at all ages relative to concrete without fly ash.

Dhir et al. [31] concluded that near surface characteristics such as absorption, intrinsic permeability and vapour diffusity are closely related with the abrasion resistance of concrete. Bilodeau and Malhotra [14] investigated the abrasion resistance of concrete incorporating high volumes of Class F fly ash. Superplasticized mixtures were developed with 55 to 60% fly ash of total cementitious materials. Test results showed that fly ash concrete had poorer abrasion resistance than concrete without fly ash.

Naik et al. [87] evaluated the abrasion resistance of concrete containing five levels of cement replacements (15, 30, 40, 50, and 70%) with one source of Class C fly ash. Test results showed that abrasion resistance of concrete having cement replacement up to 30% was comparable to the reference concrete with out fly ash, but beyond 30% cement replacement, fly ash concrete exhibited slightly lower resistance to abrasion relative to non-fly ash concretes. Naik et al. [88] reported that blending of Class C fly ash with Class F fly ash showed either comparable or better abrasion resistance results than either the control mixture with out fly ash or the unblended Class C fly ash.

Siddique [113] studied the abrasion resistance of concrete proportioned to have four levels of fine aggregate replacement (10, 20, 30 and 40%) with Class F fly ash. A Control mixture with ordinary Portland cement was designed to have 28 days compressive strength of 26 MPa. Concrete specimens of size 65 × 65 × 60 mm were made for the purpose. The abrasion resistance of concrete mixtures was determined at the ages of 28, 91, and 365 days in accordance with Indian Standard Specifications [50]. It was measured in term of depth of wear. Figure 1.21 shows the variation of depth of wear versus percentage of fine aggregate replacement with Class F fly ash, at 60 min of abrasion time. It is evident that with the increase in fly ash content, depth of wear decreased, which indicated that the abrasion resistance of concrete increased with the increase in fly ash content. This showed that for a particular percentage of fine aggregate replacement with fly ash, depth of wear decreased with increase in age, which means that abrasion resistance increased with age. This could be primarily attributed to the increase in compressive strength resulting from increased maturity of concrete with age.

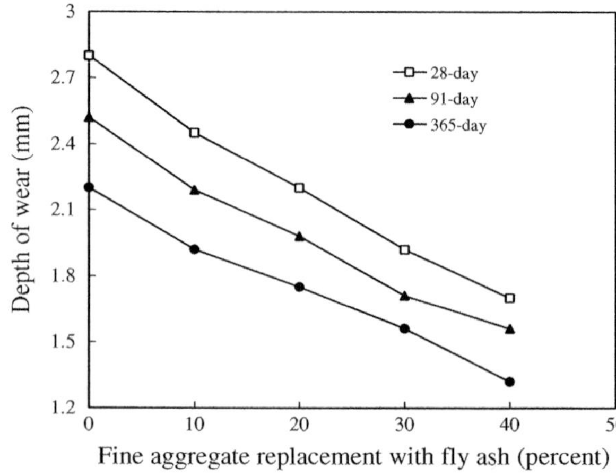

Fig. 1.21 Depth of wear at 60 min of abrasion versus fine aggregate replacement with fly ash [113]

Yazici and İnan [127] developed a relationship between mechanical properties (compressive strength and splitting tensile strength) and abrasion resistance of high strength concretes (HSC) having compressive strength between 65 and 85 MPa. They concluded that abrasion resistance of high strength concrete can be estimated from compressive and splitting tensile strength results.

Yen et al. [128] established equations based on effective compressive strength and effective water-to-binder ratios, which were modified by cement replacement and developed to predict the 28- and 91-day abrasion resistance of concretes with compressive strengths ranging from approximately 30–100 MPa. The predicted results compared favourably with the experimental results.

References

1. ACI committee 211.1.81 (1984) Standard Practice for Selecting Proportions for Normal, Heavy Weight and Mass Concrete. ACI Manual of Concrete Practice
2. ACI committee 226 3R-87: Fly ash in concrete. ACI Mater. J. **11**, 381–409 (1987)
3. Alonso, J.L., Wesche, K.: Characterization of Fly Ash. Fly Ash in Concrete, Properties and Performance, pp. 3–23, RILEM Report, E & FN Spon, New York (1992)
4. Andrade, C.: Effect of fly ash in concrete on the corrosion of steel reinforcement. ACI SP 91, pp. 609–620 (1986)
5. ASTM C 618: Standard Specification for Coal Fly Ash and Raw or Calcined Natural Pozzolan for Use as a Mineral Admixture in Concrete. Annual Book of ASTM Standards, Philadelphia (1993)
6. Atis, C.D.: Heat evolution of high-volume fly ash concrete. Cem. Concr. Res. **32**(5), 751–756 (2002)
7. Atis, C.D.: Accelerated carbonation and testing of concrete made with fly ash. Construct. Build. Mater. **17**(3), 147–152 (2003)
8. Atis, C.D., Kilic, A., Korkut, U.: Strength and shrinkage properties of mortar containing a nonstandard high-calcium fly ash. Cem. Concr. Res. **34**(1), 99–102 (2004)
9. Atis, C.D.: Strength properties of high-volume fly ash roller compacted and workable concrete, and influence of curing condition. Cem. Concr. Res. **35**(6), 1112–1121 (2005)
10. Attiogbe, E.K., Rizkalla, S.H.: Response of concrete to sulfuric acid attack. ACI Mater. J. **84**(6), 481–488 (1988)
11. Aydin, S., Yazici, H., Yiğiter, H., Baradan, B.: Sulfuric acid resistance of high-volume fly ash concrete. Build. Environ. **42**(2), 717–721 (2007)
12. Bamforth, P.B.: In situ measurement of the effect of partial cement replacement using either fly ash or ground granulated blast furnace slag on the performance of mass concrete. Proc. Instit. Civil Eng. **69**, 777–800 (1980)
13. Bamforth, P.B.: The water permeability of concrete and its relationship with strength. Mag. Concr. Res. **43**(137), 233–241 (1991)
14. Bilodeau, A., Malhotra, V.M.: Concrete incorporating high volumes of ASTM Class F fly ashes: mechanical properties and resistance to deicing salt scaling and to chloride-ion penetration. ACI Special Publication SP, 132, pp. 319–349 (1992)
15. Brown, J.H.: The strength and workability of concrete with PFA substitution. In: Proceedings International Symposium on the Use of PFA in Concrete, pp. 151–161, University of Leeds, England (1982)
16. Burns, J.S., Guarnaschelli, C., McAskill, J.: No controlling the effect of carbon in fly ash on air entrainment. In: Proceedings, Sixth International Symposium on Fly Ash Utilization, pp. 294–313, Reno, Nevada, DOE/METC (1982)

17. Buttler, F.G., Dector, M.H., Smith, G.R.: Studies on the desiccation and carbonation of systems containing Portland cement and fly ash. ACI SP 79(1), 367–383 (1983)
18. Cabrera, J.G., Lynsdale, C.J.: A new gas permeameter for measuring the permeability of mortar and concrete. Mag. Concr. Res. **40**(144), 177–182 (1988)
19. Carette, G.G., Malhotra, V.M.: Characterization of Canadian fly Ashes and their Performance in Concrete. Division report MRP/MSL 84-137, CANMET, Energy, Mines and Resources, Canada (1984)
20. Carette, G.G., Malhotra, V.M.: Characterization of Canadian fly ashes and their relative performance in concrete. Can. J. Civil Eng. **14**(5), 667–682 (1987)
21. Chalee, W., Teekavanit, M., Kiattikomol, K., Siripanichgorn, A., Jaturapitakkul, C.: Effect of W/C ratio on covering depth of fly ash concrete in marine environment. Construct. Build. Mater. **21**(5), 965–971 (2007)
22. Chindaprasirt, P., Chotithanorm, C., Cao, H.T., Sirivivatnanon, V.: Influence of fly ash fineness on the chloride penetration of concrete. Construct. Build. Mater. **21**(2), 356–361 (2007)
23. Central Electricity Generating Board (CEGB) (1967) PFA data book. London
24. Compton, F.R., Macinnis, C.: Field trial of fly ash concretes. Ontario Hydro Research News, pp. 18–21 (1952)
25. Cook, J.E.: Research and application of high strength concrete using class C fly ash. Concr. Int. **4**, 72–80 (1982)
26. Crow, R.D., Dunstan, E.R.: Properties of fly ash concrete. In: Diamond, S. (ed.) Proceedings of Symposium on Fly Ash Incorporation in Hydrated Cement Systems, pp. 214–225. Materials Research Society, Boston (1981)
27. Demirboğa, R.: Thermal conductivity and compressive strength of concrete incorporation with mineral admixtures. Build. Environ. **42**(7), 2467–2471 (2007)
28. Demirboğa, R., Türkmen, I., Karakoc, M.B.: Thermo-mechanical properties of concrete containing high-volume mineral admixtures. Build. Environ. **42**(1), 349–354 (2007)
29. Diamond, S.: Effects of two Danish fly ashes on alkali—contents of cement—fly ash pastes. Cement. Concr. Res. **11**(3), 383–394 (1981)
30. Diamond, S.: Selection and use of fly ash for high way concrete. Joint Highway Research Project, Purdue University, Indiana (1985)
31. Dhir, R.K., Hewlett, P.C., Chan, Y.N.: Near-surface characteristics of concrete: abrasion resistance. Mater. Struct. **24**(2), 122–128 (1991)
32. Dunstan, E.R.: Performance of lignite and sub-bituminous fly ash in concrete. A progress report REC-ERC-76-1, USBR (1976)
33. Dunstan, E.R.: A possible method for identifying fly ashes that—will improve sulfate resistance of concretes. ASTM Cem. Concr. Aggreg. **2**, 20–30 (1980)
34. Dunstan, E.R.: The effect of fly ash on concrete alkali–aggregate reaction. ASTM Cem. Concr. Aggreg. **3**, 101–104 (1981)
35. Electric Power Research Institute.: Classification of fly ash for use in cement and concrete. CS-5116, Project 2422-10. Palo Alto, California 94304, USA (1987)
36. Ellis, WE Jr, Rigs, E.H., Butler, W.B.: Comparative results of utilization of fly ash, silica fume and GGBS in reducing the chloride permeability of concrete. In: Proceedings of the 2nd CANMET/ACI International Conference on Durability of Concrete, Montreal, Canada. ACI SP 126(1), 443–457 (1991)
37. Erdoğdu, K., Türker, P.: Effects of fly ash particle size on strength of Portland cement fly ash mortars. Cement. Concr. Res. **28**(9), 1217–1222 (1998)
38. Fay, K.F.V., Pierce, J.S.: Sulfate resistance of concretes with various fly ashes. ASTM standardisation news, pp. 32–37 (1989)
39. Gebauer, J.: Source observations on the carbonation of fly ash concrete. Silic. Ind. **6**, 155–159 (1982)
40. Gebler, S.H., Klieger, P.: Effect of fly ash on the air void stability of concrete. In: Proceedings of the 1st International Conference on the Use of Fly Ash, Silica Fume, Slag and Other Mineral by Products in Concrete. ACI SP-79, pp. 103–142 (1983)

41. Ghosh, R.S., Timusk, J.: Creep of fly ash Concrete. ACI J. **78**(5), 351–387 (1981)
42. Gifford, P.M., Langan, B.W., Day, R.L., Joshi, R.C., Ward, M.A.: A study of fly ash concrete in curb and gutter construction under various laboratory and field curing regimes. Can. J. Civil Eng. **14**(5), 614–620 (1987)
43. Gopalan, M.K.: Sorptivity of fly ash concretes. Cem. Concr. Res. **26**(8), 1189–1197 (1996)
44. Haque, M.N., Langan, B.W., Ward, M.A.: High fly ash concretes. ACI Mater. **8**(1), 54–60 (1988)
45. Helmuth, A.R.: Fly ash in cement and concrete. Special Publication 040, p. 203. Portland Cement Association, Skokie, IL (1987)
46. Ho, D.W.S., Lewis, R.K.: Carbonation of concrete incorporating fly ash or a chemical admixture. ACI SP **79**, 333–346 (1983)
47. Hobbs, D.W.: The alkali–silica reaction: a model for predicting expansion in mortar. Mag. Concr. Res **33**(117), 208–220 (1981)
48. Hui-sheng, S., Bi-wan, X., Xiao-chen, Z.: Influence of mineral admixtures on compressive strength, gas permeability and carbonation of high performance concrete. Construct. Build. Mater. **23**(5), 1980–1985 (2009)
49. Idorn, G.M.: Concrete durability and resource economy. Concr. Int. **13**(7), 18–23 (1991)
50. IS: 1237: Method for testing abrasion resistance of concrete. Bureau of Indian Standards (BIS), New Delhi, India (1980)
51. Jawed, I., Skalny, J.: Hydration of tricalcium silicate in the presence of fly ash. Effects of fly ash incorporation in cement and concrete. In: Diamond, S. (ed.) Proceedings Symposium, pp. 60–70. Materials Research Society, (1981)
52. Jiang, L., Liu, Z., Ye, Y.: Durability of concrete incorporating large volumes of low-quality fly ash. Cem. Concr. Res. **34**(8), 1467–1469 (2004)
53. Joshi, R.C.: Experimental production of synthetic fly ash from kaolinite. MS Thesis, Iowa State University (1970)
54. Joshi, R.C.: Sources of pozzolanic activity in fly ashes—a critical review. In: Proceedings of the 5th International Fly Ash Utilization Symposium, pp. 610–623, Atlanta, GA, USA (1979)
55. Joshi, R.C.: Effect of a sub-bituminous fly ash and its properties on sulfate resistance of sand cement mortars. J. Durab. Build. Mater. **4**, 271–286 (1987)
56. Joshi, R.C., Lam, D.T.: Sources of self-hardening properties of fly ashes. Materials research proceedings, vol 86, pp. 183–184. MRS. Pittsburgh (1987)
57. Joshi, R.C., Lohtia, R.P.: Effects of premature freezing temperatures on compressive strength, elasticity and microstructure of high volume fly ash concrete. Third Canadian Symposium on Cement and Concrete, Ottawa, Canada (1993)
58. Joshi, R.C., Lohtia, R.P, Salam, M.A.: High strength concrete with high volumes of Canadian sub-bituminous coal ash. Third International Symposium on Utilization of High Strength Concrete, Lillachammer, Norway (1993)
59. Joshi, R.C., Lohtia, R.P., Salam, M.A.: Some durability related properties of concretes incorporating high volumes of sub-bituminous coal fly ash. In: Proceedings, 3rd CANMET/ACI International Conference on Durability of Concrete, Nice, France, pp. 447–464 (1994)
60. Kasai, Y., Matsui, I., Fukushima, U., Kamohara, H.: Air permeability of blended cement mortars. In: Proceedings of the 1st t International Conference on the use of Fly ash, Silica-fume, Slag and other mineral by-products in concrete. ACI SP **29**, 435–451 (1983)
61. Khan, M.I.: Rheological characteristics of HPC containing composite cementitious materials. Concr. Technol. J. Concr. Plant Int. No. 02/10, Germany, pp. 78–84 (2010)
62. Khan, M.I.: Permeation of high performance concrete. J. Mater. Civil Eng. **15**(1), 84–92 (2002)
63. Klieger, P., Perenchio, W.F.: Laboratory studies of blended cement: portland-pozzolan cements. Research and Development Bulletin RD013, Portland cement Association, USA (1972)
64. Kokubu, M., Nagataki, S.: Carbonation of concrete with fly ash and corrosion of reinforcement in 20-years tests. ACI Special publications CS- **114**, 315–329 (1989)

65. Korac, V., Ukraincik, V.: Studies into the use of fly ash in concrete for water dam structures. ACI Special Publication SP- **79**, 173–185 (1983)
66. Lane, R.O., Best, J.F.: Properties and use of fly ash in Portland cement concrete. Concr. Int. **4**(7), 81–92 (1982)
67. Langan, B.V., Joshi, R.C., Ward, M.A.: Strength and durability of concrete containing 50 percent Portland cement replacement by fly ash and other materials. Can. J. Civil Eng. **17**(1), 19–27 (1990)
68. Langley, W.S., Carette, G.G., Malhotra, V.M.: Structural concrete incorporating high volumes of ASTM class F fly ash. ACI Mater. J. **86**(5), 507–514 (1989)
69. Larsen, T.D.: Use of fly ash in structural concrete in Florida, Presented at fly ash in high way construction seminar, Altanta, GA (1985)
70. Liu, T.C.: Abrasion resistance of concrete. ACI J. Proc. **78**(5), 341–350 (1981)
71. Lohtia, R.P., Nautiyal, B.D., Jain, O.P.: Creep of fly ash concrete. ACI J. **73**(8), 469–472 (1976)
72. Lohtia, R.P., Nautiyal, B.D., Jain, K.K., Jain, O.P.: Compressive strength of plain and fly ash concrete by non-destructive testing methods. J. Instit. Eng. (India) **58a-1**:40–45 (1977)
73. Majko, R.M., Pistilli, M.F.: Optimizing the amount of Class C fly ash in concrete mixtures. Cem. Concr. Aggreg. CCAGDP **6**(2), 105–119 (1984)
74. Malhotra, V.M., Carette, G.G., Bremmer, T.W.: Durability of concrete containing granulated blast furnace slag or fly ash or both in Marine environment, report 80-18E, CANMET, EMR, Canada (1982)
75. Malhotra, V.M., Caratte, G.G., Bilodeau, A., Sivasundram, V.: Some aspects of durability of high volume ASTM class F (Low-calcium) fly ash concrete. Mineral Sciences Laboratories, Division report MSL-90-20 (OP & J) (1990)
76. McCarthy, G.J., Johansen, D.M., Steinwand, S.J.: X-ray diffraction analysis of fly ash. In: Barrett, C.S., et al. (eds.) Advances in X-Ray Analysis, vol 31. Plenum Press, New York (1988)
77. McCarthy, M.J., Dhir, R.K.: Development of high volume fly ash cements for use in concrete construction. Fuel **84**(11), 1423–1432 (2005)
78. Mehta, P.K.: Pozzolanic and cementitious by-products as mineral admixtures for concrete—a critical review. In: Proceedings of the 1st International Conference on the use of Fly Ash, Slag and Silica fume in Concrete, pp. 1–46, Montebello, Canada, ACI SP-79 (1983)
79. Mehta, P.K.: Concrete: structure, properties and materials. Prentice Hall, Englewood Cliffs (1986)
80. Mehta, P.K.: Standard specifications for mineral admixtures—an overview. ACI SP **91**, 637–658 (1988)
81. Mehta, P.K.: Sulfate Attack on concrete—a critical review. In: J. Skalny (ed.) Materials science of concrete-III. American Ceramic Society, pp 105–130 (1993)
82. Mehta, P.K.: In: Khayat, I.H., Aitcin, P.C. (eds.) Symposium on durability of concrete, pp. 99–118. Nice, France, (1994)
83. Munday, J.G.L., Ong, L.T., Wong, L.G., Dhir, R.K.: Load independent movements in OPC/PFA concrete. In: Cabreva, J.A., Cusens R.R. (eds.) Proceedings, International Symposium on the use of PFA in concrete, pp. 243–246, University of Leeds, England (1982)
84. Nagataki, S., Ohga, H., Kim, E.K.: Effect of curing conditions on the carbonation of concrete with fly ash and the corrosion of reinforcement in long term basic. ACI SP **91**, 521–540 (1986)
85. Nagataki, S., Ohga, H.: Combined effect of carbonation and chloride on corrosion of reinforcement in fly ash concrete. In: Proceedings 4th International Conference on the Use of Fly Ash, Silica Fume, Slag, and Natural Pozzolans in Concrete, vol. 32, pp. 227–244. Istanbul, Turkey, ACI SP1 (1992)
86. Naik, T.R., Singh, S.S., Hossain, M.M.: Permeability of concrete containing large amounts of fly ash. Cem. Concr. Res. **24**(5), 913–922 (1994)
87. Naik, T.R., Singh, S.S., Hossain, M.M.: Abrasion resistance of high-strength concrete made with class C fly ash. ACI Mater. J. **92**(6), 649–659 (1995)

88. Naik, T.R., Singh, S.S., Ramme, B.W.: Mechanical properties and durability of concrete made with blended fly ash. ACI Mater. J. **95**(4), 454–462 (1998)
89. Nanni, A.: Abrasion resistance of roller-compacted concrete. ACI Mater. J. **86**(53), 559–565 (1989)
90. Nasser, K.W., Marzouk, H.M.: Properties of concrete made with sulfate resisting cement and fly ash. In: Proceedings First International Conference on the use of Fly ash, Silica Fume, slag and other mineral by-products in concrete, pp. 383–395. ACI SP-79 (1983)
91. Nasser, K.W., Al-Manasser, A.A.: Shrinkage and creep of concrete containing 50 percent lignite fly ash at different stress–strength ratios. In: Proceedings of the 2nd International Conference on Fly ash, Silica fume, Slag and Natural Pozzolans in concrete, ACI SP-91(1), pp 433–448 (1986)
92. Neville, A.M.: Properties of Concrete, 2nd edn. Wiley, New York (1973). 382 p
93. Oberholster, R.E., Westra, W.B.: The effectiveness of mineral admixtures in reducing expansion due to alkali–aggregate reaction with malmesbury group aggregate. In: Proceedings of the 5th International Conference on Alkali–Aggregate Reaction in Concrete, Cape Town, South Africa (1981)
94. Ogawa, K., Uchikawa, H., Takemoto, K., Yasui, I.: The mechanism of the hydration in the system CS-pozzolana. Cem. Concr. Res. **10**(5), 683–696 (1980)
95. Ozer, B., Ozkul, M.H.: The influence of initial water curing on the strength development of ordinary Portland and pozzolanic cement concretes. Cem. Concr. Res. **34**(1), 13–18 (2004)
96. Owens, P.L.: Fly ash and its usage in concrete. Concr. Soc. J. **13**(7), 21–26 (1979)
97. Pepper, L., Mather, B.: Effectiveness of mineral admixtures in preventing excessive expansion of concrete due to alkali–aggregate reaction. Proc. ASTM **59**, 1178–1202 (1959)
98. Perencho, W.F., Klieger, P.: Further laboratory studies of Portland-pozzolan cements. Portland Cement Research and Development Bulletin RD041.01T (1976)
99. Perry, C., Day, R.L., Joshi, R.C., Langan, B.W., Gillot, J.E.: The effectiveness of twelve Canadian fly ashes in suppressing expansion due to alkali–silica reaction. In: Proceedings of the 7th International Conference on Alkali–Aggregate Reaction, Ottawa, pp. 93–97 (1987)
100. Prunsinski, J.R., Carrasquillo, R.L.: Factors affecting the sulfate resistance of concrete made with class C fly ash. In: Proceedings 11th International Symposium on Use and management of Coal Combustion By-Products (CCBs), American Coal Ash Association, Alexandria, Virginia, pp. 29-1-29-14 (1995)
101. Raba, F. Jr., Smith, S.L., Mearing, M.: Sub bituminous fly ash utilization in concrete. In: Diamond, D. (ed.) Proceedings Symposium on Fly Ash Incorporation on Hyderated Cement Systems, pp. 296–306. Materials Research Society, Boston (1981)
102. Ramakrishan, V., Coyle, W.V., Brown, J., Tluskus, A., Benkataramanyam, P.: Performance characteristics of concrete containing fly ash. In: Diamond, S. (ed.) Proceedings Symposium on Fly Ash Incorporation in Hydrated Cement Systems, pp. 233–243. Materials Research Society, Boston (1981)
103. Ravina, D.: Efficient utilization of coarse and fine fly ash in precast concrete by incorporating thermal curing. ACI J. **78**(3), 194–200 (1981)
104. RILEM CP113: Absorption of water by immersion under vacuum. Material Structures Research Testing **101**, 393–394 (1984)
105. Rodway, L.E., Fedriko, W.M.: Superplasticized high volume fly ash structural concrete. ACI SP **114**(1), 98–112 (1989)
106. Roy, D.M., Luke, K., Diamond, S.: Characterization of fly ash and its reactions in concrete. Proceedings of the Materials Research Society, Pittsburgh, Pennsylvania (1984)
107. Saraswathy, V., Muralidharan, S., Thangavel, K., Srinivasan, S.: Influence of activated fly ash on corrosion–resistance and strength of concrete. Cem. Concr. Compos. **25**(7), 673–680 (2003)
108. Schiepl, P., Hardtle, R.: Relationship between durability and fore structure properties of concretes containing fly ash. In: Khayat, I.H., Aitcin, P.C. (eds.) P.K. Mehta Symposium on Durability of Concrete, pp. 99–118, Nice, France (1994)

109. Schmidt, M.: Cement with inter-ground additives—capabilities and environmental relief. Part 2. Zement-Kalk Gips (1992)
110. Schubert, P.: Carbonation behaviors of mortars and concretes made with fly ash. ACI Special Publications SP- **100**, 1945–1962 (1987)
111. Shafiq, N., Cabrera, J.G.: Effects of initial curing condition on the fluid transport properties in OPC and fly ash blended cement concrete. Cem. Concr. Compos. **26**(4), 381–387 (2004)
112. Siddique, R.: Effect of fine aggregate replacement with class F fly ash on the mechanical properties of concrete. Cem. Concr. Res. **33**(4), 539–547 (2003)
113. Siddique, R.: Effect of fine aggregate replacement with class F fly ash on the abrasion resistance of concrete. Cem. Concr. Res. **33**(11), 877–1881 (2003)
114. Sivasundram, V., Carette, G.G., Malhotra, V.M.: Properties of concrete incorporating low quantity of cement and high volumes of low-calcium fly ash. In: Proceedings of the 3rd International Conference on Fly Ash, Silica Fume, Slag and Natural Pozzolans in Concrete, ACI SP-114, pp. 45–71 (1989)
115. Sivasundram, V., Carette, G.G., Malhotra, V.M.: Selected properties of high volume fly ash concretes. ACI Concrete International, pp. 47–50 (1990)
116. Stanton, T.E.: Expansion of concrete through reaction between cement and aggregate. Trans. ASCE Part **2**, 68–85 (1942)
117. Swamy, R.N., Mahmud, H.B.: Mix proportions and strength characteristics of concrete containing 50 percent low-calcium fly ash. In: Proceedings of the 2nd CANMET/ACI International Conference on Fly ash, Silica Fume, Slag and Natural pozzolans in Concrete, ACI SP-91, 1: 413–432 (1986)
118. Takemoto, K., Uchikawa, H.: Hydration of pozzolanic cement. 7th International Congress on the Chemistry of Cement, Paris, I IV-2/1-2/29 (1980)
119. Tattersall, G.H., Benfill, P.F.G.: The Rheology of Fresh Concrete. Pitman, London (1983)
120. Taylor, H.W.F.: Cement Chemistry. Academic Press, New York (1990)
121. Termkhajornkit, P., Nawa, T., Kurumisawa, K.: Effect of water curing conditions on the hydration degree and compressive strengths of fly ash–cement paste. Cem. Concr. Compos. **28**(9), 781–789 (2006)
122. Thomas, M.D.A., Matthews, J.D.: The permeability of fly ash concrete. Mater. Struct. **25**(151), 388–396 (1992)
123. Tikalsky, P.J., Carrasquillo, P.M., Carrasquillo, R.L.: Strength and durability considerations affecting mix proportions of concrete containing fly ash. ACI Mater. J. **85**(6), 505–511 (1988)
124. Virtanen, J.: Freeze–thaw resistance of concrete containing blast furnace slag, fly ash or condensed silica fume. In: Proceedings of the 1st International Conference on the use of Fly ash, Silica Fume, Slag and other mineral by-products, ACI SP-79: 923–942 (1983)
125. William, J.T., Owens, P.L.: The implications of a selected grade of United Kingdom pulverized fuel ash on the engineering design and use in structural concrete. In: Proceedings of the International Symposium on the Use of PFA in Concrete, pp. 301–313. University of Leeds, England (1982)
126. Yazici, H., Aydin, S., Yiğiter, H., Baradan, B.: Effect of steam curing on class C high-volume fly ash concrete mixtures. Cem. Conc. Res. **35**(6), 1122–1127 (2005)
127. Yazici, Ş.G., İnan, G.: An investigation on the wear resistance of high strength concretes. Wear **260**(6), 615–618 (2006)
128. Yen, T., Hsu, T.H., Liu, Y.W., Chen, S.h: Influence of class F fly ash on the abrasion–erosion resistance of high-strength concrete. Construct. Build. Mater. **21**(2), 458–463 (2007)
129. Yuan, R.L., Cook. J.E.: Study of class C fly ash in concrete. In: Proceeding of the 1st International Conference on the use of Fly ash, Silica Fume, Slag, and other mineral by-products in concrete, ACI SP-79, pp. 307–319 (1983)
130. Yuan, R.Z., Jin, S.X., Qian, J.C.: Effects of fly ash on rheology of fresh cement paste. Materials and Research Society Symposium Proceedings, pp. 182–191 (1982)

Chapter 2
Silica Fume

2.1 Introduction

Silica fume (SF) is a byproduct of the smelting process in the silicon and ferrosilicon industry. The reduction of high-purity quartz to silicon at temperatures up to 2,000°C produces SiO_2 vapours, which oxidizes and condense in the low-temperature zone to tiny particles consisting of non-crystalline silica. By-products of the production of *silicon metal* and the *ferrosilicon alloys* having silicon contents of 75% or more contain 85–95% non-crystalline silica. The by-product of the production of ferrosilicon alloy having 50% silicon has much lower silica content and is less pozzolanic. Therefore, SiO_2 content of the silica fume is related to the type of alloy being produced (Table 2.1).

Silica fume is also known as micro silica, condensed silica fume, volatilzed silica or silica dust.

The American concrete institute (ACI) defines silica fume as a "very fine non-crystalline silica produced in electric arc furnaces as a by product of production of elemental silicon or alloys containing silicon". It is usually a grey colored powder, somewhat similar to Portland cement or some fly ashes. It can exhibit both pozzolanic and cementitious properties.

Silica fume has been recognized as a *pozzolanic admixture* that is effective in enhancing the *mechanical properties* to a great extent. By using silica fume along with superplasticizers, it is relatively easier to obtain compressive strengths of order of 100–150 MPa in laboratory. Addition of silica fume to concrete improves the durability of concrete through reduction in the permeability, refined pore structure, leading to a reduction in the diffusion of harmful ions, reduces calcium hydroxide content which results in a higher resistance to sulfate attack. Improvement in durability will also improve the ability of silica fume concrete in protecting the embedded steel from corrosion.

Figure 2.1 shows the schematic diagram of silica fume production. The silica fume is collected in very large filters in the baghouse and then made available for use in concrete.

R. Siddique and M. Iqbal Khan, *Supplementary Cementing Materials*,
Engineering Materials, DOI: 10.1007/978-3-642-17866-5_2,

Table 2.1 SiO_2 content of silica fume produced from different alloy sources [4]

Alloy type	SiO_2 content SF (%)
50% ferrosilicon	61–84
75% ferrosilicon	84–91
Silicon metal	87–98

2.1.1 Availability and Handling

Silica fume is available in two conditions: dry and wet. Dry silica can be provided as produced or densified with or without dry admixtures and can be stored in silos and hoppers. Silica Fume slurry with low or high dosages of chemical admixtures are available. Slurried products are stored in tanks.

2.2 Properties of Silica Fume

2.2.1 Physical Properties

Silica fume particles are extremely small, with more than 95% of the particles finer than 1 μm. Its typical physical properties are given in Table 2.2. Silica fume colour is either premium white or grey (Fig. 2.2).

2.2.2 Chemical Composition

Silica fume is composed primarily of pure silica in *non-crystalline form*. X-ray diffraction analysis of different silica fumes reveals that material is essentially

Fig. 2.1 Schematic diagram of silica fume production

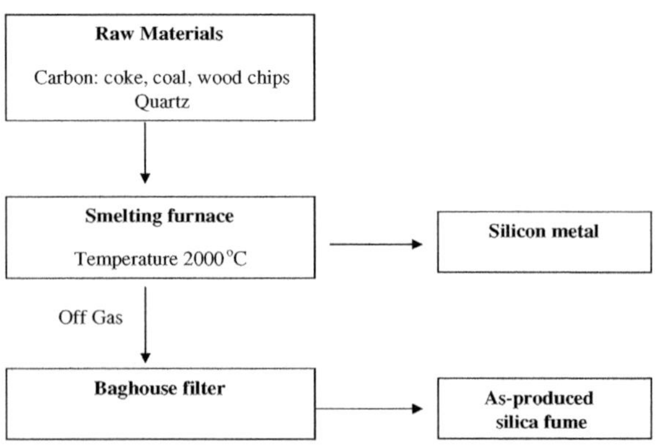

Table 2.2 Typical physical properties of silica fume [82]

Property	Value
Particle size (typical)	<1 μm
Bulk density	
As-produced	130–430 kg/m^3
Slurry	1,320–1,440 kg/m^3
Densified	480–720 kg/m^3
Specific gravity	2.22
Surface area (BET)	13,000–30,000 m^2/kg

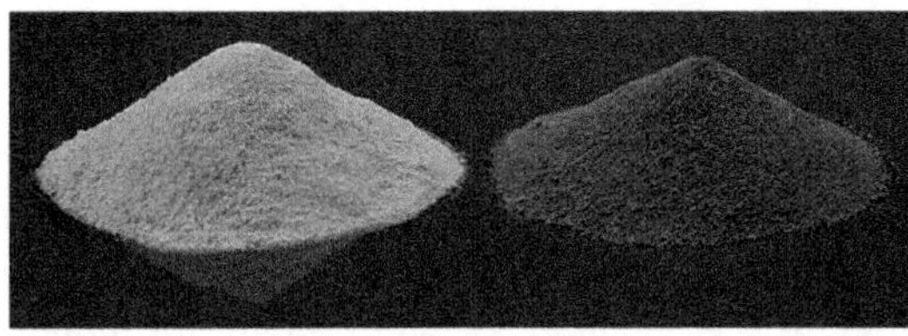

Fig. 2.2 Silica fume

vitreous silica, mainly of *cristobalite form*. Silica fume has a very high content of amorphous silicon dioxide and consists of very fine *spherical particles*. Silica fume generally contains more than 90% SiO_2. Small amounts of iron, magnesium, and alkali oxides are also found. Oxides analyses of silica fume as reported by some authors are given in Table 2.3.

2.3 Reaction Mechanism

Because of its extreme fineness and very high amorphous silicon dioxide content, silica fume is a very reactive *pozzolanic material*. As the Portland cement in concrete begins to react chemically, it releases calcium hydroxide. The silica fume reacts with this calcium hydroxide to form additional binder material called calcium silicate hydrate which is very similar to the calcium silicate hydrate formed

Table 2.3 Chemical composition of silica fume samples

Oxides	Sandvik and Gjørv [75]	Hooton and Titherington [39]	Yazici [98]
SiO_2	92.1	96.65	92.26
Al_2O_3	0.5	0.23	0.89
Fe_2O_3	1.4	0.07	1.97
CaO	0.5	0.31	0.49
MgO	0.3	0.04	0.96
K_2O	0.7	0.56	1.31
Na_2O	0.3	0.15	0.42
SO_3	–	0.17	0.33
LOI	2.8	2.27	–

from Portland cement. It is an additional binder that gives silica-fume concrete its improved properties. Mechanism of silica fume in concrete can be studied basically under three roles:

(i) *Pore-size Refinement* and Matrix Densification:

The presence of silica fume in the Portland cement concrete mixes causes considerable reduction in the volume of large pores at all ages. It basically acts as *filler* due to its *fineness* and because of which it fits into spaces between grains in the same way that sand fills the spaces between particles of coarse aggregates and cement grains fill the spaces between fine aggregates grains.

(ii) Reaction with Free-Lime (From Hydration of Cement)

CH crystals in Portland cement pastes are a source of weakness because cracks can easily propagate through or within these crystals without any significant resistance affecting the strength, durability and other properties of concrete. Silica fume which is siliceous and aluminious material reacts with CH resulting reduction in CH content in addition to forming strength contributing cementitious products which in other words can be termed as “Pozzolanic Reaction”.

(iii) Cement Paste–Aggregate Interfacial Refinement

In concrete the characteristics of the *transition zone* between the aggregate particles and cement paste plays a significant role in the *cement-aggregate bond.* Silica fume addition influences the thickness of transition phase in mortars and the degree of the orientation of the CH crystals in it. The thickness compared with mortar containing only ordinary Portland cement decreases and reduction in degree of orientation of CH crystals in transition phase with the addition of silica fume. Hence mechanical properties and *durability* is improved because of the enhancement in interfacial or bond strength. Mechanism behind is not only connected to chemical formation of C–S–H (i.e. pozzolanic reaction) at interface, but also to the microstructure modification (i.e. CH) orientation, porosity and transition zone thickness) as well.

2.4 Heat of Hydration

Silica fume is amorphous in nature and may contain some crystalline silica in the form of quartz or cristobalite. The higher surface area and amorphous nature of silica fume make it highly reactive. The hydration of C_3S, C_2S, and C_4AF are accelerated in the presence of silica fume [55, 94]. Grutzeck et al. [34] concluded that silica fume experiences rapid dissolution in the presence of $Ca(OH)_2$ and a supersaturation of silica with respect to a silica-rich phase. This unstable silica-rich phase forms a layer on the surface of the silica fume particles. The layer is then partly dissolved and the remainder acts as a substitute on which conventional C–S–H is formed.

Uchikawa and Uchida [94] reported that addition of silica fume accelerates the hydration of ordinary Portland cement at all stages of hydration. Immediately after mixing, the saturation factor of $Ca(OH)_2$ indicative of the concentrations of Ca^{2+} and OH^- ions, in the paste containing silica fume was reduced compared to that made of ordinary Portland cement. However, the saturation factor sharply increased to its maximum earlier than for ordinary Portland cement paste. During the course of hydration, the cumulative heat evolved due to hydration of ordinary Portland cement containing silica fume was always higher than from ordinary Portland cement paste. However, this trend may be reversed if water-reducing admixture is added to the mixing water. In the presence of melamine based water-reducer, the major hydration peak was accelerated in a silica fume cement paste. The cumulative heat evolved also increased in the presence of silica fume in the paste, and the higher the amount of silica fume in the paste, the greater the heat evolved and the shorter the hydration time [40].

Meland [63] observed that cumulative heat evolved is lower when paste containing silica fume and *lignosulfonate*. In addition, the higher the amount of silica fume, the smaller the amount of heat evolved. In the presence of lignosulfonate, the *hydration reaction* was retarded and less heat was evolved from paste containing silica fume.

Uchikawa [93] mentioned that use of excessive superplasticizer may cause substantial delays in setting times of cement paste containing silica fume.

Lohtia and Joshi [58] concluded that partial replacement of cement by silica fume results in reduction of *heat of hydration* without any reduction in strength. For a high strength concrete having 540 kg/m^3 cement and 10% cement replacement with silica fume, heat was 9% less compared to the mix without silica fume. Addition of silica fume may accelerate the temperature rise during the first 2–3 days, but a net decrease in temperature rise of silica fume concrete was observed at later stages (7–28 days) when compared to corresponding plain concrete. At early age, due to fast pozzolanic reaction of silica fume, a greater amount of heat is liberated compared to Portland cement. Ratio of heat liberated by pozzolanic activities of silica fume during the first 2–3 days per gram of silica fume to that of Portland cement is reported to be of the order of 1–2.

Langan et al. [56] studied the effect of silica fume on the heat of hydration of Portland cement. Silica fume was added as a partial replacement of cement at 10% by weight of the total *cementitious material*. Calorimeter tests were performed on these mixtures at water/cementitious ratios (w/cm) of 0.35, 0.40 and 0.50, up to a period of 24 h. However, several were carried on for 72 h to observe any later reactions. Effect of silica fume on the accumulative heat of hydration is shown in Table 2.4. It is evident that the presence of silica fume increased heat evolution during the first 30 min of hydration, and during the period from 8 to 24 h regardless of the w/cm ratio. Heat evolved during the dormant period remained almost constant for all mixtures, while the heat during the period from 2 to 8 h was reduced. Total heat evolved at 1 and 3 days was not changed by the presence of silica fume at w/c ratio of 0.35. Total heat at 1 day does increase with an increase in w/cm.

Table 2.4 Effect of silica fume on heat evolution of Portland cement hydration [56]

Mix type	Heat of hydration (Kcal/kg)						
	w/cm	0.0–0.5 h	0.5–2.0 h	2.0–8.0 h	8.0–24 h	Total at 1 day	Total at 3 days
0% SF	0.35	2.6	0.4	11.7	31.3	45.9	56.9
10% SF	0.35	3.1	0.4	8.7	34.5	46.6	56.1
0% SF	0.40	2.6	0.5	11.8	31.8	46.7	–
10% SF	0.40	3.2	0.4	10.3	33.6	47.4	–
0% SF	0.50	2.6	0.4	10.2	33.3	46.4	–
10% SF	0.50	3.2	0.5	9.7	35.3	48.7	–

Kadri and Duval [45] investigated the influence of silica fume on the hydration heat of concrete. Portland cement was replaced by *silica fume* (10–30% by mass) in concrete with w/(c + sf) ratios varying between 0.25 and 0.45. The heat of hydration was monitored continuously by a semi-adiabatic calorimetric method for 10 days at 20°C. They concluded that (i) hydration rate of silica fume concretes mainly depends on two parameters: the w/(c + sf) ratio and the silica fume content. Regardless of silica fume in concrete, it is evident that hydration rate decreased with an increase in w/(c + sf) ratio. The reduction in the hydration rate for the lower w/(c + sf) ratios is due to the lack of water available for cement hydration, whereas increased superplasticizer addition may be accounted for extension for the dormant period. On the contrary, for the 0.45 w/(c + sf) ratio, the amount of water is enough to fill the voids and coat the cement grains which enable a greater hydration of cement; (ii) addition of silica fume altered the hydration process at very early ages. For three lower w/(c + sf) ratios, the hydration heat increased with silica fume in the first hour; (iii) difference in the rate of heat evolution between a 30% silica fume concrete and the plain concrete exceeds 50% for the 0.25 and 0.35 w/(c + sf) ratios. On the contrary for the 0.45 ratio, the heat evolution rate of the reference concrete is higher than that of the silica fume concrete; and (iv) total heat of hydration (Fig. 2.3) in the ten first days depends on the silica fume content. With the increase in silica fume content, quantity of cement reduces. The reduction in the cement content tends to diminish the total heat released, while the pozzolanic reaction tends to increase it. A 10% substitution of Portland cement increased the cumulative heat of hydration as compared to control concrete (0% silica fume) due to the pozzolanic effect. At 30% silica fume content, total heat of hydration decreased since at a later stage the hydration rate of cement slows down and forms less $Ca(OH)_2$. The pozzolanic reaction is controlled by the $Ca(OH)_2$ formation and depends on the available amount of $Ca(OH)_2$.

2.5 Silica Fume Efficiency

Silica fume efficiency [8] in concrete is not constant at all percentages of replacement. The "overall efficiency factor" of silica fume can be assessed in two

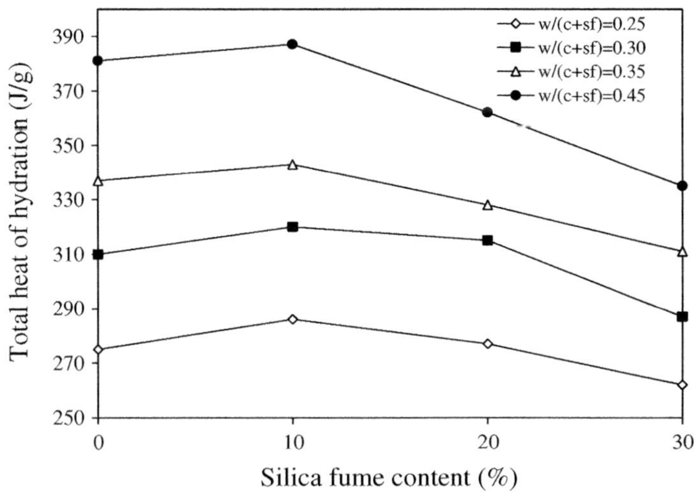

Fig. 2.3 Total heat of hydration over 10 days as a function of silica fume content [45]

separate parts; "general efficiency" which is constant at all percentages of replacement and the "percentage efficiency factor" which varies with the replacement percentage. The activity of silica fume in concrete is obtained in terms of the amount of cement replaced through its "cementing efficiency factor" (K). Efficiency factor for silica fume in concrete can be defined as the number of parts of cement that may be replaced by one part of the silica fume, without changing the property being investigated generally the compressive strength.

$K = (K_e) \times K_p$
K = Overall Efficiency Factor
K_p = Percentage Efficiency Factor (K_p) $= 0.0015\ \mathrm{pr}^2 - 0.3671\ \mathrm{pr} + 2.8502$
K_e = General Efficiency Factor (K_e). It is taken as 3, usually kept constant for all the percentages of replacement.
pr = the percentage of silica fume in the total cementitious materials

2.6 Advantages of Using Silica Fume

- High early *compressive strength*
- High tensile, *flexural strength*, and *modulus of elasticity*
- Very low *permeability* to chloride and water intrusion
- Enhanced durability
- Increased *toughness*
- Increased *abrasion resistance* on decks, floors, overlays and marine structures

- Superior resistance to chemical attack from chlorides, acids, nitrates and sulfates and life-cycle cost efficiencies.
- Higher *bond strength*
- High *electrical resistivity* and low permeability

2.7 Applications of Silica Fume

- *High Performance Concrete* (HPC) *containing silica fume*—for highway bridges, parking decks, marine structures and bridge deck overlays which are subjected to constant deterioration caused by rebar corrosion current, abrasion and chemical attack. Silica fume will protect concrete against deicing salts, seawater, road traffic and freeze/thaw cycles. Rebar corrosion activity and concrete deterioration are virtually eliminated, which minimizes maintenance expense.
- *High-strength concrete enhanced with silica fume*—provides architects and engineers with greater design flexibility. Traditionally used in high-rise buildings for the benefit of smaller columns (increasing the usable space) high-strength concrete containing silica fume is often used in precast and prestressed girders allowing longer spans in structural bridge designs.
- *Silica-fume Shotcrete*—delivers greater economy, greater time savings and more efficient use of *sprayed concrete*. Silica fume produces superior shotcrete for use in rock stabilization; mine tunnel linings, and rehabilitation of deteriorating bridge and marine columns and piles. Greater bonding strength assures outstanding performance of both wet and dry process shotcreting with less rebound loss and thicker applications with each pass of the shotcrete nozzle.
- *Oil Well Grouting*—whether used for primary (placement of grout as a hydraulic seal in the well-bore) or secondary applications (remedial operations including leak repairs, splits, closing of depleted zones); the addition of silica fume enables a well to achieve full production potential. Besides producing a blocking effect in the oil well grout that prevents gas migration, it provides these advantages such as (i) Improved flow, for easier, more effective application; (ii) dramatically decrease permeability, for better control of gas leakage; and (iii) lightweight
- *Repair Products*—silica fume is used in a variety of cementitious repair products. Mortars or grouts modified with silica fume can be tailored to perform in many different applications—overhead and vertical mortars benefit from silica fume's ability to increase surface adhesion. Silica fume significantly improves cohesiveness making it ideal for use in underwater grouts, decreases permeability in grouts used for post-tensioning applications and increases the resistance to aggressive chemicals.

- *Refractory and Ceramics*—the use of silica fume in refractory castables provides better particle packing. It allows for less water to be used while maintaining the same flow characteristics. It also promotes low temperature sintering and the formation of mullite in the matrix of the castable. This produces a castable that has a low permeability to avoid gas, slag and metal penetration. Castables incorporating silica fume are stronger than non-silica fume containing castables especially at high temperatures with higher density they attain lower porosity and are more volume stable.

2.8 Effect of Silica Fume on Fresh Properties of Cement/Mortar/Concrete

Rheological properties of a fresh cement paste play an important role in determining the *workability* of concrete. The water requirement for flow, hydration behaviour, and properties of the hardened state largely depends upon the degree of dispersion of cement in water. Properties such as fineness, particle size distribution, and mixing intensity are important in determining the rheological properties of cement paste. Due to the charges that develop on the surface, cement particles tend to agglomerate in the paste and form flocs that trap some of the mixing water. Factors such as water content, early hydration, water reducing admixtures and mineral admixtures like silica fume determine the degree of flocculation in a cement paste.

Fresh concrete containing silica fume is more cohesive and less prone to *segregation* than concrete without silica fume. Concrete containing silica fume shows substantial reduced *bleeding*. Additionally silica fume reduces bleeding by physically blocking the pores in the fresh concrete. Use of silica fume does not significantly change the unit weight of concrete.

2.8.1 Consistency

Rao [72] determined the influence of silica fume on the *consistency* of cement pastes and mortars. *Specific gravity* and *specific surface* of the silica fume were 2.05 and 16,000 m^2/kg, respectively. Silica fume was varied from 0 to 30% at a constant increment of 2.5/5% by weight of cement. Since the SF is finer than the cement, the specific surface increased with increase in SF content. The standard consistency of pure cement paste was found out to be 31.50%; while at 30% SF, it was 44.25%. It was observed that the consistency of cement increased with the increase in SF content. As much as 40% of additional water requirement was observed for cement pastes containing 20–30% SF. Figure 2.4 shows the variation of consistency of cement at different silica fume contents.

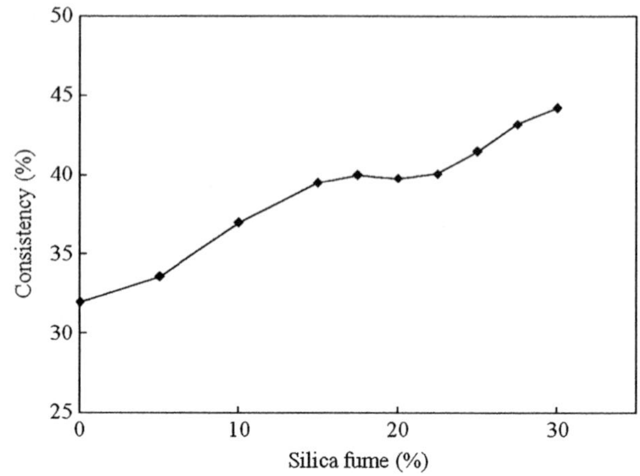

Fig. 2.4 Variation of consistency of cement pastes containing different percentages of silica fume [72]

Qing et al. [71] examined the influence of nano-SiO_2 (NS) addition on consistency of cement paste incorporating NS or silica fume. The influence of NS or silica fume addition on consistency and setting time of fresh pastes is given in Table 2.5. It was found that with increasing the NS content, fresh pastes for sample A-series grew thicker gradually and their penetration depths (consistency value) decreased gently as compared with that of control sample CO. While with increasing the silica fume content, the pastes for sample B-series grew thinner and their depths increased. They concluded that silica fume makes cement paste thinner as compared with NS.

2.8.2 Setting Times

Alshamsi et al. [3] reported that addition of micro-silica lengthened the *setting time* of pastes. This was expected since micro-silica replaces part of the OPC, reducing the early stiffening potential. While the addition of *micro-silica* (10%)

Table 2.5 Mix proportions, consistency of pastes made of cement and NS and silica fume [71]

Sample	Mix proportion in mass					Consistency (mm)
	Cement	Nano silica	Silica fume	Water	SM	
CO	100	0	0	22	2.5	34
A1	99	1	0	22	2.5	34
A2	98	2	0	22	2.5	33
A3	97	3	0	22	2.5	33
A5	95	5	0	22	2.5	32
B2	98	0	2	22	2.5	35
B3	97	0	3	22	2.5	35
B5	95	0	5	22	2.5	36

had little effect on setting times, higher percentages produced significant influences. There was 6–20% increase in setting times when OPC was replaced with 20% micro-silica.

Lohtia and Joshi [58] concluded that the addition of silica fume to concrete in the absence of water-reducer or superplasticizer causes delay in setting time, compared to non-silica fume concrete of equal strength, especially when the silica fume content was high. The additions of 5–10% *silica fume* to either superplasticized or non-superplasticized concrete with W/(C + SF) ratio of 0.40 did not exhibit any significant increase in setting time. However, when 15% silica fume was added with superplasticizer, both the initial and final setting times were delayed by approximately 1 and 2 h, respectively. The observed delay was attributed to the relatively high dose of superplasticizers needed for the high amount of silica fume added to concrete.

Rao [72] studied the influence of silica fume on the setting time of cement paste. Specific gravity and specific surface of the silica fume were 2.05 and 16,000 m^2/kg, respectively. Figure 2.5 shows the variation of setting times with the addition of silica fume in cement pastes. It was observed that initial setting time decreased with the increase in silica fume content. At smaller contents, the setting time of cement paste did not affect much. However, at higher silica fume contents, the initial setting time was significantly decreased. At 30% silica fume, the initial setting time had been only 30 min. The final setting time seem to be not influenced by the silica fume. The pozzolanic action of silica fume seems to be very active at early hours of hydration. Therefore, he concluded that silica fume contents result in quick setting of cement.

Qing et al. [71] investigated the influence of nano-SiO_2 (NS) on the setting time of cement paste incorporating NS or silica fume. The influence of NS or silica fume addition on consistency and setting time of fresh pastes is presented in

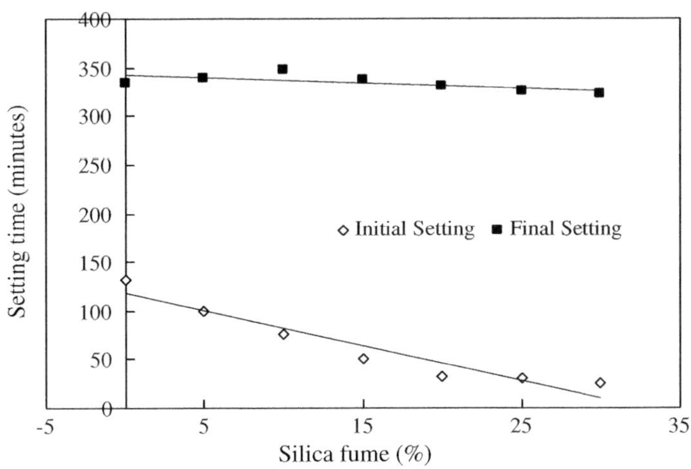

Fig. 2.5 Variation of setting time with different percentages of silica fume addition [72]

Table 2.6. It was observed that the setting of fresh pastes (sample A-series) was slightly accelerated but the difference between initial and final setting time decreased with increase in NS content. While the setting of fresh pastes (sample B-series) was obviously retarded and the difference was also decreased with increasing the silica fume content. They concluded that silica fume makes cement paste thinner and retards the cement setting process as compared with NS.

2.8.3 Workability

Sellevold and Redjy [80] reported that there is net decrease in water requirements in concretes containing high concentration of silica fume and *water-reducer* or superplasticizers. The addition of water-reducer or a superplasticizer causes the dispersion of cement and silica fume particles and reduces the concentration of contact points between the different grains; resulting in less water requirement to achieve a given consistency.

Alshamsi et al. [3] highlighted that addition of micro-silica to cement pastes or concretes leads to lower *workability*. Such effect can result in higher water demand to maintain a constant slump. Hence water-reducing admixtures or superplasticizers should be dosaged by weight of micro-silica in order to keep water demand similar to that of control. Table 2.7 shows that higher the replacement level, the larger the water demand and larger the superplasticizer dosage required to maintain the standard *consistency*. The physical properties of micro-silica are known to reduce workability mainly due to small particle size that leads to higher water demand. The workability of concrete mix (9) containing mineral admixture is considerably improved by using chemical admixture. The combination of a superplasticizer and a mineral admixture (silica-fume) is desirable, since silica fume in the amount exceeding 5% from the mass of cement considerably increases the fine fraction volume and hence the water requirement of the binder.

Table 2.6 Mix proportions, setting time of pastes made out of cement and NS and silica fume [71]

Sample	Mix proportion in mass					Setting time	
	Cement	Nano silica	SF	Water	SM	Initial	Final
CO	100	0	0	22	2.5	2 h 57 m	4 h 23 m
A1	99	1	0	22	2.5	2 h 57 m	4 h 05 m
A2	98	2	0	22	2.5	2 h 55 m	3 h 50 m
A3	97	3	0	22	2.5	2 h 48 m	3 h 40 m
A5	95	5	0	22	2.5	2 h 16 m	3 h 06 m
B2	98	0	2	22	2.5	3 h 50 m	4 h 45 m
B3	97	0	3	22	2.5	4 h 35 m	5 h 20 m
B5	95	0	5	22	2.5	4 h 45 m	5 h 28 m

Table 2.7 Mix proportions for standard consistency [3]

Mix #	OPC	Micro-silica	Water	Admixture (ml)
1	100	0	27.5	0
2	95	5	30	0
3	90	10	32	0
4	85	15	37.5	0
5	80	20	43	0
6	95	4	27.5	3
7	90	10	27.5	5
8	85	15	27.5	6.5
9	80	20	27.5	8

Khayat and Aitcin [50] reported that addition of 10% silica fume in a lean concrete (100 kg/m^3) of cement reduced the water demand. However, it exhibited poor durability against *freeze–thaw attack*. In normal structure concrete, even with 5% silica fume addition, the water demand is increased to maintain constant *slump*. For producing very high strength and durable concrete, silica fume up to 10% is added as an admixture and use of superplasticizer to maintain specified slump is found necessary. When no plasticizers are used an additional 1 l/m^3 of water should be used for every 1 kg/m^3 of silica fume addition to maintain constant level of fluidity.

Wong and Razak [97] studied the *cementing efficiency factor* (k) of silica fume. Specific gravity of silica fume was 2.22. Three water-to-cementitious material ratios (w/cm) of 0.27, 0.30 and 0.33 were used in concrete mixtures. At each w/c ratio, cement was replaced with 0, 5, 10, and 15% silica fume. Slump and *Vebe time* results are shown in Table 2.8. It could be seen from this table that mixtures achieved slump values ranging from 30 to 260 mm, while Vebe time was in the range of 1–15 s. The large variation of workability across mixtures was due to the constant superplasticizer dosage used for mixtures with the same w/cm ratio.

Mazloom et al. [62] made *high-performance concrete* containing silica fume. The silica fume content was 0, 6, 10, and 15%, and water–cementitious ratio being 0.35. The water/cement ratio and the slump of control high-strength concrete were 0.35 and 100 ± 10 mm, respectively. The same water/binder ratio of 0.35 was used for the other concrete mixes with the same slump. Consequently, the dosage of superplasticizer changed due to the effect of the different levels of silica fume. Details of mix proportions for concrete containing different levels of silica fume

Table 2.8 Workability characteristics [97]

Mixture	w/cm	Slump (mm)	Vebe (s)	w/cm	Slump (mm)	Vebe (s)	w/cm	Slump (mm)	Vebe (s)
C	0.27	165	8	0.30	225	3	0.33	240	1
SF 5		100	8		215	3		180	3
SF 10		50	12		117	5		100	6
SF 15		35	15		30	16		35	16

Table 2.9 Mix proportions of concrete containing different levels of silica fume [62]

Mix components	Concrete mixes			
	OPC	SF 6	SF 10	SF 15
Cement (kg/m^3)	500	470	450	425
Silica fume (kg/m^3)	–	30	50	75
Superplasticizer (kg/m^3)	8.17	9.78	11.71	13.34

Gravel: 1,203 kg/m^3, sand: 647 kg/m^3, water: 175 kg/m^3, w/cm = 0.35

are given in Table 2.9. It was observed that mixes incorporating higher silica fume content tended to require higher dosages of superplasticizer. The higher demand of superplasticizer with the concrete containing silica fume was attributed to the very fine particle size of silica fume that causes some of the superplasticizer being adsorbed on its surface. Also mixes incorporating more silica fume were more cohesive, which is in agreement with the findings of Khatri et al. [47].

Rao [72] studied the influence of silica fume on the workability (*flow table* test) of mortars. Specific gravity and specific surface of the silica fume were 2.05 and 16,000 m^2/kg, respectively. The *specific gravity* and the *bulk density* of sand were 2.68 and 1,584 kg/m^3, respectively. The cement–sand ratio in the mortars was 1:3, and w/b ratios were 0.45 and 0.50. Figure 2.6 demonstrates the variation of workability as percentage flow at w/cm ratios 0.45 and 0.50 with different silica fume contents. It is evident from these results that workability of mortar slightly decreased as the silica fume content increased. This has been due to the higher specific surface of silica fume, which needs more water for complete hydration and for workability. When very fine particles of silica fume are added to the mortar, the size of flow channels further reduced because these fine particles are able to adjust

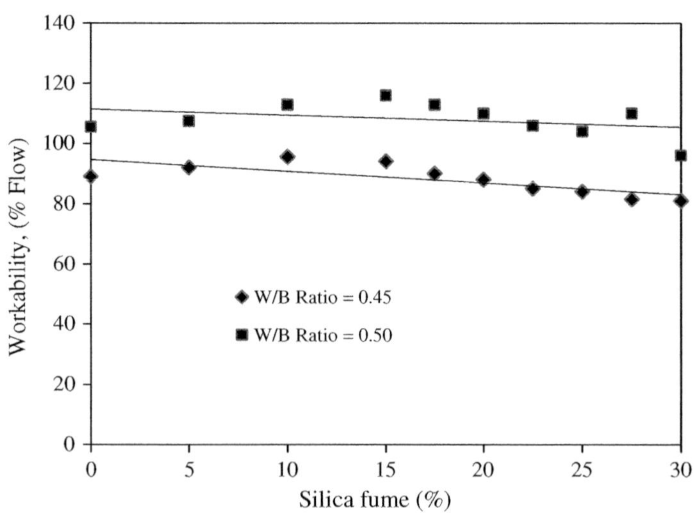

Fig. 2.6 Variation of workability of cement mortars with silica fume at different contents [72]

their positions to occupy the empty spaces between cement particles. Due to increase in the number of contact points between solid particles, the cohesiveness of mortar mixture greatly improves when silica fume is present. In fact, the presence of too much silica fume in mortar (>10% by weight of cement) tends to make the mixture stiff. The addition of small amounts of silica fume does not require the use of extra water or super plasticizers. However, with higher dosages of silica fume, the workability of mortar was found to be reduced.

2.9 Effect of Silica Fume on the Hardened Properties of Cement/Mortar/Concete

2.9.1 Compressive Strength

When silica fume is added to concrete, it results in a significant change in the *compressive strength* of the mix. This is mainly due to the aggregate-paste bond improvement and enhanced *microstructure*.

2.9.1.1 Compressive Strength of Cement Paste/Mortar

Huang and Feldman [41] found that mortar without silica fume has lower strength than cement paste with the same water–cement ratio, while mortar with 30% of cement replaced with silica fume has a higher strength than cement–silica fume paste with the same water–cementitious ratio. They concluded that the addition of silica fume to mortar resulted in an improved bond between the hydrated cement matrix and sand in the mix, hence increasing strength. This improved bond is due to the conversion of the calcium hydroxide, which tends to form on the surface of aggregate particles, into calcium silicate hydrate due to the presence of reactive silica.

Cong et al. [20] observed that the replacement of cement by silica fume (up to 18%) and the addition of superplasticizer increased the strength of cement paste. Concrete containing silica fume as a partial replacement of cement exhibited an increased compressive strength largely because of the improved strength of cement paste matrix. But, changes in paste aggregate interface caused by the incorporation of silica fume had little effect on the compressive strength of concrete.

Gleize et al. [30] determined *compressive strength* of *silica fume* mortar having proportion 1:1:6(Cement + silica fume: lime: sand). The results are given in Table 2.10. They concluded that in Portland cement mortars, silica fume acts mainly at the interface paste-aggregate, where there is a higher concentration of calcium hydroxide and greater porosity than in paste. In Portland cement mortars with silica fume, lime is better suited in the paste and there is no evidence of concentration of silica fume at the interface paste aggregate.

Table 2.10 Mortar compressive strength [30]

Age (days)	0% Silica fume	10% Silica fume
7	3.26 ± 0.12	2.93 ± 0.13
28	6.58 ± 0.19	7.11 ± 0.25

Gutiérrez et al. [36] studied the effect of silica fume on the compressive strength of *fibre reinforced mortar*. Different types of natural and synthetic fibres were also used. These were embedded in the blended cement mortars in the proportion of 2.5% by weight of the cement. The effect of addition of silica fume in plain mortar is shown in Fig. 2.7. Incorporation of silica fume increased the average compressive strength by 23%. But fibres in the plain mortar caused a reduction in its compressive strength. However this loss was compensated by addition of silica fume into the matrix. The compressive strength of the matrix reinforced with glass fibres gained an increment of up to 68% when silica fume was included as a part of the cementitious material.

2.9.1.2 Compressive Strength of Normal Strength/High-Performance Concrete

Bentur et al. [11] reported that the strength of *silica fume concrete* is greater than that of silica fume paste which they attributed to the change in the role of the aggregate in concrete. In cement concrete, the aggregate functions as inert filler but due to the presence of weak *interfacial zone*, composite concrete is weaker then cement paste. But, in silica fume concrete, the presence of silica fume eliminates this weak link by strengthening the cement paste aggregate bond and forming a less porous and more homogenous microstructure in the interfacial

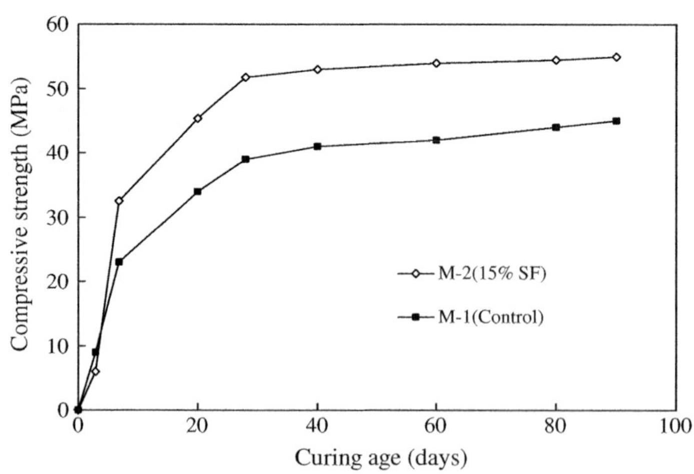

Fig. 2.7 Effects of addition of silica fume in plain mortar [36]

Table 2.11 Development of compressive strength with age (MPa) [62]

Concrete mixes	Silica fume (%)	Compressive strengths (MPa)						
		7 days	14 days	28 days	42 days	90 days	365 days	400 days
OPC	0	46	52	58	62	64	73	74
SF 6	6	50.5	58	65	69	71	73	73
SF 10	10	52	61	67.5	71	74	73	73
SF 15	15	53	63	70	73	76	75	76

region. Thus, silica fume concrete is stronger than silica fume cement paste, taking into account that the strength of aggregate exceeds the strength of cement paste.

Mazloom et al. [62] investigated the compressive strength of high performance concrete containing silica fume. The silica fume content was 0, 6, 10, and 15%, and water–cementitious ratio being 0.35. The results are given in Table 2.11. From the results it can be seen that (i) at the age of 28 days, the silica fume concrete was 21% stronger than control concrete; (ii) *compressive strength* development of concrete mixtures containing silica fume was negligible after the age of 90 days; however, there was 26% and 14% strength increase in the control concrete after 1 year compared to its 28 and 90 days strength, respectively. Also the tests showed that at the age of 400 days, the compressive strength of control concrete and concrete mixes containing different proportions of silica fume were the same. According to Wild et al. [96], this difference in strength development in OPC concrete and silica fume concrete can be attributed to the rapid formation of an inhibiting layer of reaction product preventing further reaction of silica fume with calcium hydroxide beyond 90 days.

Sobolev [84] studied the compressive strength of high performance concretes. The compressive results of HPC mixture are shown in Table 2.12. It was observed that (i) increase in superplasticizer dosage from 8 to 18% led to a reduction of w/c from 0.31 to 0.26 and improved the concrete compressive strength from 86 to 97 MPa; (ii) maximum compressive strength of 91 MPa was obtained at 15% silica fume.; (iii) lower strength value of 90 MPa occurred at 10 and 20% silica fume; and (iii) reduction of w/c 0.32–0.19 increased the compressive strength of cement concrete and resulted in super high strength concrete having strength up to 135 MPa.

Table 2.12 Details of HPC mixtures [84]

Proportions (kg/m^3)	SF (5%)	SF (10%)	SF (15%)	SF (20%)
Cement	426	449	468	478
Silica fume	22	50	83	120
Age	Compressive strength (MPa)			
1 day	16.8	24.1	34.4	45.1
3 days	28.6	42.2	63.0	84.9
7 days	50.1	67.2	84.8	102.5
28 days	60.0	80.0	100.0	120.0

Wong and Razak [97] studied the compressive strength of concrete containing silica fume. Concrete mixtures with w/cm *ratios* of 0.27, 0.30 and 0.33 were prepared. At each w/c ratio, cement was replaced with 0, 5, 10, and 15% silica fume. The results are given in Table 2.13. They observed that (i) silica fume did not produce an immediate strength enhancement; instead, the blended mixtures only achieved higher strength than the control from 7 days onwards. Strength loss in the early ages, which was proportional to the cement replacement level, was probably due to the dilution effect of the *pozzolan* and as well as the slow nature of pozzolanic reaction; and (ii) after 90 days of curing, the average strength enhancement with 10% silica fume achieved 17% increment. It was also found that reducing the w/c ratio from 0.30 to 0.27 did not trigger a significant strength enhancement as anticipated.

Poon et al. [69] reported the results of *compressive strength* of *high-performance concrete* with silica fume. Two series of concrete mixes were prepared at the w/b ratios of 0.30 and 0.50. Each series included 5 and 10% silica fume. Compressive strength results are given in Table 2.14. It is clear from the results that the incorporation of silica fume did not result in any strength increase for the concrete at 3 days, although it increased the strength at the ages of or after 7 days.

Behnood and Ziari [10] designed concrete mixtures to evaluate the effect of silica fume on the compressive strength of the *heated and unheated concrete* specimens. Three mixtures were made with a constant water-to-cement ratio (w/c) 0.30. The dosages of replacing cement by silica fume were 0% (W30OPC), 6% (W30SF6) and 10% (W30SF10). One mixture was prepared with w/c of 0.40 without silica fume (W40OPC), whereas other concrete was produced with w/c of 0.35 containing 6% silica fume (W35SF6). The results of the *compressive strength* are given in Table 2.15. As was expected, the replacement of cement by 6 and 10% silica fume increased the 28-day compressive strength approximately by 19

Table 2.13 Cube compressive strength [97]

Mixture	Compressive strength (MPa)						
	1 day	3 days	7 days	28 days	56 days	90 days	180 days
w/cm 0.27	39	68	72.5	84	86.5	87.5	90
SF 5	35	63	75.5	88.5	93	96.5	97.5
SF 10	25	61	79	95.5	100	104	107
SF 15	24.5	59.5	76.5	101	103.5	106	109
w/cm 0.30	48	63.5	72	83.5	84.5	85.5	87.5
SF 5	46	62	81	91	95.5	95.5	97
SF 10	42	61.5	78.5	95	97	99	103
SF 15	38	57.5	74.5	98.5	101.5	104	106.5
w/cm 0.33	41.0	58.0	62.5	75	78	79	81.5
SF 5	35.0	55.0	69.5	83.0	85.0	90.0	90.0
SF 10	32.0	53.0	70.5	89.5	90.5	92.0	93.5
SF 15	31.0	47.5	70.5	88.5	93.0	95.5	100.5

Table 2.14 Compressive strength of control and blended concrete [69]

Series	w/b	Mix	Compressive strength			
			3 days	7 days	28 days	90 days
1	0.30	Control	68.5	81.1	96.5	102.5
		5% SF	67.0	79.3	106.5	110.2
		10% SF	63.2	76.9	107.9	115.6
2	0.50	Control	28.6	41.2	52.1	60.4
		5% SF	27.4	47.0	54.3	67.5
		10% SF	25.8	47.4	58.4	69.1

Table 2.15 Results of compressive strength at different temperatures [10]

Mixture name	SF (%)	w/c	Compressive strength (MPa)					
			20°C		100°C	200°C	300°C	600°C
			7-day	28-day				
W40OPC	0	0.40	48.3	61.8	53.3	55.5	46.5	20.6
W35SF6	6	0.35	61.5	73.9	62.8	64.7	56.5	21.8
W30OPC	0	0.30	55.3	67.4	57.6	59.7	49.0	21.0
W30SF6	6	0.30	69.1	80.3	68.0	69.0	56.5	23.4
W30SF10	10	0.30	74.1	84.2	70.8	71.7	57.9	22.6

and 25% respectively. This was due to the reaction of silica fume with calcium hydroxide formed during the hydration of cement that caused the formation of *calcium silicate hydrate* (C–S–H). It was also due to the filler role of very fine particles of silica fume. Furthermore, concretes containing different levels of silica fume showed lower rates of compressive strength gain in early ages. They concluded that (i) Concrete containing silica fume had significantly higher strength than that of OPC concrete at room temperature. After exposure to 100°C, significant reductions occurred in the compressive strength of concrete with and without silica fume; (ii) In the range 300–600°C, severe strength losses occurred in all three concretes, which were 68.8, 70.9 and 73.2% of the initial values for W30OPC, W30SF6 and W30SF10 concretes respectively. This was because during exposure to high temperatures, cement paste contracts, whereas aggregates expand. Thus, the transition zone and bonding between aggregates and paste are weakened. As a result, this process as well as chemical decomposition of hydration products causes severe deteriorations and strength loses in concrete after subjecting to high temperatures; and (iii) After heating to 600°C, the residual compressive strength of all three concretes were approximately same, whereas the relative residual compressive strengths of concretes containing 6 and 10% silica fume were 6.7 and 14.1% lower than those of the OPC concretes, respectively, after exposure to 600°C. Therefore, the rate of strength loss was significantly higher in silica fume concretes. This was attributed to the presence and amount of silica fume in concretes that produced very dense transition zone between aggregate and paste due to ultra fine particles as filler.

Köksal et al. [53] studied the compressive strength of *steel fibre reinforced concrete* with silica fume. Cold drawn steel fibres with hooked ends were used. Aspect ratios (l/d) of fibres were 65 and 80 and volume fractions (Vf) of steel fibres were 0.5 and 1%. Silica fume content was 0, 5, 10, and 15% weight of cement. The test results are given in Table 2.16. They observed that (i) a considerable increase in the compressive strength of the concretes without steel fibres by increasing the *silica fume* content. The increases were 12, 73.4 and 85.5% for 5, 10 and 15% silica fume, respectively. These result were clearly dependant on increasing *bond strength* of *cement paste–aggregate interface* by means of filling effect of silica fume; and (ii) compressive strengths of concretes produced by additions of both steel fibre and silica fume had higher than the ones containing silica fume only.

2.9.1.3 Effect of Curing on the Compressive Strength of Concrete

Bentur and Goldman [12] studied the effect of water and *air-curing* in mild environmental conditions on the compressive strength at the age of 90 days. The air curing resulted in a somewhat lower strength compared to continuous water curing. This was attributed to the observations that the strengthening influence of the silica fume takes place quite early during the period 1–28 days and possibly slower rate of drying from within the silica fume concrete, which apparently developed a tight micro-structure after 7 days of water curing. Similar trends were

Table 2.16 Test results of concrete steel fibre concrete [53]

Series	Silica fume (%)	Steel fibre content (%)	Compressive strength (N/mm^2)
A	0	0	32.4
		0.5	33.4
		1.0	37.4
		0.5	34.1
		1.0	38.5
B	5	0	36.4
		0.5	38.3
		1.0	48.1
		0.5	41.4
		1.0	45.7
C	10	0	56.2
		0.5	60.4
		1.0	66.9
		0.5	59.7
		1.0	63.7
D	15	0	60.1
		0.5	66.5
		1.0	69.3
		0.5	63.2
		1.0	70.5

observed for 28 days except that the strength values were usually lower by about 10%.

Hooton [38] determined the compressive strengths of concretes containing 0, 10, 15, and 20% silica fume up to the age of 5 years. Results are presented in Table 2.17. While strengths of the concretes containing silica fume were higher at ages between 7 and 91 days, the Portland cement concrete continued to gain strength at later ages; with a 55% increase between 28 days and 5 years. In contrast, the long-term strength gain of silica fume concretes were very low and 5-year strengths were ±12% of the day strengths. Reductions in strengths were noted to be within the normal variation of strengths observed in long-term studies with Portland cement concretes.

2.9.1.4 Compressive Strength of Recycled Aggregate Silica Fume Concrete

González-Fonteboa and Martínez-Abella [32] studied the properties of concrete using *recycled aggregates* from Spanish demolition debris (RC mixes) and the impact of the addition of silica fume on the properties of recycled concrete (RCS mixes). A comparison was made between both these materials and standard conventional concrete (CC mixes), which was also modified by adding silica fume (CCS mixes). It also aimed to study the effect of addition of silica fume on the basic properties of recycled concrete. For the test four series (10 CC, 7CCS, 10 RC and 6 RCS) of mixes were made. Compressive strength results are shown in Fig. 2.8. They reported that (i) *Pozzolanic effect* of silica fume was seen between 7 and 21 days which tends to increase the compressive strength of the concrete; and (ii) concrete containing 8% silica fume displayed greater compressive strength than concretes that did not contain this admixture, at all ages.

Almusallam et al. [6] investigated the effects of silica fume on the compressive strength of concrete made with *low-quality coarse aggregates*. Four types of low quality coarse aggregates, namely calcareous, dolomitic, and quartzitic limestone

Table 2.17 Moist-cured concrete compressive strength development [38]

Testing age	Concrete mixture details				
	Control	10% SF	15% SF	20% SF	10% SF in slurried product
1 day	25.6	25.2	28.0	27.2	33.2
7 days	44.6	59.8	63.2	64.5	69.9
28 days	55.6	70.7	75.2	74.2	85.0
56 days	63.5	74.0	76.4	72.9	82.6
91 days	63.4	77.6	72.9	74.0	83.7
182 days	72.5	73.2	71.4	77.8	82.7
365 days	79.0	76.7	70.0	80.3	80.2
2 years	86.1	81.5	71.4	82.0	–
3 years	88.2	89.7	85.1	88.3	–
5 years	86.0	79.6	68.6	70.3	–

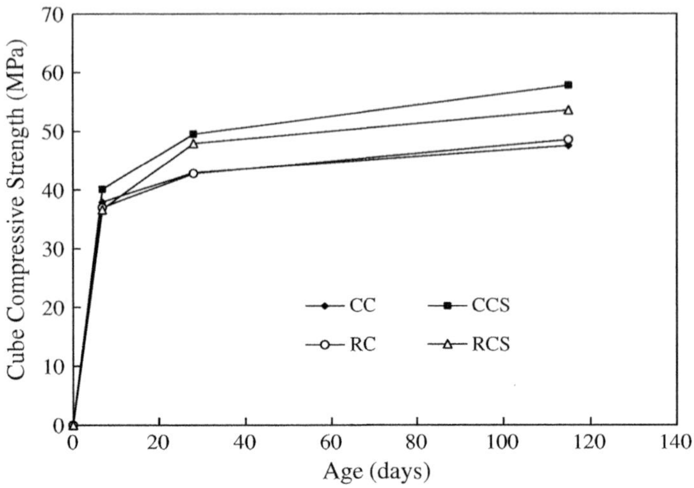

Fig. 2.8 Development of cube compressive strength [32]

and steel slag were used, and silica fume content was 10 and 15% as partial replacement of cement. The concrete specimens had a w/c ratio of 0.35 and a coarse aggregate to fine aggregate ratio of 1.63. The influence of aggregate quality on the compressive strength of 15% silica fume cement concretes is shown in Fig. 2.9. They observed that compressive strength increased with age in all the concrete specimens. After 180 days of curing, highest compressive strength was noted in the 15% silica fume cement concrete specimens (54 MPa) followed by

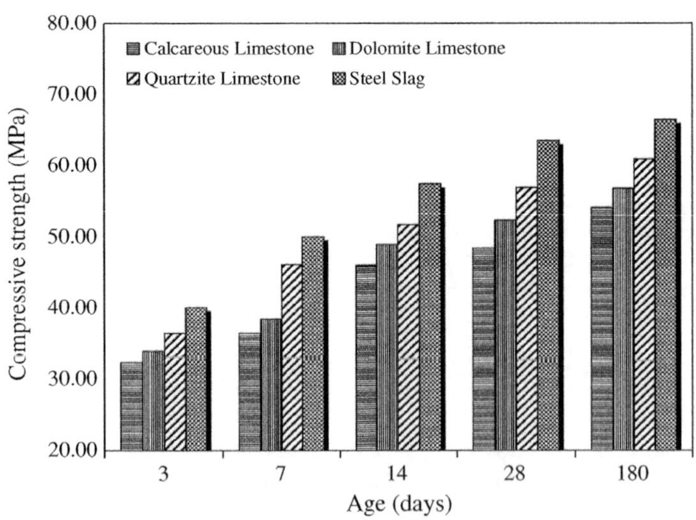

Fig. 2.9 Effect of aggregate type on the compressive strength of 15% silica fume cement concrete [6]

those prepared with 10% silica fume (52 MPa), and plain cement concrete (49 MPa). The higher compressive strength noted in the silica fume cement concrete, compared to plain cement concrete, may be attributed to the reaction of the silica fume with calcium hydroxide liberated during the hydration of cement. Khatri et al. [47] stated that it results in formation of secondary calcium silicate hydrate that fills up the pores due to the hydration of the initial calcium silicate hydrate.

Babu and Babu [9] studied the use of *expanded polystyrene* (EPS) beads as *lightweight aggregate* both in concrete and mortars containing silica fume as a supplementary cementitious material. Three percentages of silica fume—3, 5 and 9% (by weight of the total cementitious materials) were used. They concluded that the rate of strength development was greater initially and decreased as the age increased. A comparison of strengths at 7 days reveals that concretes with 3% silica fume developed almost 75% of its 28-day strength, while that with 5 and 9% silica fume developed almost 85 and 95% of the corresponding 28-day strength. They concluded that rate of strength gain was increasing with an increasing percentage of silica fume.

2.9.2 Tensile Strength

Hooton [38] reported the *splitting tensile strength* of silica fume concretes up to the age of 182 days (Table 2.18). It can be seen that except at 28 days, the splitting tensile strength was not improved for silica fume concrete mixes. Also it was observed that with increasing replacement of silica fume split tensile strength decreased.

Bhanja and Sengupta [14] studied the isolated contribution of silica fume on the tensile strengths of *high-performance concrete*. Five concrete mixes, at w/cm ratios of 0.26, 0.30, 0.34, 0.38 and 0.42 were prepared by partial replacement of cement by equal weight of silica fume. The dosage of silica fumes were 0% (control mix), 5, 10, 15, 20 and 25% of the total cementitious materials. For all the mixes, tensile strengths were determined at the end of 28 days. Studies clearly exhibited that very high percentages of silica fume did not significantly increase the splitting tensile strength and increase was insignificant beyond 15% (Fig. 2.10).

Table 2.18 Splitting tensile strength of concrete [38]

Test age (days)	Concrete mixes			
	Control	10% SF	15% SF	20% SF
28	5.2	6.3	6.2	4.6
91	6.8	6.7	6.2	5.6
182	7.1	6.2	6.5	5.6

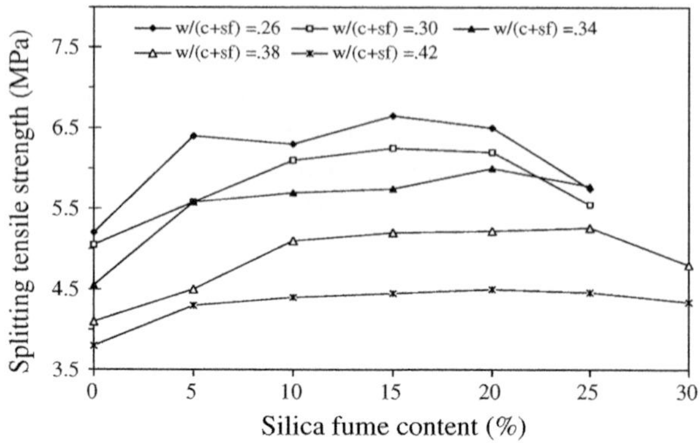

Fig. 2.10 Relationship between 28-day split tensile strength and percentage replacement of silica fume [14]

Yazıcı [98] conducted tests on *self-compacting concrete* investigating various properties like *freezing and thawing resistance, chloride penetration resistance* along with mechanical properties. He also found that silica fume addition improves the tensile strength at all fly ash replacement levels.

Almusallam et al. [6] investigated the effects of silica fume on the splitting tensile strength of concrete made with low-quality coarse aggregates. Four types of low quality coarse aggregates, namely calcareous, dolomitic, and quartzitic limestone and steel slag were used, and silica fume content was 10 and 15% as partial replacement of cement. The concrete specimens had a w/c ratio of 0.35 and a coarse aggregate to fine aggregate ratio of 1.63. The influence of aggregate quality on the compressive strength of 15% silica fume cement concretes is shown Fig. 2.11. The tests showed that the splitting tensile strength increased with age in

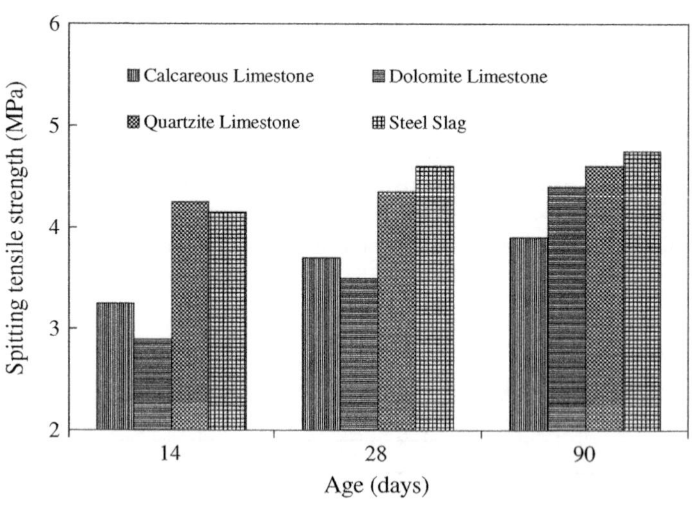

Fig. 2.11 Effect of aggregate type on split tensile strength of 15% silica fume cement concrete [6]

all the concrete specimens. The highest splitting tensile strength was observed in 15% silica fume cement concrete specimens followed by those prepared with 10% silica fume. They found that the splitting tensile strength of the silica fume cement concrete was more than that of plain concrete. After 90 days of curing, the splitting tensile strength of 15% silica fume concrete was maximum being 4.39 MPa. Also, after 90 days, the split tensile strength of 10 and 15% silica fume cement concrete specimens was nearly the same, being 4.54 and 4.59 MPa, respectively.

Tanyildizi and Coskun [88] studied the effect of silica fume on tensile strength of lightweight concrete exposed to *high temperature*. In mixtures containing silica fume, 0, 10, 20 and 30% of Portland cement by weight was replaced with silica fume. They reported that the tensile strength started to drop with temperature starting from 200°C. The reduction in splitting tensile strength of lightweight concrete containing 10% silica fume was 3.11, 11.46 and 80.15% at the 200, 400 and 800°C, respectively. The reduction in splitting tensile strength of lightweight concrete containing 20% silica fume was 4.69, 12.91 and 78.87% at the 200, 400 and 800°C, respectively. And the reduction in splitting tensile strength of lightweight concrete containing 30% silica fume was 5.8, 40.62 and 75.08% at the 200, 400 and 800°C, respectively. They concluded that addition of admixture silica fume prevented the decrease in the tensile strength of concrete.

González-Fonteboa and Martínez-Abella [32] studied the properties of concrete using recycled aggregates from Spanish demolition debris (RC mixes) and the impact of the addition of silica fume on the properties of recycled concrete (RCS mixes). A comparison was made between both these materials and standard conventional concrete (CC mixes), which was also modified by adding silica fume (CCS mixes). For the test, they made 10 CC, 7CCS, 10RC and 6RCS mixes. The results of *splitting tensile strength* are shown in Table 2.19. They stated that neither the addition of recycled aggregates nor the addition of silica fume had any particular impact on the tensile strength of the concrete types. The reason behind this was that silica fume mainly affected mechanical properties like compressive strength of high strength concretes.

Köksal et al. [53] studied the splitting tensile of *steel fibre reinforced concrete* with silica fume. Cold drawn steel fibres with hooked ends were used. Aspect ratios (l/d) of fibres were 65 and 80 and volume fractions (Vf) of steel fibres were 0.5 and 1%. Silica fume content was 0, 5, 10, and 15% weight of cement. The test results are given in Table 2.20 They found that increasing the silica fume and steel fibre contents, a considerable increase in the splitting tensile strength of the

Table 2.19 Average tensile splitting tensile strength (MPa) for the various test ages [32]

Cube	Average splitting tensile strength (MPa)		
	7 days	28 days	115 days
CC	3.12	3.15	3.32
CCS	3.04	3.15	3.31
RC	3.17	3.00	3.37
RCS	3.24	3.36	3.35

Table 2.20 Test results of concrete mechanical properties [53]

Series	Silica fume (%)	Steel fibre content (%)	Splitting tensile strength (N/mm^2)
A	0	0	3.48
		0.5	3.75
		1.0	4.59
		0.5	3.7
		1.0	6.6
B	5	0	3.82
		0.5	4.05
		1.0	8.98
		0.5	4.4
		1.0	6.9
C	10	0	5.36
		0.5	6.91
		1.0	9.56
		0.5	7.3
		1.0	9.7
D	15	0	6.54
		0.5	8.4
		1.0	10.01
		0.5	7.5
		1.0	10.0

concrete occurred. The increases in the splitting tensile strengths of the concretes without steel fibres were determined as 9.7, 54 and 87.9% for the 5, 10 and 15% silica fume, respectively.

Babu and Babu [9] studied the use of *expanded polystyrene* (EPS) beads as lightweight aggregate both in concrete and mortars containing silica fume as a supplementary cementitious material. Three percentages of silica fume 3, 5 and 9% (by weight of the total cementitious materials) were used. Split tensile strength test was conducted at 28 days. It was found that tensile strength increased with an increase in compressive strength. And failure observed was more gradual and the specimens did no separate into two, as was earlier reported for plastic shredded aggregate concretes [5].

Sata et al. [76] also observed that tensile strength of concrete slightly increased with the increase in the compressive strength.

2.9.3 Flexural Tensile Strength

Bhanja and Sengupta [14] studied the contribution of silica fume on the *flexural strength* of high performance concrete (HPC). Five series of concrete mixes, at w/cm ratios of 0.26, 0.30, 0.34, 0.38 and 0.42 were made with partial replacement of cement by equal weight of silica fume. The dosages of silica fumes were 0, 5, 10, 15, 20 and 25% of the total cementitious materials. The variations of flexural tensile strength with silica fume replacement percentage at different w/cm ratios in shown in Fig. 2.12. They stated that silica fume seemed to have a pronounced effect on flexural strength in comparison with splitting tensile strength. For flexural strengths, even very high percentages of silica fume significantly improve the strengths. Also it was found that there was a steady increase in the flexural strength with increase in the silica fume replacement percentage.

Köksal et al. [53] evaluated the flexural strength of concrete incorporating hooked steel fibres and silica fume. Aspect ratios (l/d) of fibres were 65 and 80 and volume fractions (Vf) of steel fibres were 0.5 and 1%. Silica fume was added to concrete directly as the percentages of 0, 5, 10 and 15% by weight of cement. Table 2.21 gives the flexural strength results. Significant increases in the flexural strengths of the concretes were observed by adding silica fume and steel fibres. The increases in the flexural strengths of the concretes without steel fibres were 7, 42.1 and 64.9% for the 5, 10 and 15% silica fume, respectively. Also they found that the flexural strengths of concretes containing 1% steel fibre were found to be greater than that of the concrete with 0.5% steel fibre for each of the silica fume content.

Kılıç et al. [52] examined the influence of aggregate type on the flexural strength characteristics of high-strength silica fume concrete. Five different aggregate types (gabbro, basalt, quartsite, limestone and sandstone) were used to produce high strength concrete containing silica fume. Silica fume replacement ratio with cement was 15% on a mass basis. Water-binder ratio was 0.35. The amount of hyperplasticizer was 4% of the binder content by mass. The flexural tensile strengths of concretes were measured at 3, 7, 28 days, and 3 months. The results are given in Table 2.22. They showed that (i) flexural tensile strength increased with the increase in curing time; and (ii) sandstone concrete showed the

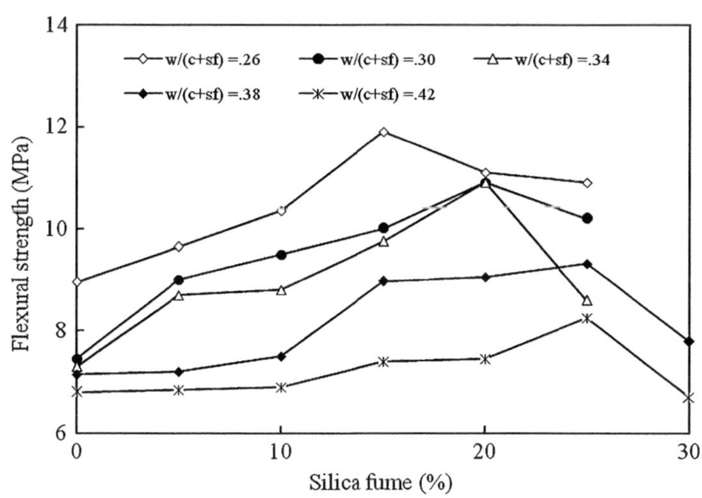

Fig. 2.12 Relationship between 28-day flexural strength and percentage replacement of silica fume [14]

Table 2.21 Test results of concrete mechanical properties [53]

Series	Silica fume (%)	Steel fibre content (%)	Flexural tensile strength (N/mm^2)
A	0	0	5.7
		0.5	5.9
		1.0	6.69
		0.5	6.10
		1.0	10.10
B	5	0	6.1
		0.5	7.2
		1.0	8.7
		0.5	7.6
		1.0	10.3
C	10	0	8.08
		0.5	8.5
		1.0	9.6
		0.5	8.98
		1.0	11.3
D	15	0	9.35
		0.5	9.52
		1.0	10.28
		0.5	9.58
		1.0	12.8

lowest flexural tensile strength, while Gabbro concrete showed the highest flexural tensile strength.

2.9.4 Modulus of Elasticity

Hooton [38] reported the *modulus of elasticity* of silica fume concretes up to the age of 365 days (Table 2.23). It can be seen that elastic modulus of the Portland cement concrete was approximately equal to silica fume concretes at 28 days but continued to increase at later ages.

Mazloom et al. [62] investigated the effect of silica fume on the *secant modulus of elasticity* of high performance concrete. The percentages of silica fume were: 0,

Table 2.22 Flexural strengths of concrete at different curing times [52]

Curing time (days)	Flexural tensile strength (MPa)				
	Gabbro (247)	Basalt (132)	Quartsite (160)	Limestone (110)	Sandstone (52)
3	12.6	11.4	12.9	7.9	3.2
7	16.1	15.4	14.9	12.5	4.5
28	17.3	16.7	16.2	12.8	5.2
90	18.4	17.9	16.9	13.9	5.6

Table 2.23 Modulus of elasticity of silica fume concrete [38]

Testing age (days)	Concrete mix			
	Control	10% SF	15% SF	20% SF
28	43.2	43.7	42.8	43.4
91	48.0	46.2	45.0	45.7
182	49.2	46.7	46.1	46.1
385	51.8	48.4	48.1	48.1

6, 10 and 15%. The results of secant modulus of elasticity of concrete specimens containing different levels of silica fume are given in Table 2.24. From the results it was observed that increasing the silica fume replacement level increased the secant modulus of concrete.

Almusallam et al. [6] determined the elastic modulus of concretes prepared with four types of low quality coarse aggregates, namely calcareous, dolomitic, and quartzitic limestone and steel slag, and 10 and 15% silica fume The results are given in Table 2.25. The type of coarse aggregate had a significant effect on the modulus of elasticity of concrete. After 28 days of curing, the modulus of elasticity of plain cement concrete prepared with calcareous, dolomitic, and quartzitic limestone and steel-slag aggregates was 22.0, 25.0, 29.0 and 30.0 GPa, respectively. The modulus of elasticity of steel-slag aggregate concrete was the highest while the modulus of elasticity of calcareous limestone aggregate concrete was the lowest. On average, the increase in the modulus of elasticity was 16 and 32% due to the incorporation of 10 and 15% silica fume, respectively. Moreover, the modulus of elasticity of concrete specimens prepared with steel-slag aggregate was more than that of concrete specimens prepared with limestone aggregate.

González-Fonteboa and Martínez-Abella [32] concluded by saying that addition of silica fume did not improve the elastic modulus of concrete made from demolition waste.

Table 2.24 Compressive strength and secant modulus of elasticity [62]

Kind and age of concrete	Compressive strength (MPa)	Measured modulus (GPa)
OPC		
7 days	46	28.8
28 days	58	34.4
SF 6		
7 days	50.5	31
28 days	65	35.5
SF 10		
7 days	52	31.1
28 days	67.5	37
SF 15		
7 days	53	31.5
28 days	70	38.1

Table 2.25 Modulus of elasticity of concrete after 28 days of curing [6]

Aggregate	Modulus of elasticity (GPa)		
	0% SF	10% SF	15% SF
Calcareous limestone	21.6	26	29.3
Dolomitic limestone	24.5	25.9	32.8
Quartzitic limestone	28.8	36.2	38
Steel slag aggregates	29.6	32.9	40.4

Güneyisi et al. [35] studied the modulus of elasticity of *rubberized concretes* with and without silica fume. Two types of *tire rubber*, *crumb rubber* and *tire chips*, were used as fine and coarse aggregate, respectively. Two control mixtures were designed at w/cm ratios of 0.60 and 0.40, and silica fume content varied between 5 and 20%. The results showed that the moduli of elasticity of the plain concretes were about 33 and 46 GPa at 0.60 and 0.40 w/cm ratios, respectively. However, the silica fume concretes had slightly greater elastic modulus values which were about 36 and 47 GPa for high and low w/cm ratios, respectively, irrespective of the amount of silica fume used.

2.9.5 Toughness

Köksal et al. [53] studied the effect of silica fume (0, 5, 10, and 15%) on the *steel fibre reinforced concrete*. Steel fibres with hooked ends were used. Aspect ratios (l/d) of fibres were 65 and 80 and volume fractions (Vf) of steel fibres were 0.5 and 1%. Figure 2.13 shows the relations between toughness of concrete, evaluated up to a 10 mm deflection, and silica fume content for each aspect ratio. It was concluded that steel fibres in matrixes with a high strength can exhibit a broken fracture down behavior without being pulled-out from matrix due to since strong bond between fibres and matrix. However, for low silica fume content or low matrix strength, the common failure type at the fracture plane appeared as the pulling-out of fibres from matrix, demonstrating the adverse effect of relatively resulting in a weaker bond.

2.9.6 Absorption

Demirboğa and Gül [25] studied high strength concretes using *blast furnace slag aggregates* (BFSA). Silica fume and a superplasticizer were used to improve BFSA concretes. They concluded that water absorption values were somewhat less than those of control specimens. Silica fume and BFSA were considered responsible for this behavior.

Gonen and Yazicioglu [31] studied the *capillary absorption* performance of concrete by adding *mineral admixtures*, silica fume and fly ash in the concrete

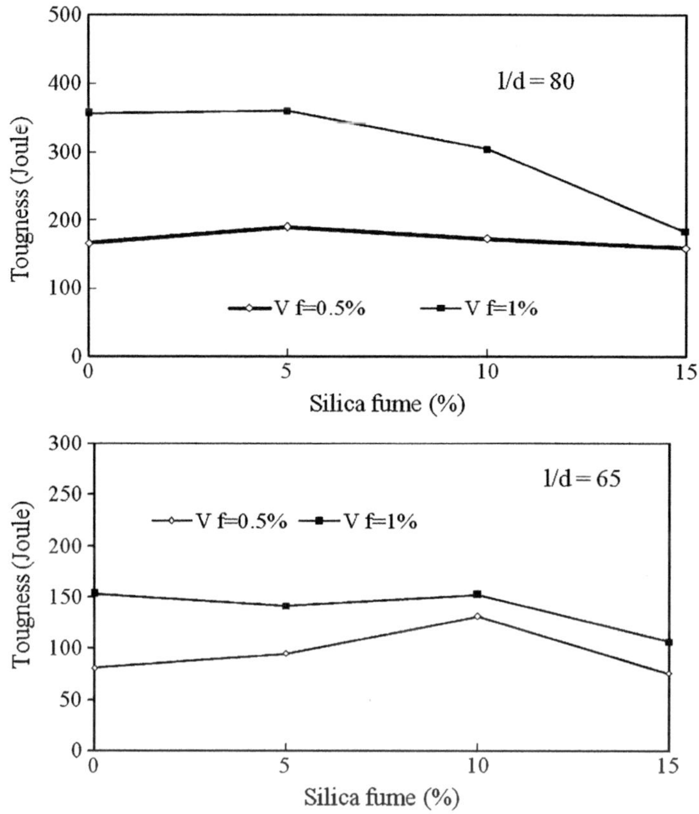

Fig. 2.13 Toughness versus silica fume content for different aspect ratios [53]

mixes, the replacement of fly ash and silica fume were kept at the level of 15 and 10% as the weight of cement, respectively. Test results are given in Table 2.26. It can be seen that the capillary absorption of concrete sample with FA was increased by as much as 47%; however, this increasing trend was reversed in specimens with fly ash and silica fume (double adding). Since silica fume is very fine, pores in the bulk paste or in the interfaces between aggregate and cement paste is filled by these mineral admixtures, hence, the Capillary pores are reduced.

Table 2.26 Mix proportions [31]

Mixes		Composite of binder pastes (%)			Capillary absorption coefficient (cm/s½)
		Cement	Fly ash	Silica fume	
I	NC	100	–	–	0.66
II	SFC	90	–	10	0.41
III	FAC	85	15	–	0.97
IV	SFAC	70	20	10	0.46
V	SSFC	90	–	10	0.41

González-Fonteboa and Martínez-Abella [32] indicated that recycled concretes showed higher water absorption ratios than conventional concretes. Babu and Babu [9] studied the use of expanded polystyrene (EPS) beads as *lightweight aggregate* both in concrete and mortars containing silica fume as a supplementary cementitious material. Three percentages of silica fume—3, 5 and 9% (by weight of the total cementitious materials) were used. The total *absorption* values of EPS concretes, ranging from 3 to 6%, decreased as the silica fume percentage increased. This could be attributed to the effect of silica fume and the advantage of the nonabsorbent nature of the EPS aggregate.

Gutiérrez et al. [36] found that the steel fibre reinforced material with the inclusion of silica fume showed the lowest percentage of water absorption. The incorporation of silica fume improved the *water absorption* of the material because of the reduction of permeable voids.

Krishnamoorthy et al. [54] did investigations on the cementitious *grouts* containing *supplementary cementitious materials* (SCM), in which he found that water absorption of the specimens having SCM's (like SF, FA, GGBS) was lower than that of control specimens having neat cement.

2.9.7 Porosity

Gleize et al. [30] investigated the effect of silica fume on the porosity of mortar. 10% of Portland cement was replaced with silica fume in a 1:1:16(cement/lime/sand mix proportion by volume) masonry mortar. The *porosity* results are given in Table 2.27. They found that the silica fume lowered the porosity only at 28 days and the pore structure of mortar with silica fume was found to be finer than that of non-silica fume mortar. But this refinement in pore size was more pronounced at 28 days than 2 days due to silica fume pozzolanic reaction.

Igarashi et al. [42] evaluated the capillary porosity and *pore size distribution* in *high-strength concrete* containing 10% silica fume at early ages. They concluded that silica-fume-containing concretes were found to have fewer coarse pores than the ordinary concretes, even at early ages of 12 and 24 h. The threshold diameter at which porosity starts to steeply increase with decreasing pore diameter was smaller in silica-fume-containing concretes than in ordinary concretes at 12 h.

Table 2.27 Total porosity of mortars [30]

Silica fume content (%)	Age (days)	Total porosity (%)
0	7	30.57
10	7	32.31
0	28	28.53
10	28	27.92

This smaller threshold diameter in silica-fume-containing concretes indicated higher packing density of binder grains in these concretes.

Khan [49] observed that the inclusion of silica fume (0–15% as partial replacement of cement) resulted in more significant reductions in porosity in mixtures. However, the reduction in the porosity was greater when silica fume was incorporated at up to 10% replacement level, beyond which the reduction was marginal or reversed.

Gonen and Yazicioglu [31] studied the performance of concrete by adding mineral admixtures, silica fume and fly ash. In the concrete mixes, the replacement of fly ash and silica fume were kept at the level of 15 and 10% as the weight of cement, respectively. The porosity results are given in Table 2.28. Porosity of mixtures varied between 6 and 16%. The porosity values indicated the effect of mineral admixtures on the porosity of concrete. When FA as a single mineral admixture was used in mix, porosity value was higher compared to NC. However, the porosity of concrete improved when two types of mineral admixtures (fly ash and silica fume) were added at the same time (double approach). Their results showed that the porosity of concrete with double mineral admixtures is smaller than the other series.

Poon et al. [69] examined the *porosity* of concrete mixtures using MIP. Two series of concrete mixes were prepared at the w/b ratios of 0.3 and 0.5. Each mixture included two silica fume contents (10 and 15%). Porosity results are given in Table 2.29. The results showed the decrease in porosity in with age due to addition of silica fume .

Cwirzen and Penttala [22] did investigations on eight non-air-entrained concretes having water-to-binder (w/b) ratios of 0.3, 0.35 and 0.42 and different additions of condensed silica fume. Their results from the MIP investigation showed that the concrete having a w/b ratio of 0.3 showed decrease in capillary porosity with silica fume. However for w/b ratios of 0.35 and 0.42, capillary and total porosities appeared to be quite similar.

Rossignolo [74] wrote a paper which deals with the effect of silica fume and styrene-butadiene latex (SBR) on the microstructure of the interfacial transition zone (ITZ) between Portland cement pastes and aggregates (basalt). It was observed that the usage of 10% of silica fume, in relation to the cement paste, caused a reduction of 36% in the thickness of the matrix-aggregate ITZ, in relation to the reference concrete.

Table 2.28 Mix Proportions [31]

Mixes	Composite of binder pastes (%)			Porosity (%)
	Cement	Fly ash	Silica fume	
NC	100	–	–	10.47
SFC	90	–	10	6.82
FAC	85	15	–	15.82
SFAC	70	20	10	6.41
SSFC	90	–	10	8.37

Table 2.29 MIP measured total porosity of series 1 concrete (w/b = 0.3) [69]

Mix	MIP measured total porosity (% v/v)			
	3 days	7 days	28 days	90 days
Control	8.69 ± 0.11	8.44 ± 0.13	7.92 ± 0.12	6.97 ± 0.28
5% SF	7.53 ± 0.16	7.24 ± 0.14	6.31 ± 0.11	5.85 ± 0.02
10% SF	7.64 ± 0.15	6.14 ± 0.13	5.66 ± 0.12	5.11 ± 0.41

2.9.8 Thermal Properties

Demirboğa [24] studied the effect of silica fume on *thermal conductivity* (TC) of concrete. *Density* decreased with the replacement of silica fume. The variation of TC of concrete is shown in Fig. 2.14. It can be seen that the highest value of TC of concrete was obtained for specimens produced with 100% PC. Further, the graph declines largely with increasing silica fume replacement for PC. For 7.5 and 15% silica fume replacement, keeping other conditions constant, the reductions were 5 and 14%, respectively, compared to the corresponding control specimens.

Demirboğa [23] reported that silica fume decreased thermal conductivity of mortar up to 40 and 33% at 30% replacement of PC, respectively. Chen and Chung [19] and Postacioğlu and Maddeler [70] had reasoned that the reduction in thermal conductivity was primarily due to the low density of LWAC (*Lightweight Aggregate concrete*) with silica fume and fly ash content, and may be partly due to the amorphous silica content of silica fume and fly ash

2.9.9 Creep

According to Mindess and Young [64], there are number of factors that determine the amount of creep a concrete will undergo. First, *creep* is approximately

Fig. 2.14 Relationship between thermal conductivity and silica fume [24]

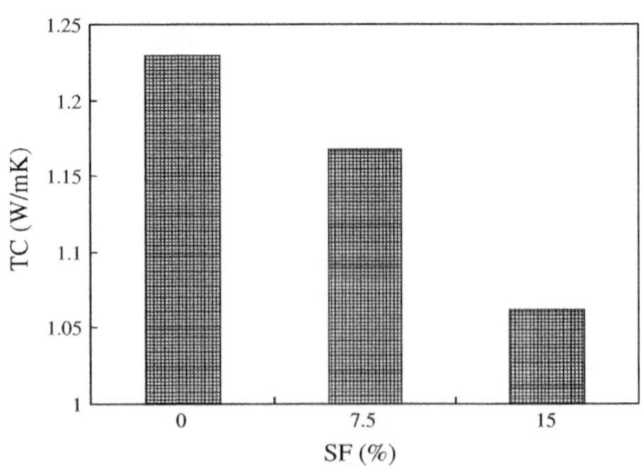

proportional to the level of applied stress as a percent of ultimate strength of the concrete within the normal range of long-term loading (up to 50% of ultimate strength). In fact, his relationship is only a gross estimate because the concrete continues to hydrate and gain strength and stiffness over time. The compressive strength of the concrete is inversely related to the specific creep.

Khatri et al. [47] studied the behaviour of concretes containing silica fume having a constant water/solids ratio of 0.35 and a total cementitious materials content of 430 kg/m^3. They observed that silica fume reduced the strain due to creep compared with Portland cement concrete. Adding silica fume to concrete containing 65% slag did not affect the creep. Ternary mixes containing 15 or 25% fly ash and 10% silica fume experienced greater creep than control concrete.

Mazloom et al. [62] studied the *creep* of high performance concrete having silica fume. The control mix was made with OPC, while the other mixes were prepared by replacing part of the cement with silica fume at four different (0, 6, 10 and 15%) replacement levels by mass. The w/c ratio was 0.35. It was found that silica fume had a significant influence on the long-term creep. As the proportion of silica fume increased to 15%, the creep of concrete decreased by 20–30% (Table 2.30).

Tao and Weizu [89] carried out an experimental study on the early-age *tensile creep behavior* of *high-strength* concrete (HSC) comprising of silica fume concrete under uniaxial restraining stresses. The experiments were performed with three 0.35 w/b mixtures, including plain concrete OPC, double-blended concrete silica fume (6% replacement of OPC by silica fume). The compressive creep strain for silica fume and OPC concretes during the temperature rising period is shown in Fig. 2.15. It was found that about 70% of free expansion deformation was compensated by compressive creep within the first day. After this period, the compressive creep was replaced by tensile creep due to high tensile stress development in specimens.

2.9.10 Shrinkage

Taylor [92] identified four effects contributing to *drying shrinkage*; *capillary stress*, surface free energy, disjoining pressure, and movement of interlayer water. Capillary stress describes the phenomenon of transfer of the tension from the

Table 2.30 Values of creep of 80 × 270 mm high specimens on completion of the tests (microstrain) [62]

Age of loading (days)	Concrete mixes			
	OPC	SF 6	SF 10	SF 15
7	595	510	459	417
28	413	407	381	328

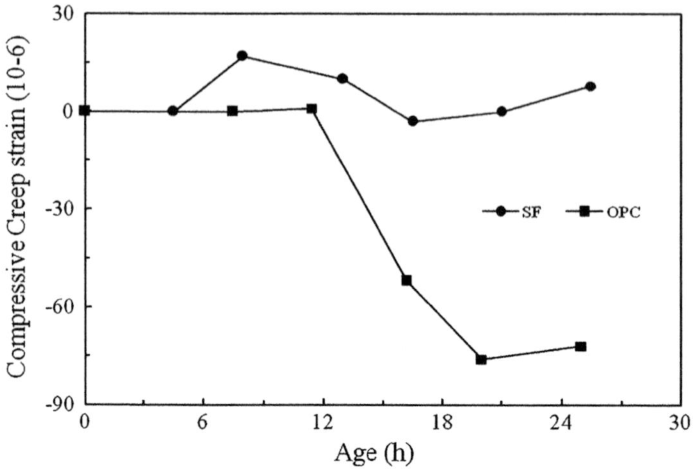

Fig. 2.15 Compressive creep strain under semi-adiabatic condition [89]

meniscus of capillary pore water to the walls of the pore as water evaporates. The pore shrinks and may even collapse, in which case it will not expand on rehydration. The surface tension of solid particles is reduced by the adsorption of molecules. When they are removed, the particles tend to contract. Disjoining pressure is analogous to the phenomenon that occurs in the swelling of clays as water is drawn between adjacent particles forcing them apart. As the water is removed, the particles come back together

Tazawa et al. [90] investigated the effects of silica fume addition on the drying shrinkage of mortar. They observed that magnitude of drying shrinkage is determined by the balancing of two factors; the shrinkage stress due to capillary tension in the pores, and rigidity of the structure as determined by compressive strength.

Mazloom et al. [62] conducted tests for total, autogenous and drying shrinkage of high performance concrete having silica fume. The percentages of silica fume were: 0, 6, 10 and 15% with w/c ratio being 0.35. The results showed that (i) silica fume did not have much effect on drying specimens (total shrinkage); (ii) silica fume considerably affected the shrinkage of sealed specimens. It is clear that the general effect of increasing the silica fume inclusion is to increase autogenous shrinkage; and (iii) there was significant increase in autogenous shrinkage at high levels of silica fume. In fact, inclusion of 10 and 15% silica fume increased the autogenous shrinkage of concrete by 33 and 50%, respectively. The effect of silica fume on autogenous shrinkage was explained by its influence on the pore structure and pore size distribution of concrete as well as its pozzolanic reaction.

According to Sellevold [78] the inclusion of silica fume at high replacement levels significantly increased the *autogenous shrinkage* of concrete due to the refinement of pore size distribution that leads to a further increase in capillary tension and more contraction of the cement paste.

Table 2.31 Maximum plastic shrinkage strain in plain and blended cement concretes, exposed to a wind velocity of 15 km/h, temperature of 45°C and RH of 35% [2]

Silica fume type	Replacement (%)	Maximum plastic shrinkage strain (μm)
1	5	1,322
	7.5	1,645
	10	2,348
2	5	1,724
	7.5	2,794
	10	2,924
3	5	1,038
	7.5	1,370
	10	1,656
4	5	1,122
	7.5	1,183
	10	1,224
5	5	783
	7.5	939
	10	1,119
Plain cement	0	716

Al-Amoudi et al. [1] found that the maximum *plastic shrinkage strain* was observed in silica fume (undensified) cement concrete. This was attributed to the undensified nature of this silica fume. On the other hand, the lowest plastic shrinkage strain was noted in the plain cement concrete.

Al-Amoudi et al. [2] varied the dosage of silica fume to investigate its effect on the plastic shrinkage of concrete exposed to *hot-weather conditions*. A summary of the maximum plastic shrinkage strains attained during the 24-h exposure is presented in Table 2.31. The plastic shrinkage strain increased with increasing dosage of silica fume in the parent cement. This trend was noted in all the concrete specimens prepared with the selected silica fume cements.

Zhang et al. [99] did an experimental study on the *autogenous shrinkage* of Portland cement concrete (OPC) and concrete incorporating silica fume. The water-to-cementitious materials (w/c) ratios of the concrete were in the range of 0.26–0.35, and silica fume content was between 0 and 10% by weight of cement. Autogenous shrinkage up to 98 days is presented in Table 2.32. It appeared that both the w/c ratio and the incorporation of silica fume had significant effect on the autogenous shrinkage strain of the concrete. The autogenous shrinkage increased with decreasing w/c ratio and with increasing silica fume content. This is in agreement with those reported by Tazawa and Miyazawa [91], Brooks et al. [16], Mak et al. [60], and Persson [68]. The results indicated that the concrete with low w/c ratio and with silica fume induce autogenous shrinkage rapidly even at early ages. Particularly at w/c ratio of 0.26, the autogenous shrinkage strains of the silica fume concrete at the age of 2 days were more than 100 micro strain.

2.10 Effect of Silica Fume on the Durability Properties of Concrete

2.10.1 Permeability

Perraton et al. [67] studied the effect of silica fume on the *chloride permeability* of concretes. Concretes were made with water–cementitious ratios of 0.4 and 0.5. Silica fume dosage varied from 5 to 20% by weight of cement. Concretes were moist cured for 7 days before drying in air at normal and low temperatures for 6 month. They observed significant reduction in the chloride-ion diffusion in silica fume concretes which further decreased with increasing addition of silica fume as shown in Fig. 2.16. Main reason that could be attributed to reduced permeability is that addition of *silica fume* cause considerable pore refinement i.e. transformation of bigger pores into smaller one due to their pozzolanic reaction concurrent with cement hydration. By this process the permeability of hydrated cement paste as well as porosity of the transition zone between cement paste and aggregate are reduced.

Gjrov [29] illustrated that silica fume can greatly reduce the *water permeability* in a lean concrete, but it can have a small influence on the permeability of rich mixture. It was found that water permeability co-efficient of a concrete containing 100 kg/m^3 of cement can decrease from 1.6×10^{-7} to 4×10^{-10} m/s when 10 kg/m^3 of silica fume is used. The latter permeability value was comparable to that obtained from non-silica fume concrete containing 250 kg/m^3 of cement.

Hooton [38] studied the permeability of silica fume concretes and stated that permeability decreased with addition of silica fume. Water-permeability of control mix was 1.8×10^{-14} m/s; whereas it was less than 1×10^{-17} m/s for concrete mix made with 10% silica fume. Permeability of silica fume concretes with higher dosage of silica fume could not be measured.

Ozyildirim and Halstead [66] found that ternary mixes containing fly ash and silica fume performed better in chloride resistance than comparable concretes containing only Portland cement.

Kayali and Zhu [46] concluded that reinforced concrete slabs whose concrete included silica fume as 10% by mass of cement and whose strength was around 70 MPa, showed extremely low value of corrosion current density and half-cell

Table 2.32 Autogenous shrinkage of concrete [99]

w/c	Autogenous shrinkage (micro strain)/% of 98-day shrinkage											
	SF 0%				SF 5%				SF 10%			
	2 days	7 days	14 days	98 days	2 days	7 days	14 days	98 days	2 days	7 days	14 days	98 days
0.26	49/25	100/51	129/65	197	101/38	170/64	194/73	266	101/36	174/62	221/78	282
0.30	36/20	87/48	115/64	180	77/35	149/68	174/80	218	74/27	161/59	213/64	251
0.35	25/63	34/85	40/100	40	49/22	101/47	128/60	215	41/16	115/46	160/64	251

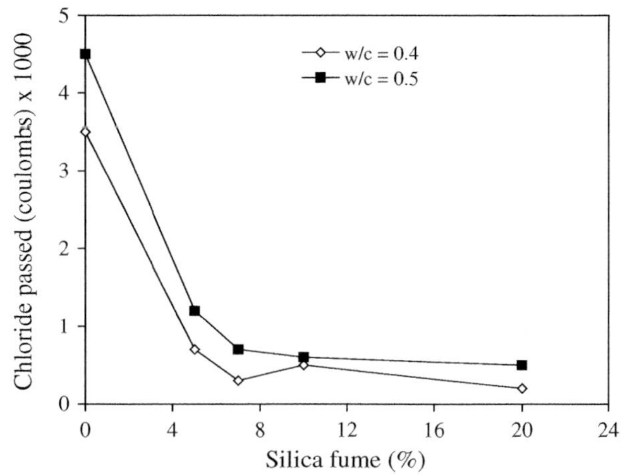

Fig. 2.16 Changes in chloride ion permeability of concrete with amount of silica fume replacement [67]

potentials. These values remained very low even after long exposure to chloride ion solution.

Song et al. [85] presented a procedure for predicting the *diffusivity* of high strength silica fume concrete, developed by considering water-to-binder ratio, silica fume replacement ratio, and degree of hydration as major influencing factors. Relative diffusivity of bulk paste and ITZ decreases as silica fume replacement ratio increases; however, if silica fume replacement is about 10%, it would be marginal (Figs. 2.17 and 2.18).

Gutiérrez et al. [36] determined the effect of incorporation of various *supplementary cementitious materials* on various properties of Portland cement fibre-reinforced

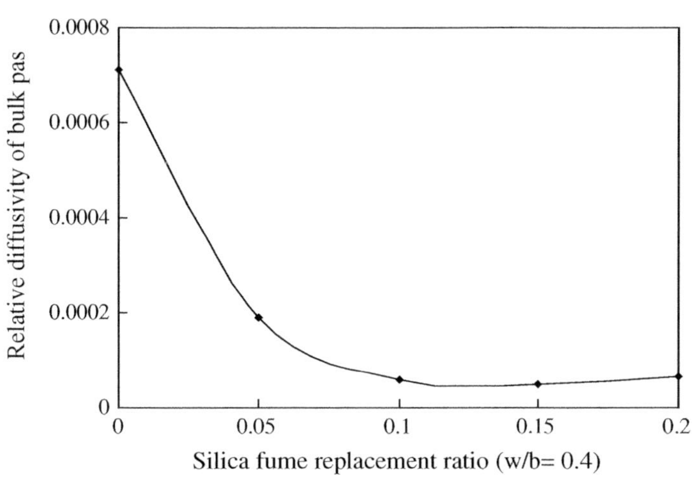

Fig. 2.17 Relative diffusivity of ITZ versus silica fume replacement ratio [85]

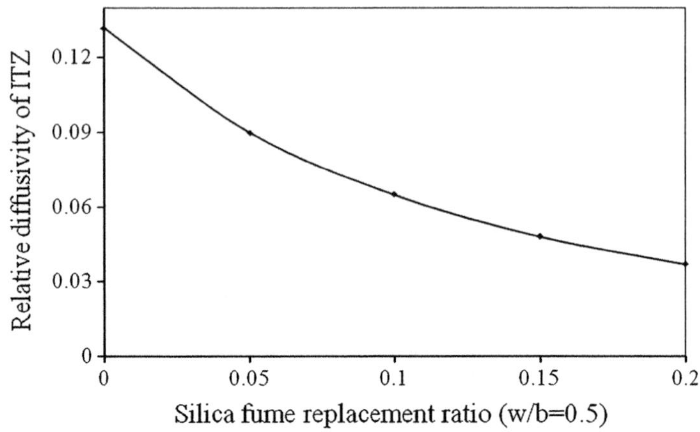

Fig. 2.18 Relative diffusivity of ITZ vs. silica fume replacement ratio [85]

mortars. It was found that the pozzolanic materials and the slag were effective in reducing the permeation of chloride ion. The greatest effect on the plain matrix was produced by the silica fume addition followed by the additions of MK, GGBS, and FA. It was also noted that the fibre inclusion negatively affects the penetration of chlorides because of the increase in capillary porosity. This was observed for both synthetic and natural fibre reinforced specimens.

Soroushian et al. [87] reported a 75% reduction in the permeability to chloride ions when polypropylene fibres were used along with silica fume in a Portland cement matrix. The decrease in the coefficient of chloride diffusion was 98% when silica fume was added to the glass fibre reinforced mortar. The positive effects of silica fume additions were attributed to the increase in density and reduction in capillary porosity caused by reaction products such as calcium silicates and calcium aluminates, which change the material microstructure.

Babu and Babu [9] studied chloride resistance of concrete and mortar made with expanded polystyrene (EPS) beads as lightweight aggregate. It also contained silica fume as a supplementary cementitious material. Three percentages of silica fume were 3, 5 and 9% (by weight of the total cementitious materials). They concluded that as per the assessment criteria, all the EPS concretes containing silica fume showed a low chloride permeability of <1,000 C. These concretes also exhibited much lower corrosion rates compared to the normal concrete.

2.10.2 Freezing and Thawing

Sørensen [86] studied the effect of silica fume on *salt-scaling* of concrete. He found that drying-rewetting history of concrete prior to *freezing and thawing* has a significant effect on conventional concrete, whereas silica fume concrete is relatively unaffected. *Air entrainment* has a beneficial effect on both types of concrete, but frost-resistant silica fume concretes can be made with out entrained air.

Feldman [27] investigated the effect of silica fume and sand/cement ratio on pore structure and frost resistance of Portland cement mortars. Silica fume-Portland cement blend mortars fabricated with 0, 10 and 30% silica fume at a water/binder ratio of 0.60 and a sand/cement ratio of 2.25 were monitored by mercury porosimetry while being cured for 1–180 days. The threshold value for pore intrusion increased with pore size and becomes less abrupt with silica fume addition; it was in the 0.5 to 20 $\times$ 10^3 nm regions. Mortars were also made with and without 10% silica fume at a water/cement ratio of 0.60 and sand/cement ratios of 0, 1.0, 1.5, 1.8, 2.0, 2.25 and 3.0. Mercury intrusion measurements were carried out after 14 days of curing. In the presence of silica fume pore volume in the 0.5 to 20 $\times$ 103 nm pore diameter range increased with sand/cement ratio. Mortar prisms were subjected to freezing and thawing cycles (two cycles in 24 h) according to ASTM standard test method C 666, Procedure B. Results indicated that if the sand/cement ratio was 2.25 or over, expansion was less than 0.02% after 500 cycles. At lower sand/cement ratios 10% silica fume gives little protection.

Hooton [38] investigated the *frost resistance* of concretes containing 0. 10, 15, and 20% silica fume by mass of cement. Water–cementitious materials ratio of concretes was between 0.360 and 0.369. Concretes contained 1% air. Test results showed that Portland cement concrete failed ASTM C 666 (Procedure A), after 58 cycles, while all silica fume concretes had durability factor in excess of 90% after 300 cycles. Concretes with 10, 15 and 20% silica fume had durability factor of 97.5, 93.4, and 92.8, respectively. He attributed this excellent performance to a low degree of saturation due to self-desiccation during hydration.

Johnston [44] observed that concretes containing 10 and 15% silica fume and having the maximum permissible water–cementitious materials ratio of 0.45 had barely acceptable resistance to salt-scaling. Resistance to salt scaling correlated well with the water/cement ratio (not water–cementitious material ratio)

Cwirzen and Penttala [22] studied the influence of the cement paste–aggregate *interfacial transition zone* (ITZ) on the frost durability of high-performance silica fume concrete (HPSFC). Investigation was carried out on eight non-air-entrained concretes having water to binder (w/b) ratios of 0.3, 0.35 and 0.42 and different additions of condensed silica fume. Results of the freeze–thaw tests are presented in Fig. 2.19. It was found that none of the concretes having w/b ratio of 0.35 and 0.42 had surface scaling results under 1,500 g/m^2 and the dynamic modulus of elasticity was less than 60% after 56 cycles. Scaling decreased with increasing silica fume amount and decreasing w/b. The weakest concrete with respect to surface scaling appeared to be concrete having w/b ratio of 0.42 without any silica fume addition. Internal damage was in line with the surface scaling except for the mix 0.42-3SF (w/b ratio 0.42 and 3% silica fume of cement weight). They said that neither internal damage nor surface scaling was observed for mixes having a w/b ratio of 0.3. The results also showed that the transition zone initiates and accelerates damaging mechanisms by enhancing movement of the pore solution within the concrete during freezing and thawing cycles. They concluded that moderate additions of silica fume seemed to densify the microstructure of the ITZ.

Yazıcı [98] conducted tests on *compressive strength* and *splitting tensile strength* of self-compacting concrete after freezing and thawing cycles. Test results shown in Fig. 2.20 indicated that, the residual compressive strength ratio of control mixture after 90 freeze–thaw cycles was 93%. It was also found that the compressive strength of control mixture had been exceeded by both H (without silica fume) and HS (with silica fume) series at all FA replacement level after freezing and thawing. For H series, the residual strength of H30 was 108%, which meant that freeze–thaw cycling caused increase in compressive strength. For HS series, at 30 and 40% fly ash content, gain in compressive strength was clear.

2.10.3 Corrosion

Berke [13] used electrochemical tests on concrete samples monitored for 2 years, and found that using silica fume (up to 15% addition to cement) improved the long-term *corrosion resistance*. Rasheeduzzafar and Al-Gahtani [73] reported that blending of plain cements with 10 or 20% silica fume significantly improved the corrosion resistance. They found hardly any tangible advantage in corrosion-initiation time by increasing the silica-fume content from 10 to 20%.

Khayat and Aitcin [50] observed that iron oxide layer on conventional steel reinforcing bars becomes unstable when the pH of surrounding concrete dropped to approximately 10–11 or when this layer comes in contact with chloride ions. When silica fume was used as cement replacement, the pH of concrete decreased

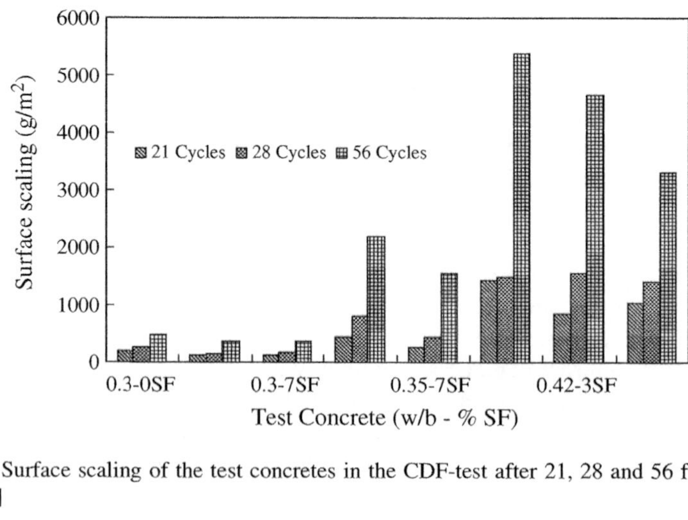

Fig. 2.19 Surface scaling of the test concretes in the CDF-test after 21, 28 and 56 freeze–thaw cycles [22]

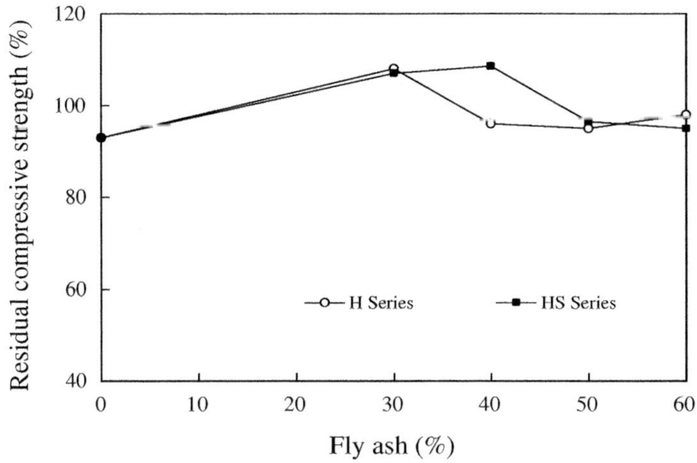

Fig. 2.20 Residual compressive strength after 90 freeze–thaw cycles [98]

because cement content is less. Also decrease in $Ca(OH)_2$ content due to pozzolanic reaction of silica fume and reduction in alkali-pore water concentration further reduces the pH. But these factors have small effects in destabilizing the passive iron oxide layer since pH of concrete does not fall below 12 even when 30% silica fume was used. Diffusion coefficient of chloride and chlorides content in concrete are reduced significantly in presence of silica fume. Also the use of silica fume substantially increased the electrical resistivity of concrete hence slowing the rate of corrosion.

Khedr and Idriss [51] studied the effectiveness of silica fume concrete in resisting damage caused by corrosion of embedded steel using an accelerated impressed voltage-testing setup. Silica fume concrete included 0, 10, 15, 20, and 25% silica as equal replacement of ordinary Portland cement. Concrete samples were treated in either fresh water or 4% NaCl saline water for 7 and 28 days. STC (Susceptibility to corrosion) was used as an index of resistance of concrete to corrosion-related damage. Blending of plain concrete with 10–20% silica fume significantly improved the corrosion resistance. At 7 days of curing, STC index of control concrete (0% silica fume) was 16.61×10^{-4}/W h^2 with fresh water and 29.32×10^{-4}/W h^2 with saline water. STC index was almost constant (same as with 0% silica fume) up to 15% silica fume content, but, at higher dosages of silica fume (20–25%) significantly reduced the STC index ($2.5–9.8 \times 10^{-4}$/W h^2). At 28 days curing, control concrete (0% silica fume) achieved STC index of 6×10^{-4}/W h^2 with fresh water and 8.1×10^{-4}/W h^2 with saline water. An optimal effect at silica fume replacement dosage of 15% was observed. STC was always lower for longer curing periods. For control mix the STC values of 28-day samples were 28–36% of those of 7-day samples. This percentage was significantly lower (0.8–17%) for Silica Fume concrete.

Dotto et al. [26] studied the influence of silica fume on the *corrosion behaviour* of reinforcement bars. Concretes with different water–binder ratio (cement + silica fume) 0.50, 0.65 and 0.80 were used. Silica fume additions were 0, 6 and 12% by weight of cement. Tests were conducted for electrical resistivity, and polarization curves. The results showed that the addition of 6% silica fume increases the electrical resistivity of concrete by 2.5 times and 12% silica fume increases it by 5 times. This suggests that the addition of silica fume can be effectively used in protecting steel reinforcement against corrosion.

Kayali and Zhu [46] did tests on high-strength reinforced silica fume–cement concrete slabs with a compressive strength of 70 MPa for chloride diffusion and corrosion activity after partial immersion in a 2% chloride solution. It was found that high-strength concrete containing 10% silica fume possessed exceedingly high corrosion resistance.

Civjan et al. [18] carried a long-term corrosion study was conducted to determine the effectiveness of calcium nitrite, silica fume, fly ash, ground granulated blast furnace slag, and disodium tetrapropenyl succinate (DSS) in reducing corrosion of reinforcing steel in concrete. Mixture proportions included single, double, and triple combinations of these admixtures. They concluded that for optimal protection against corrosion in structural concrete, a triple combination of CN, SF, and FA (or a double combination of CN and BFS), all at moderate dosages, was recommended.

2.10.4 Sulfate Resistance

According to ACI Committee 234 [4], the effect of silica fume on *sulfate resistance* is due more to the reduction in permeability than to dilution of the C_3A content because of the relatively low doses of silica fume used in practice.

Sellevold and Nilsen [79] reported field studies of concretes with and with out 15% silica fume. After 20 years' exposure to ground water containing 4 g/L sulfate and 2.5–7.0 pH, the performance of the silica fume concrete was found equal to that of the concretes made with sulfate-resisting Portland cement, even though the water/cementitious materials ratio was higher for silica fume concrete (0.62) than for control (0.50).

Cohen and Bentur [21] studied the effect of 15% silica fume replacement of Types I and V Portland cement on the resistance to sulfate attack in magnesium and sodium sulfate solutions. The water–cementitious materials ratio was 0.3. In the sodium sulfate solutions, the silica fume concrete specimens were resistant to sulfate attack. In the magnesium sulfate solutions, all the specimens expanded, with the Type I cement specimens (with or without silica fume) expanding more than Type V cement specimens (with or without silica fume). Since specimens were thin (6 mm), the authors attributed the effect of silica fume on sulfate resistance more to chemical effects than to reduced permeability.

Hooton [38] used a 10.7% C_3A cement for *mortar bars* tested according to ASTM C 1012 [7]. Cement was replaced with 0, 10, and 20% of silica fume by mass. 10% silica fume bars were made both with and with out superplasticizer to maintain constant water content at constant slump. The 20% silica fume bars were made only with superplasticizer. A control mortar bars were made using Type V cement. After 1-year, results indicated that all of the silica fume mortar had less expansion than Type V mortar, and all of the mixes except the Type I cement easily passed the proposed ASTM failure criterion of 0.10% expansion. The Type V cement mortar exceeded the 0.10% expansion limit after 1.7 years, while none of the silica fume mortars had exceeded the expansion limit after 5 years.

Mangat and Khatib [61] investigated the influence of silica fume (0, 5, 9, and 15%) on the sulfate resistance of concrete containing under different *curing conditions* (initial air curing at 45°C and 55% RH; initial wt/air curing at 45°C and 25% RH; initial air curing at 20°C and 25% RH) up to the age of 512 days. They concluded that (i) replacement level between 5 and 15% of cement with silica fume increased considerably the sulfate resistance of concrete; (ii) under initial air-curing at 45°C, 25% RH, an expansion of 0.275% was obtained for the control mix (0% silica fume), compared with only 0.04% for the 9% silica fume after 502 days of exposure to sulfate solution; (ii) under initial wet/air-curing at 45°C, 25% RH, the control mix disintegrated before 207 days in sulfate solution where as silica fume concrete did not disintegrated until after 502 days; and (iii) air-curing at 20°C, 55% RH did not show large expansions and was effective in improving the sulfate resistance of normal concrete as with replacement of cement by Silica Fume in optimum quantity. Reasons that could be attributed to the increase in sulfate resistance was probably to refined pore structure of silica fume incorporated mixes or to the reduction in calcium hydroxide cement in the presence of silica fume which reduces the extent of gypsum formation and hence, increase sulfate resistance.

Irassar et al. [43] investigated the sulfate resistance of concrete made with *silica fume*. Concrete specimens were half-buried in sulfate soil for 5 years. Mineral admixtures were used as a partial replacement for ordinary Portland cement ($C_3A = 8.5\%$), and the progress of sulfate attack was evaluated by several methods (visual rating, loss in mass, dynamic modulus, strength, X-ray analysis). Results showed that silica fume improved the sulfate resistance when the concrete was buried in the soil. However, concretes with high content of silica fume exhibited a greater surface scaling over soil level due to the sulfate salt crystallization.

Hekal et al. [37] reported that partial replacement of Portland cement by silica fume (10–15%) did not show a significant improvement in sulfate resistance of hardened cement pastes.

Ganjian and Pouya [28] studied the effect of silica fume on deterioration resistance to sulfate attack in *seawater* within tidal zone and simulated wetting–drying conditions. The performance of pastes and concrete specimens with silica fume exposed to simulation ponds and site tidal zone were inferior to those without silica fume replacement.

Lee et al. [57] studied the effectiveness of silica fume in controlling the damage arising from sulfate attack. The water/cementitious materials ratios (w/cm) of the mortar mixtures were 0.35, 0.45 and 0.55. Under this sulfate environment, the incorporation of 10% silica fume in OPC matrix showed no evidence of spalling and cracking up to about 1 year of exposure, and strength loss (Fig. 2.21) increased as the w/cm ratio increased; and the total strength loss as well as that between different w/cm ratio levels was greater in mortar specimens without silica fume compared to those with silica fume.

Wee et al. [95] also showed that silica fume, at replacement levels of 5 and 10% by mass of OPC plays a key role in resisting sodium sulfate attack, indicating no signs of *spalling* after about 1 year of exposure in 5% sodium sulfate solution.

Shannag and Shaia [81] prepared high-performance concrete mixes containing various proportions of natural pozzolan and silica fume (up to 15% by weight of cement). They were stored in sodium and magnesium sulfate solutions, in Dead Sea and Red Sea waters. After 1 year immersion in sulfate solution and sea water, the concrete mix containing a combination of 15% silica fume, and 15% natural pozzolan (by weight of cement) showed a maximum protection against sulfate attack compared to those investigated in the study. This mix retained more than 65% of its strength after 1 year of storage in sulfates solutions and sea waters. The superior resistance of that mix against sulfate attack was attributed to the pore refinement process and further densification of the transition zone occurring due to the conversion of lime forming from the hydration of cement into additional binding material through lime-pozzolan reaction. The results also showed that magnesium sulfates had a more damaging effect than sodium sulfates; which was consistent with the data available in the literature [65]

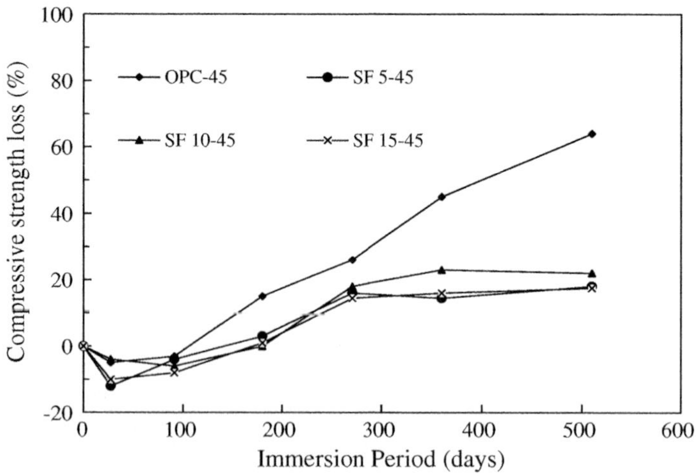

Fig. 2.21 Compressive strength loss of mortars subjected to sulfate attack in 5% sodium sulfate solution (w/cm¼0.45) [57]

2.10.5 Carbonation

Skjolsvold [83] investigated *carbonation* depths of field concrete with or with out silica fume. The results were normalized to correct the differences in compressive strength and length exposure to the atmosphere. The mean carbonation depth was greater for silica fume concretes under theses conditions, but the variation was quite high. Laboratory study showed that for a given compressive strength, silica fume concrete had greater carbonation rates than concretes with out silica fume. Schubert [77] believed that the consumption of $Ca(OH)_2$ in the pozzolanic reaction acts to increase the rate of carbonation, while the blocking of capillary pores acts to decrease it. Grimaldi et al. [33] found that the carbonation depth was greater in mortars containing silica fume than in controls. They attributed this result to the reduction of pH caused by the pozzolanic reaction.

Khan and Lynsdale [48] did investigations which aimed at developing high-performance concrete. Binary and ternary blended cementitious systems based on ordinary Portland cement, pulverised fuel ash and silica fume were investigated. PFA up to 40% was used, and to these blends, 0, 5, 10 and 15% silica fume were incorporated as partial cement replacements. Carbonation measurements were carried out for concrete cubes of 100 mm after 2 years of exposure in a constant temperature room at 20 ± 3°C and 65 ± 5% RH in normal atmospheric conditions. The carbonation study was limited to concrete prepared with w/b ratio of 0.27. The samples (100 mm cubes) were broken into two halves at the age of 2 years. The results (Fig. 2.22) clearly demonstrated that there was an increase in carbonation with an increase in PFA content, whilst silica fume inclusion of its own did not exhibit significant influence on the carbonation; this was in good agreement with earlier findings [17, 59].

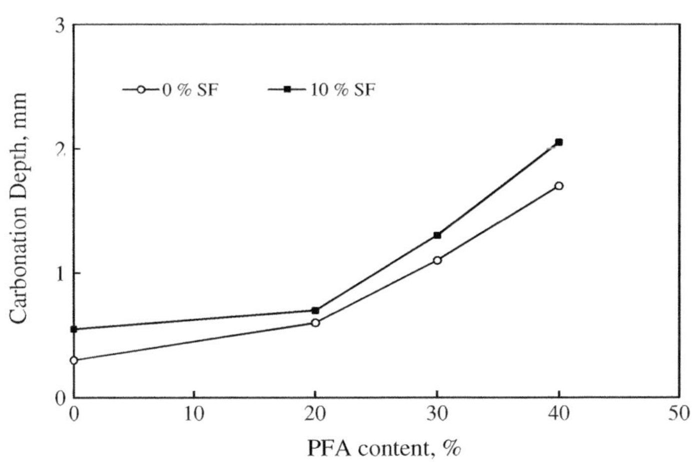

Fig. 2.22 Carbonation depth of concrete at the age of 2 years, w/b ratio of 0.27 [48]

Byfors [17] investigated the *carbonation* of silica fume and pulverized fuel ash blended cement concrete (compared on equal w/b ratio) and found that the incorporation of 10–20% silica fume has no effect on carbonation as compared to OPC control, whilst 15–40% PFA exhibited higher rate of carbonation. So Khan and Lynsdale [48] concluded that still silica fume inclusion slightly increases *carbonation depth* as compared to the OPC control and PFA added mix.

Gonen and Yazicioglu [31] studied the performance of concrete by adding *mineral admixtures*, silica fume and fly ash. In the concrete mixes, the replacement of fly ash and silica fume were kept at the level of 15 and 10% as the weight of cement, respectively. They concluded that the depth of carbonation in concrete mixtures containing FA was slightly higher than that of control concrete. In concrete mixtures containing silica fume and fly ash at the same time, depth of carbonation was lower compared to the results of other concrete mixtures, where silica fume had little effect on carbonation. The lower depth of carbonation in SFAC was attributed to the lower porosity.

2.10.6 Alkali–Silica Reaction

Concrete prepared with sand or aggregates containing amorphous silica or siliceous aggregates (gneiss, schist) can deteriorate by the formation of expansive alkali silicate gels. The alkalis (Na_2O and K_2O) from the cement and other sources, with hydroxyl ions and certain siliceous constituents leading to formation of distinctive gelatinous hydrates which expand as water is imbibed and exert pressure on surrounding matrix. Pressure generated by the swelling gel ruptures the aggregate particles and causes cracks to extend into the surrounding concrete. Typically, alkali–silica reaction results in the formation of map-pattern cracking of the concrete.

The optimum method for minimizing the potential for expansion due to *alkali–silica reaction* in concrete is to replace a portion of the Portland cement with a supplementary cementing material. Low-lime fly ash, ground granulated blast furnace slag, silica fume, metakaolin and natural pozzolans used in the appropriate quantities have been found to be an effective antidote for alkali–silica reaction. Mixtures of two supplementary cementing materials with Portland cement (so-called ternary mixtures) are also very effective in preventing deterioration due to alkali–silica reaction.

Hooton [38] studied the influence of silica fume on the *expansion of mortars* made with *high-alkali cement* up to the age of 365 days. Expansions were reduced with increasing replacement of silica fume. It was concluded that mortars made with 10, 15 and 20% silica fume met the ASTM expansion limit of 0.020% at the age of 14 days.

Boddy et al. [15] investigated the possible use and effectiveness of "lower grade" silica fume with SiO_2 contents less than 85% as such materials do not meet current ASTM standards for silica fume. The performance of two silica fumes with

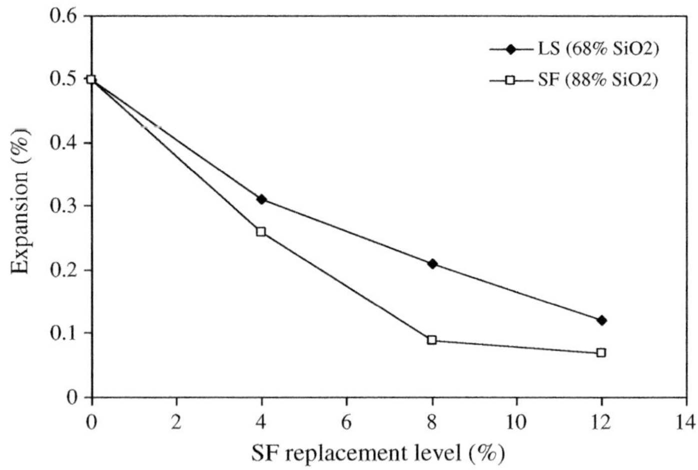

Fig. 2.23 Effect of silica fume level of replacement and SiO_2 content on mortar bar expansion at 14 days [15]

different silica contents; LS (68% SiO_2) and silica fume (88% SiO_2) were compared by examining the effect of the materials on the expansion due to alkali–silica reaction (ASR) and the composition of the pore solution. The concrete mixtures included 0, 4, 8, and 12% silica fume replacement by mass of cement. The accelerated mortar bar test was used for determining alkali–silica reactivity. Test results are shown in Fig. 2.23. Mortar bar mixtures containing LS (68% SiO_2) silica fume at levels of replacement up to 12% failed to control ASR expansion of Spratt aggregate. An NS (88% SiO_2) silica fume was effective at 12% replacement, and just below the 0.10% expansion limit criterion at 8% replacement.

References

1. Al-Amoudi, O.S.B., Abiola, T.O., Maslehuddin, M.: Effect of superplasticizer on plastic shrinkage of plain and silica fume cement concretes. Construct. Build. Mater. **20**(9), 642–647 (2006)
2. Al-Amoudi, O.S.B., Abiola, T.O., Maslehuddin, M.: Effect of type and dosage of silica fume on plastic shrinkage in concrete exposed to hot weather. Construct. Build. Mater. **18**(10), 737–743 (2004)
3. Alshamsi, A.M., Sabouni, A.R., Bushlaibi, A.H.: Influence of set retarding superplasticizers and microsilica on setting time of pastes at various temperatures. Cem. Concr. Res. **23**(3), 592–598 (1993)
4. ACI Committee 234: Guide for the use of silica fume in concrete (ACI 234R). ACI Mater. J. **92**(4), 437–440 (1995)
5. Al-Manaseer, A.A., Dalal, T.R.: Concrete containing plastic aggregates. Concr. Intern. **19**(8), 47–52 (1997)
6. Almusallam, A.A., Beshr, H., Maslehuddin, M., Al-Amoudi, O.S.B.: Effect of silica fume on the mechanical properties of low quality coarse aggregate concrete. Cem. Concr. Compos. **26**(7), 891–900 (2004)

7. ASTM C 1012-89: Standard test method for length change of hydraulic cement mortars exposed to mixed sodium and magnesium sulfate solutions. Annual Book of ASTM Standards 4.01, pp. 442–446 (1994)
8. Babu, K.G., Prakash, P.V.S.: Efficiency of silica fume in concrete. Cem. Concr. Res. **25**(6), 1273–1283 (1995)
9. Babu, K.G., Babu, D.S.: Behaviour of lightweight expanded polystyrene concrete containing silica fume. Cem. Concr. Res. **33**(5), 755–762 (2003)
10. Behnood, A., Ziari, H.: Effects of silica fume addition and water to cement ratio on the properties of high-strength concrete after exposure to high temperatures. Cem. Concr. Compos. **30**(2), 106–112 (2008)
11. Bentur, A., Goldman, A., Cohen, M.D.: Contribution of transition zone to the strength of high quality silica fume concretes. Proc. Mater. Res. Soc. Symp. **114**, 97–103 (1987)
12. Bentur, A., Goldman, A.: Curing effects, strength and physical properties of high strength silica fume concretes. J. Mater. Civil Eng. **1**(1), 46–58 (1989)
13. Berke, N.S.: Resistance of micro-silica concrete to steel corrosion, erosion and chemical attack. ACI Special Publications SP 114, pp. 861–886 (1989)
14. Bhanja, S., Sengupta, B.: Influence of silica fume on the tensile strength of concrete. Cem. Concr. Res. **35**(4), 743–747 (2005)
15. Boddy, A.M., Hooton, R.D., Thomas, M.D.A.: The effect of the silica content of silica fume on its ability to control alkali–silica reaction. Cem. Concr. Res. **33**(8), 1263–1268 (2003)
16. Brooks, J.J., Cabrera, J.G., Megat, J.M.A.: Factors affecting the autogenous shrinkage of silica fume high-strength concrete. In: Proceedings of the International Workshop on Autogenous Shrinkage of Concrete, pp. 185–92. Japan Concrete Institute, Hiroshima (1998)
17. Byfors, K.: Carbonation of concrete with silica fume and fly ash. J. Nordic Concr. Res. **4**, 26–35 (1985)
18. Civjan, S.A., LaFave, J.M., Trybulski, J., Lovett, D., Lima, J., Pfeifer, D.W.: Effectiveness of corrosion inhibiting admixture combinations in structural concrete. Cem. Concr. Compos. **27**(6), 688–703 (2005)
19. Chen, P., Chung, D.D.L.: Effect of polymer addition on thermal stability and thermal expansion of cement. Cem. Concr. Res. **25**(3), 465–469 (1995)
20. Cong, X., Gong, S., Darwin, D., McCabe, S.L.: Role of silica fume in compressive strength of cement paste, mortar and concrete. ACI Mater. J. **89**(4), 375–387 (1992)
21. Cohen, M.D., Bentur, A.: Durability of Portland-silica fume pastes in magnesium sulfate and sodium sulfate solutions. ACI Mater. J. **85**(3), 148–157 (1988)
22. Cwirzen, A., Penttala, V.: Aggregate-cement paste transition zone properties affecting the salt-frost damage of high-performance concretes. Cem. Concr. Res. **35**(4), 671–679 (2005)
23. Demirboğa, R.: Influence of mineral admixtures on thermal conductivity and compressive strength of mortar. Energy Build. **35**(2), 189–192 (2003)
24. Demirboğa, R.: Thermal conductivity and compressive strength of concrete incorporation with mineral admixtures. Build. Environ. **42**(7), 2467–2471 (2007)
25. Demirboğa, R., Gül, R.: Production of high strength concrete by use of industrial by-products. Build. Environ. **41**(8), 1124–1127 (2006)
26. Dotto, J.M.R., Abreu, A.G.D., Molin, D.C.C.D., Müller, I.L.: Influence of silica fume addition on concretes physical properties and on corrosion behaviour of reinforcement bars. Cem. Concr. Compos. **26**, 31–39 (2004)
27. Feldman, R.F.: Influence of condensed silica fume and sand/cement ratio on pore structure and frost resistance of Portland cement mortars. ACI Special Publications SP-91, pp. 973–990 (1986)
28. Ganjian, E., Pouya, H.S.: Effect of magnesium and sulfate ions on durability of silica fume blended mixes exposed to the seawater tidal zone. Cem. Concr. Res. **35**(7), 1332–1343 (2005)
29. Gjrov, O.E.: Durability of concrete containing condensed silica fume. ACI Special Publications SP-79, pp. 695–708 (1993)

30. Gleize, P.J.P., Müller, A., Roman, H.R.: Microstructural investigation of a silica fume–cement–lime mortar. Cem. Concr. Compos. **25**(2), 171–175 (2003)
31. Gonen, T., Yazicioglu, S.: The influence of mineral admixtures on the short and long-term performance of concrete. Build. Environ. **42**, 3080–3085 (2007)
32. González-Fonteboa, B., Martínez-Abella, F.: Concretes with aggregates from demolition waste and silica fume. Materials and mechanical properties. Build. Environ. **43**(4), 429–437 (2008)
33. Grimaldi, G., Carpio, J., Raharinaivo, A.: Effect of silica fume on carbonation and chloride penetration in mortars. In: Alsali, M. (ed.) Third CANMET/ACI International Conference on Fly Ash, Silica Fume, Slag and Naural Pozzolans in Concrete, pp. 320–334 (1989)
34. Grutzeck, M., Atkinson, S., Roy, D.M.: Mechanism of hydration of condensed silica fume in calcium hydroxide solutions. ACI Special Publications SP-79 (2), pp. 643–664 (1983)
35. Güneyisi, E., Gesoĝlu, M., Özturan, T.: Properties of rubberized concretes containing silica fume. Cem. Concr. Res. **34**(12), 2309–2317 (2004)
36. Gutiérrez, R.M.D., Díaz, Ln, Delvasto, S.: Effect of pozzolans on the performance of fibre-reinforced mortars. Cem. Concr. Compos. **27**(5), 593–598 (2005)
37. Hekal, E.E., Kishar, E., Mostafa, H.: Magnesium sulfate attack on hardened blended cement pastes under different circumstances. Cem. Concr. Res. **32**(9), 1421–1427 (2002)
38. Hooton, R.D.: Influence of silica fume replacement of cement on physical properties and resistance to sulfate attack freezing and thawing, and alkali–silica reactivity. ACI Mater. J. **90**(2), 143–152 (1993)
39. Hooton, R.D., Titherington, M.P.: Chloride resistance of high-performance concretes subjected to accelerated curing. Cem. Concr. Res. **34**(9), 1561–1567 (2004)
40. Huang, C.Y., Feldman, R.F.: Influence of silica fume on the micro-structural development in cement mortars. Cem. Concr. Res. **15**(2), 285–294 (1985)
41. Huang, C.Y., Feldman, R.F.: Hydration reactions in Portland cement–silica fume blends. Cem. Concr. Res. **15**(4), 585–592 (1985)
42. Igarashi, S.I., Kawamura, A., Watanabe, M.: Evaluation of capillary pore size characteristics in high-strength concrete at early ages. Cem. Concr. Res. **35**(3), 513–519 (2005)
43. Irassar, E.F., Maio, A.D., Batic, O.R.: Sulphate attack on concrete with mineral admixtures. Cem. Concr. Res. **26**(1), 113–123 (1996)
44. Johnston, C.D.: Deice salt scaling resistance and chloride permeability. Concr. Int. **16**(8), 48–55 (1994)
45. Kadri, E.H., Duval, R.: Hydration heat kinetics of concrete with silica fume. Construct. Build. Mater. **23**(11), 3388–3392 (2009)
46. Kayali, O., Zhu, B.: Corrosion performance of medium-strength and silica fume high-strength reinforced concrete in a chloride solution. Cem. Concr. Compos. **27**(1), 117–124 (2005)
47. Khatri, R.P., Sirivivatnanon, V., Gross, W.: Effect of different supplementary cementitious materials on mechanical properties of high performance concrete. Cem. Concr. Res. **25**(1), 209–220 (1995)
48. Khan, M.I., Lynsdale, C.J.: Strength, permeability, and carbonation of high-performance concrete. Cem. Concr. Res. **32**(1), 123–131 (2002)
49. Khan, M.I.: Isoresponses for strength, permeability and porosity of high performance Mortar. Build. Environ. **38**(8), 1051–1056 (2003)
50. Khayat, K.H., Aitcin, P.C.: Silica fume: a unique supplementary cementitious material. In: Ghosh, S.N. (ed.) Mineral Admixtures in Cement and Concrete, vol. 4, pp. 227–265. ABI Books Private Limited, Delhi (1993)
51. Khedr, S.A., Idriss, A.F.: Resistance of silica fume concrete to corrosion-related damage. J. Mater. Civil Eng. **7**(2), 102–107 (1995)
52. Kılıç, A., Atiş, C.D., Teymen, A., Karahan, O., Özcan, F., Bilim, C., Özdemir, M.: The influence of aggregate type on the strength and abrasion resistance of high strength concrete. Cem. Concr. Compos. **30**(4), 290–296 (2008)
53. Köksal, F., Altun, F., Yiĝit, I., Sùahin, Y.: Combined effect of silica fume and steel fibre on the mechanical properties of high strength concretes. Construct. Build. Mater. **22**(8), 1874–1880 (2008)

54. Krishnamoorthy, T.S., Gopalakrishnan, S., Balasubramanian, K., Bharatkumar, B.H., Rao, P.R.M.: Investigations on the cementitious grouts containing supplementary cementitious materials. Cem. Concr. Res. **32**(9), 1395–1405 (2002)
55. Kurdowski, W., Nocun-Wczelik, W.: The tricalcium silicate hydration in the presence of active silica. Cem. Concr. Res. **13**(3), 341–348 (1983)
56. Langan, B.W., Weng, K., Ward, M.A.: Effect of silica fume and fly ash on heat of hydration of Portland Cement. Cem. Concr. Res. **32**(7), 1045–1051 (2002)
57. Lee, S.T., Moon, H.Y., Swamy, R.N.: Sulfate attack and role of silica fume in resisting strength loss. Cem. Concr. Compos. **27**(1), 65–76 (2005)
58. Lohtia, R.P., Joshi, R.C.: Mineral admixtures. In: Ramachandran, V.S. (ed.) Concrete Admixture Handbook, 1153pp. Noyes Publications, USA (1996)
59. Maage, M.: Carbonation in concrete made of blended cements. In: Symposium of Materials Research Society, Boston, USA, vol. 65, pp. 193–197 (1986)
60. Mak, S.L., Ritchie, D., Taylor, A., Diggins, R.: Temperature effects on early age autogenous shrinkage in high performance concretes. In: Proceedings of International Workshop on Autogenous Shrinkage of Concrete, Hiroshima, Japan, pp. 155–165. E & FN Spon, London (1998)
61. Mangat, P.S., Khatib, J.M.: Influence of fly ash, silica fume and slag on sulphate resistance of concrete. ACI Mater. J. **92**(5), 542–552 (1993)
62. Mazloom, M., Ramezanianpour, A.A., Brooks, J.J.: Effect of silica fume on mechanical properties of high-strength concrete. Cem. Concr. Compos. **26**(4), 347–357 (2004)
63. Meland, I.: Influence of condensed silica fume and fly ash on the heat evolution in cement pastes. ACI Special Publications SP-79 (2), pp. 665–676 (1983)
64. Mindess, S., Young, J.F.: Concrete. Prentice Hall, Englewood Cliffs (1981)
65. Neville, A.M.: Properties of Concrete, 4th and final edn., pp. 271–729. Wiley, New York (1997)
66. Ozyildirim, C., Halstead, W.J.: Improved concrete quality with combinations of fly ash and silica fume. ACI Mater. J. **91**(6), 587–594 (1994)
67. Perraton, D., Aitcin, P.C., Vezina, D.: Permeabilities of silica fume concretes. ACI Special Publications SP-108, pp. 63–84 (1988)
68. Persson, B.S.M.: Shrinkage of high-performance concrete. In: Proceedings of International Workshop on Autogenous Shrinkage of Concrete, Hiroshima, Japan, pp. 105–115. E & FN Spon, London (1998)
69. Poon, C.S., Kou, S.C., Lam, L.: Compressive strength, chloride diffusivity and pore structure of high performance metakaolin and silica fume concrete. Construct. Build. Mater. **20**(10), 858–865 (2006)
70. Postacioğlu, B., Maddeler, B.: Cementing Materials. Matbaa Teknisyenleri Basımevi, Istanbul, Turkey, vol. 1, pp. 63–66 (in Turkish) (1986)
71. Qing, Y., Zenan, Z., Deyu, K., Rongshen, C.: Influence of nano-SiO_2 addition on properties of hardened cement paste as compared with silica fume. Construct. Build. Mater. **21**(3), 539–545 (2007)
72. Rao, G.A.: Investigations on the performance of silica fume-incorporated cement pastes and mortars. Cem. Concr. Res. **33**(11), 1765–1770 (2003)
73. Rasheeduzzafar, S.S.A., Al-Gahtani, A.S.: Reinforcement corrosion-resisting characteristics of silica-fume blended-cement concrete. ACI Mater. J. **89**(4), 337–344 (1993)
74. Rossignolo, J.A.: Interfacial interactions in concretes with silica fume and SBR latex. Construct. Build. Mater. **23**(2), 817–821 (2008)
75. Sandvik, M., Gjørv, O.E.: Prediction of strength development for silica fume concrete. Proceedings of 4th International Conference on the Use of Fly Ash, Silica Fume, Slag and Natural Pozzolans in Concrete, Istanbul, Turkey. ACI Special Publication **132**, 987–996 (1992)
76. Sata, V., Jaturapitakkul, C., Kiattikomol, K.: Influence of pozzolan from various by-product materials on mechanical properties of high-strength concrete. Construct. Build. Mater. **21**(7), 1589–1598 (2007)

77. Schubert, P.: Carbonation behaviour of mortars and concrete made with fly ash. ACI Special Publications SP-100, pp. 1945–1962 (1987)
78. Sellevold, E.J.: The function of silica fume in high strength concrete. In: Proceedings of International Conference on Utilization of High Strength Concrete, Stavanger, Norway, pp. 39–50 (1987)
79. Sellevold, E.J., Nilsen, T.: Condensed silica fume in concrete: a world review. In: Malhotra, V.M. (ed) Supplementary Cementing Materials for Concrete, Ottawa, pp. 165–243 (1987)
80. Sellevold, E.J., Redjy, F.F.: Condensed silica fume (microsilica) in concrete: water demand and strength development. In: Malhotra, V.M. (ed.) The Use of Fly Ash, Silica Fume, Slag and Other Mineral By-Products in Concrete, ACI SP-79, pp. 677–694 (1983)
81. Shannag, M.J., Shaia, H.A.: Sulfate resistance of high-performance concrete. Cem. Concr. Compos. **25**(3), 363–369 (2003)
82. Silica Fume Association: Silica fume manual.38860 Sierra Lane, Lovettsville, VA 20180, USA (2005)
83. Skjolsvold, O.: Carbonation depths of concrete with and with out condensed silica fume. ACI Special Publications SP-91, pp. 1031–1048 (1986)
84. Sobolev, K.: The development of a new method for the proportioning of high-performance concrete mixtures. Cem. Concr. Compos. **26**(7), 901–907 (2004)
85. Song, H.W., Jang, J.W., Saraswathy, V., Byun, K.J.: An estimation of the diffusivity of silica fume concrete. Build. Environ. **42**(3), 1358–1367 (2007)
86. Sørensen, E.V.: Freezing and thawing resistance of condensed silica fume concrete exposed to deicing chemicals. ACI Special Publications SP-79 (2), pp. 709–718 (1983)
87. Soroushian, P., Mirza, F., Alhozaimy, A.: Permeability characteristics of polypropylene fibre reinforced concrete. ACI Mater. J. **92**(3), 291–295 (1995)
88. Tanyildizi, H., Coskun, A.: Performance of lightweight concrete with silica fume after high temperature. Construct. Build. Mater. **22**(10), 2124–2129 (2008)
89. Tao, Z., Weizu, Q.: Tensile creep due to restraining stresses in high-strength concrete at early ages. Cem. Concr. Res. **36**(3), 584–591 (2006)
90. Tazawa, E., Yonekura, A., Tanaka, S.: Rate of hydration and drying shrinkage of condensed silica fume mortar prepared by double mixing. In: Alsali, M. (ed.) Third CANMET/ACI International Conference on Fly Ash, Silica Fume, Slag and Natural Pozzolans in Concrete, pp. 350–364 (1989)
91. Tazawa, E., Miyazawa, S.: Autogenous shrinkage of concrete and its importance in concrete technology. In: Proceedings of the 5th International RILEM Symposium on Creep and Shrinkage of Concrete, pp. 159–168. E & FN Spon, London (1993)
92. Taylor, H.W.F.: Cement Chemistry. Academic Press, New York (1990)
93. Uchikawa, H.: Effect of blending components on hydration and structure formation. In: Eighth International Congress on the Chemistry of Cement, Rio de Janeiro, pp. 249–280 (1986)
94. Uchikawa, H., Uchida, S.: Influence of pozzolans on the hydration of C_3A. In: Seventh International Congress on the Chemistry of Cement, Paris, pp. IV-23–IV-29 (1980)
95. Wee, T.H., Suryavanshi, A.K., Wong, S.F., Anisur Rahman, K.M.: Sulfate resistance of concrete containing mineral admixture. ACI Mater. J. **97**(5), 536–549 (2000)
96. Wild, S., Sabir, B.B., Khatib, J.M.: Factors influencing strength development of concrete containing silica fume. Cem. Concr. Res. **25**(7), 1567–1580 (1995)
97. Wong, H.S., Razak, H.A.: Efficiency of calcined kaolin and silica fume as cement replacement material for strength performance. Cem. Concr. Res. **35**(4), 696–702 (2005)
98. Yazıcı, H.: The effect of silica fume and high-volume Class C fly ash on mechanical properties, chloride penetration and freeze–thaw resistance of self-compacting concrete. Construct. Build. Mater. **22**(4), 456–462 (2008)
99. Zhang, M.H., Tam, C.T., Leow, M.P.: Effect of water-to-cementitious materials ratio and silica fume on the autogenous shrinkage of concrete. Cem. Concr. Res. **33**(10), 1687–1694 (2003)

Chapter 3
Ground Granulated Blast Furnace Slag

3.1 Introduction

Ground granulated blast furnace slag (GGBS) is a by-product from the blast-furnaces used to make iron. Blast-furnaces are fed with controlled mixture of iron-ore, coke and limestone, and operated at a temperature of about 1,500°C. When iron-ore, coke and limestone melt in the blast furnace, two products are produced—molten iron, and molten slag. The molten slag is lighter and floats on the top of the molten iron. The molten slag comprises mostly silicates and alumina from the original iron ore, combined with some oxides from the limestone. The process of granulating the slag involves cooling of molten slag through high-pressure water jets. This rapidly quenches the slag and forms granular particles generally not bigger than 5 mm. The rapid cooling prevents the formation of larger crystals, and the resulting granular material comprises around 95% non-crystalline calcium-aluminosilicates. The granulated slag is further processed by drying and then grinding in a rotating ball mill to a very fine powder, which is GGBS.

GGBS can be used as a direct replacement for ordinary cement on one-to-one basis by weight. Replacement rates for GGBS vary from 30% to up to 85%. Generally 50% is used in most applications. Higher replacement rates up to 85% are used in specialist applications such as in aggressive environments and to reduce heat of hydration. GGBS can be used at replacement levels of 70% in lean mix concrete.

3.1.1 *Storage and Handling of GGBS*

Bulk GGBS is stored and handled in conditions identical to that of Portland cement. Bulk storage is in watertight silos. Transportation is by bulk tankers, as for Portland cement. GGBS can also be moved by air-slides, cement screws and

R. Siddique and M. Iqbal Khan, *Supplementary Cementing Materials*, Engineering Materials, DOI: 10.1007/978-3-642-17866-5_3,

bucket elevators. Dust control is the same as that required for Portland cement. GGBS dust does not present any fire or explosion hazard.

3.1.2 Environmental Benefits of Using GGBS

The use of GGBS as partial cement replacements with lower environmental burdens offers opportunities for significant reductions in energy use and carbon dioxide emissions. Proportions of up to 70 or even 80% can be used with advantage in suitable situations. The use of GGBS in concrete results in following environmental benefits:

- Saves energy.
- Reduces emission of carbon dioxide.
- Conserves natural resources.

3.2 Characteristics of GGBS

3.2.1 Physical Properties

Ground granulated blast-furnace slag is a glassy material. The colour of GGBS varies from beige to dark to off-white depending on moisture content, chemistry and efficiency of granulation. When it is ground it has usually white colour. It imparts a lighter, brighter color to concrete. Figure 3.1 shows the colour of GGBS.

GGBS is a fine glassy material. Its specific gravity is less than that of cement, but has more fineness. Its physical properties, reported by some authors are given in Table 3.1.

Fig. 3.1 Granulated blast furnace slag

Table 3.1 Physical properties of GGBS

Property	Tasong et al. [70]	Oner and Akyuz [57]	Hui-sheng et al. [38]
Specific gravity	2.9	2.87	2.89
Specific surface (m^2/kg)	425–470	425	371
Bulk density (kg/m^3)	1,200	–	

Table 3.2 Surface area and average diameter of GGBS [75]

Samples	Ball-mill (A)	Vibromill (B)	Airflow mill (C)	Ball-mill (D)
Surface area (m^2/kg)	510	685	515	512
Diameter (μm)	13.69	9.12	11.72	13.15

3.2.2 Particle Morphology

Wan et al. [75] reported the diameter and surface area of GGBS samples (A, B, C, and D) prepared by using four processing approaches, a ball mill, steel balls as grinding medium, an airflow mill and a vibro-mill, respectively. Surface area and diameter of GGBS are given in Table 3.2. It can be seen that diameter of GGBS particles varied from 9.12 to 13.69 μm whereas surface area was between 510 and 685 m^2/kg.

The shape of GGBS processed by a vibro-mill was predominantly spherical with a smooth surface, while that by a ball mill and an airflow mill appeared to have similar edges.

Wan et al. [75] observed the morphology of four GGBS samples (A, B, C, and D) prepared by using four techniques a ball mill, steel balls as grinding medium, an airflow mill and a vibro-mill, respectively. SEM images of samples A–D are shown in Fig. 3.2. It can be seen that the shape of GGBS is not spherical; it varies according to different grinding techniques. GGBS is mainly composed of glassy phases, which are in a continuous network structure, and there is no stress concentration in the interfacial area; thus, grinding GGBS leads to the disconnection of bonds between molecules and atoms. When slag particles are broken, the shape of the broken surface is not fixed. Fig. 3.2a and d is the appearance of GGBS processed by a ball mill, which is predominately in anomalous shape with clear edges and angles. This is due to inter-impacting and inter-rubbing between steel balls in the ball mill. Figure 3.2b shows that the shape of GGBS is mostly in sphericity and its surface is relatively smooth. The reason for this phenomenon is that GBBS is crushed by the interaction between a steel cylinder and a steel ring and that between a steel ring and a vessel wall; in addition, this increases the probability of inter-rubbing between particles and leads to smooth edges and angles of particles. Figure 3.2c shows that the GGBS sample has the most uniform particle size, and the particle appearance is similar to sample A. This has much to do with the working mechanism of airflow mills. When an airflow mill is working with a high-speed air current, particles collide severely with fixed manganese-steel boards, and this leads to pulverization and thus decreases the probability of contacts between particles.

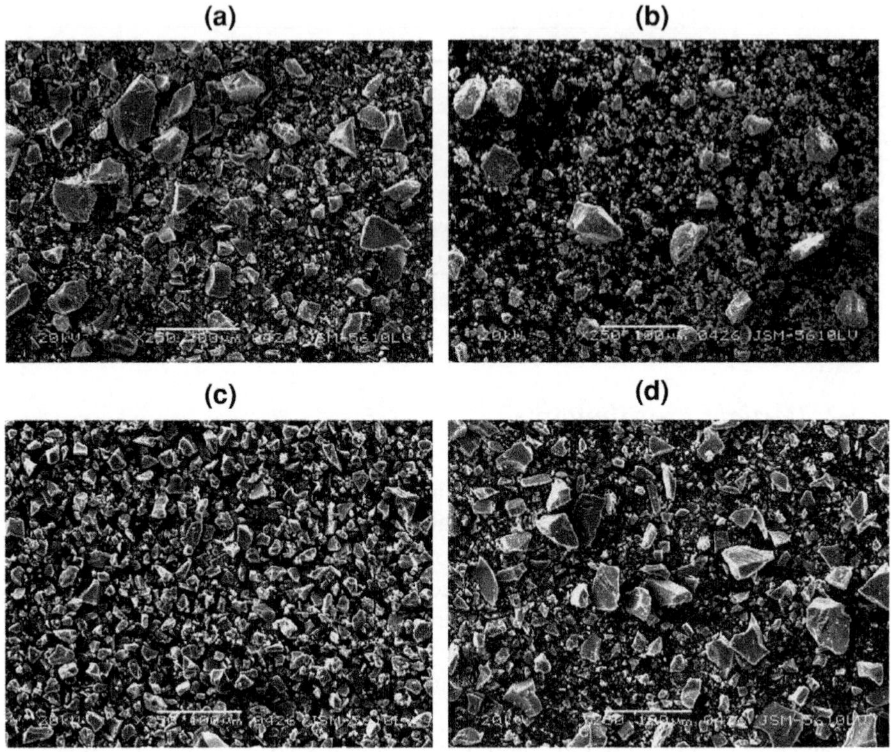

Fig. 3.2 SEM of granulated blast furnace slag samples: **a** SEM of sample A (Ball-mill), **b** SEM of sample B (Vibromill), **c** SEM of sample C (Airflow mill) **d** SEM of sample D (Ball-mill) [75]

3.2.3 Chemical Composition

Blast furnace slag is a non-metallic product, consisting essentially of silicates and alumino-silicates of calcium and other bases. Slag is made up of both glassy and crystalline phases. The glassy nature is responsible for its cementitious properties. In GGBS, glass content is between 85 and 90%. GGBS comprises mainly of CaO, SiO_2, Al_2O_3, MgO. It has the same main chemical constituents as ordinary Portland cement, but in different proportions. Typical Chemical Composition of GGBS, reported by some authors is given in Table 3.3.

3.2.4 Reactivity

ASTM C 989 [6] is the standard for evaluating the cementitious potential of a slag. The factors that determine the cementitious properties of a slag are:

- Chemical composition of the slag.
- Alkali concentration of the reacting system.

Table 3.3 Chemical composition of GGBS

Composition (%)	Tasong et al. [70]	Oner and Akyuz [57]	Hui-sheng et al. [38]
SiO_2	35.34	39.18	36.39
Al_2O_3	11.59	10.18	13.76
Fe_2O_3	0.35	2.02	2.44
CaO	41.99	32.82	30.13
MgO	8.04	8.52	9.36
MnO	0.45	–	
S_2	1.18	–	
SO_3	0.23	–	1.30

- Glass content of the slag.
- Fineness of both slag and cement.
- Temperature variations during the early phases of hydration process.

Daube and Bakker [22] defined two basic parameters that determine the hydraulic properties of a slag (1) chemical composition; and (2) vitreous state. Cheron and Lardinois [19] have defined the hydraulic activity index of the slag as below. Slag indices between 1.65 and 1.85 are considered normal:

$$\text{Hydraulic activity index} = (\text{CaO} + 1.4\ \text{MgO} + 0.56\text{Al}_2\text{O}_3)/\text{SiO}_2$$

3.2.4.1 Hydration Reaction

The hydration mechanism of a combination of GGBS and Portland cement is slightly more complex than that of a Portland cement. This reaction involves activation of the GGBS by alkalis and sulfates to form its own hydration products. Some of these combine with the Portland cement products to form further hydrates which have a pore blocking effect. The result is a hardened cement paste with more of very small gel pores and fewer of the much larger capillary pores for the same total pore volume. Generally, the rate of strength development is slower than for a Portland cement mortar.

The resulting hardened cement paste using GGBS is also more chemically stable. It contains much less free lime, which in concrete made with Portland cement leads to the formation of further reaction products such as ettringite or efflorescence. In addition, GGBS contains no C_3A, making GGBS concrete much less reactive to sulfates.

When GGBS is used in concrete, resulting hardened cement paste has more smaller gel pores and fewer larger capillary pores than is the case with concrete made with normal Portland cement. This finer pore structure gives GGBS concrete a much lower permeability, and makes an important contribution to the greater durability of this concrete.

Table 3.4 Assessment of the peak heat rate characteristics of GGBS concretes [70]

Binder composition	Time to q_{max}, t_{20} (h)	q_{max} (W/kg)	q_{max} normalized to 100% CEM I (W/kg)
100% CEM I	17.6	2.80	2.80
20% GGBS	15.3	2.39	2.99
40% GGBS	15.7	1.81	3.02
60% GGBS	14.2	1.30	3.25
80% GGBS	10.7	0.73	3.65

Ballim and Graham [15] studied the effect of GGBS on the heat of hydration of concrete. Concretes were prepared with GGBS, blended with Portland cement in proportions ranging from 20 to 80%. These concretes were subjected to heat of hydration tests under adiabatic conditions. Heat rate performances of materials are expressed in terms of maturity or t_{20} (h), which refer to the equivalent time of hydration at 20°C. This form of expression of heat rate function and the justification for its use is described by [14]. Table 3.4 shows that the time to reach this peak hydration rate is marginally reduced as the proportion of GGBS is increased to 60%. However, the reduction in time to peak heat rate is more significant as the proportion of GGBS was increased from 60 to 80%. It is also evident that normalized peak heat rates for the GGBS concretes are higher than that of the plain CEM I concrete. This indicates that the hydration of the GGBS is contributing to the generation of heat in the concrete, even at these relatively early ages.

Gao et al. [28] studied the hydration of concrete made with GGBS. Specific surface area of cement and GGBS were 361, 425, 600 m^2/kg, respectively. Cement was replaced with 40% GGBS. The self-hydration of cement and GGBS produces $Ca(OH)_2$. In a saturated solution of $Ca(OH)_2$, the pozzolanic reaction of GGBS consumes $Ca(OH)_2$. Therefore, the quantity of $Ca(OH)_2$ crystals depends on its formation and reaction rates in a $Ca(OH)_2$ saturated solution. When the formation rate is faster than reaction rate, then, amount of $Ca(OH)_2$ crystal and height of the XRD peak increases.

Figure 3.3 shows the XRD diagram of paste containing 40% GGBS (specific surface area 425 m^2/kg) as partial replacement of Portland cement. The peak height in the XRD curve, corresponding to $Ca(OH)_2$ increased when the self-hydration rate of the $Ca(OH)_2$ crystal XRD peak height. The above analysis demonstrates that XRD diagrams can indicate the pozzolanic reaction rate of GGBS indirectly.

Figure 3.4 is an XRD diagram of paste containing GGBS (specific surface area 600 m^2/kg) partially replacing (40%) of Portland cement. It was indicated in Fig. 3.4 that the CH peak height at 7 days was not as high as that in Fig. 3.3; this means that the pozzolanic reaction of GGBS in the sample starts to consume CH at 7 days. It means that the pozzolanic reaction of GGBS (specific surface area 600 m^2/kg and a weight fraction of 40%) starts earlier and proceeds at a fast rate.

Zang et al. [80] studied the hydration mechanism and mineral phase structures by waterglass activation of granulated blast furnace slag (GBFS). The GBFS used

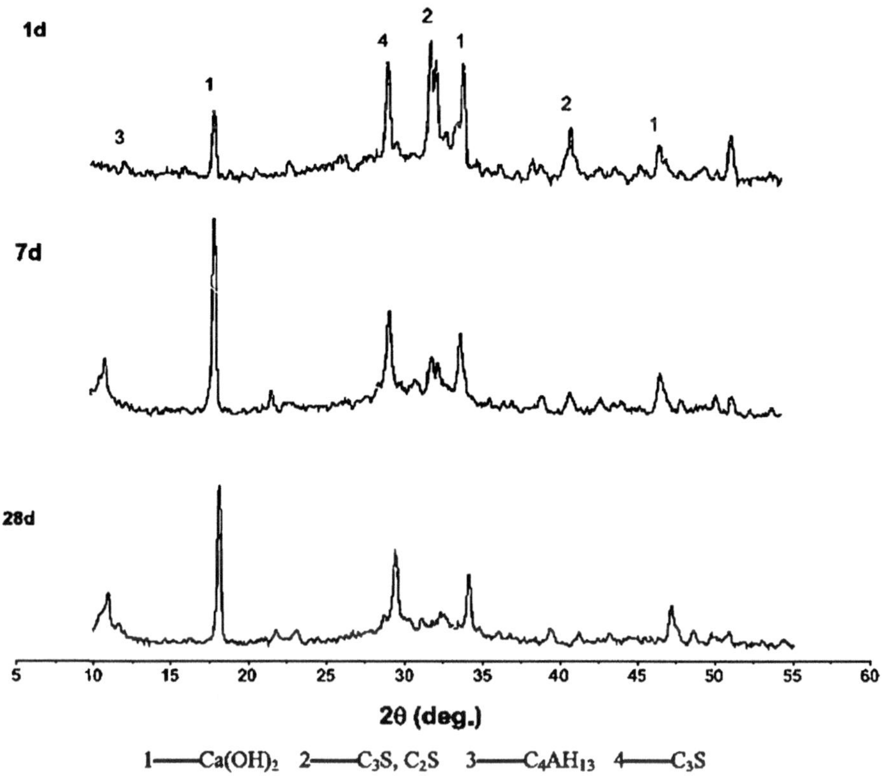

Fig. 3.3 XRD diagram of paste with GGBS (specific surface area 425 m^2/kg) replacing 40% Portland cement [28]

for synthesis of geopolymer had Blaine specific surface area of 701 m^2/kg. Chemical reagent (sodium metasilicate $Na_2SiO_3 \cdot 9H_2O$) was used as alkali activator with waterglass modulus as 1.0. Raw material was blended in the ratio of waterglass:GBFS:water = 1:9:2.5 (in weight). The paste was cast into 50 × 31.6 × 31.6 mm^3 metal model. After demolding, a triplicate set of samples were put in a curing box at 20°C with 99% relative humidity and 0.5 h initial curing time, and final setting times of 1-day (1d), 3 days (3d), 7 days (7d), and 28 days (28d), respectively. The X-ray diffraction (XRD) patterns of geopolymers were carried out on an X'Pert PRO MPD diffractometer.

Figure 3.5 shows the XRD patterns of samples by waterglass activation of GBFS. From the pattern of GBFS it can be observed that there is a broad diffuse hump peak in the region 20–38° 2θ suggesting that the GBFS predominantly consists of glassy phases. Besides, there are four kinds of mineral phases, akermanite, gehlenite, calcium silicate, and merwinite. In comparison to the pattern of GBFS, the relative intensities of diffraction peaks had no distinct changes and have no new peak appearance from setting time for 0.5 h to final setting time

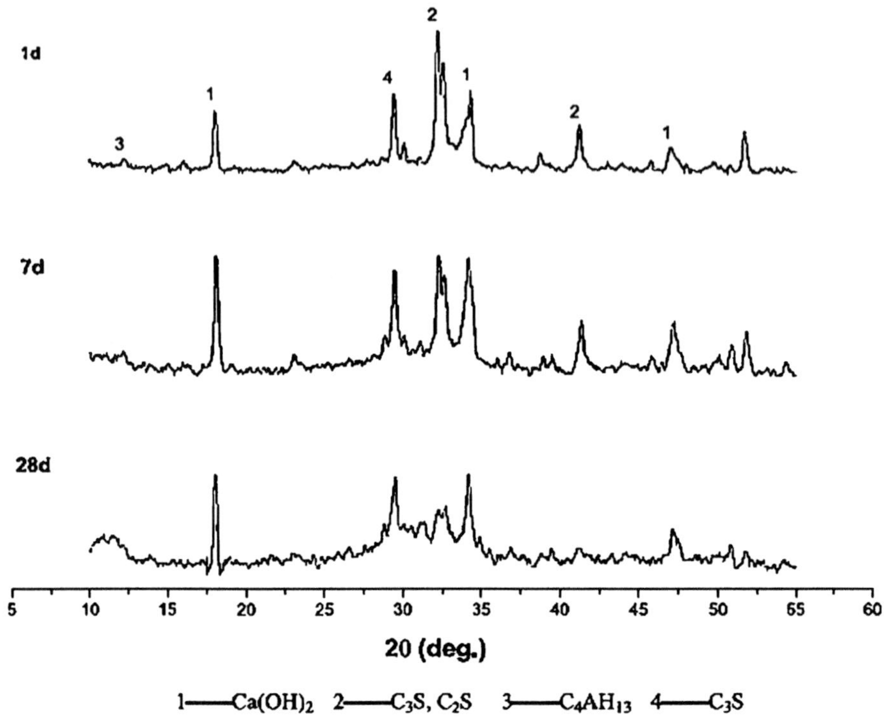

Fig. 3.4 XRD diagram of paste with GGBS (specific surface area 600 m^2/kg) replacing 40% Portland cement [28]

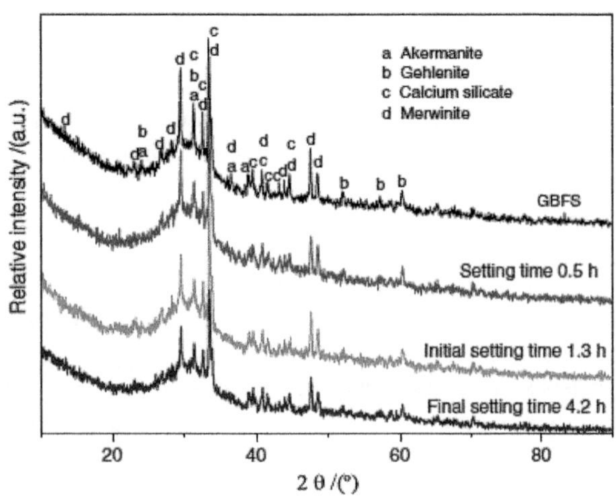

Fig. 3.5 XRD patterns of samples by waterglass activation of GBFS. The curve of GBFS, setting time for 0.5 h, initial time for 1.3 h and final setting time for 4.2 h, respectively [80]

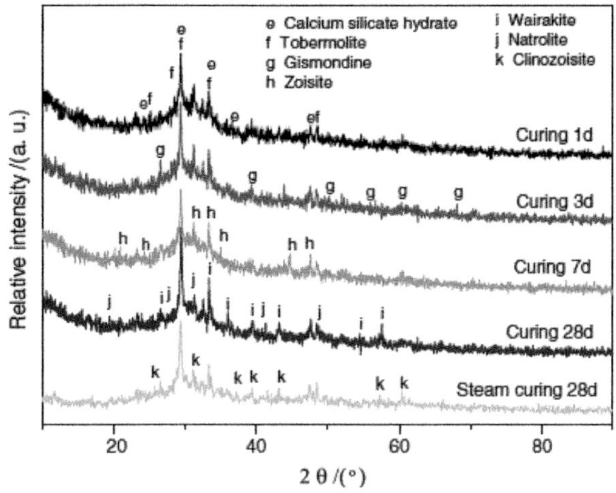

Fig. 3.6 XRD patterns of samples by waterglass activation of GBFS in the period of curing times 1–28 days [80]

for 4.2 h, implying that sodium hydroxide deriving from the hydrolysis of waterglass promotes some vitreous dissolution in the periods of curing ages.

The XRD patterns of samples for curing ages 1d, 3d, 7d, 28d and steam curing time of 28d are displayed in Fig. 3.6. After the curing time of 1d, the intensities of some diffraction peaks remarkably decreased and some peaks disappeared in comparison to final setting time in Fig. 3.3 (a) demonstrating that those crystal phases, either partial or whole, dissolved into aqueous sodium hydroxide solution derived from the hydrolysis of waterglass. Some new phases, tobermolite and calcium silicate hydrate (CSH) gel, were generated during curing time of 1d.

3.2.5 Specifications of GGBS

ASTM C 989 [6] provides for three strength grades of slag, depending upon their respective mortar strengths when blended with an equal mass of Portland cement. The classifications are; Grades 120, 100, and 80, based on the slag-activity index expressed as SP/P $\times$ 100, where SP is the average compressive strength of slag mortar cubes and P is the compressive strength of reference cement mortar cubes (without slag). Grades 100 and 120 are the most commonly used as admixtures in concrete. Table 3.5 gives criteria for ASTM C 989 classification.

3.2.6 Advantages of Using GGBS

Use of GGBS in cement and concrete results in

- Improved workability and compaction characteristics.
- Increased pumpability.

Table 3.5 GGBS-Activity index standards for various grades [6]

Age and grade	Average of last five consecutive samples	Any individual sample
7-day index, minimum		
Grade 80	–	–
Grade 100	75	70
Grade 120	95	90
28-day index, minimum		
Grade 80	75	70
Grade 100	95	90
Grade 120	115	110

- Increased strength.
- Enhanced durability.
- Reduced permeability.
- High resistance to chloride penetration.
- High resistance to sulfate attack.
- High resistance to ASR.
- Low heat of hydration.
- Improved surface finish.
- Enhanced architectural appearance
- Suppresses efflorescence
- Enhancement of the life cycle of concrete structures.
- Reduction in maintenance and repair costs.
- Slashes lifetime construction costs.
- Production of GGBS involves virtually zero CO_2 emissions, and no emissions of SO_2 and NOx.

3.3 Fresh Properties of Mortar/Paste/Concrete Containing GGBS

3.3.1 Bleeding Characteristics

Bleeding is the term used to describe the movement of water to the surface of the freshly placed concrete. All concrete bleeds to some extent but bleed water is only observed on the surface when the rate of bleeding exceeds the rate of evaporation. Immediately after compaction, there is a short dormant period which is followed by a period in which the rate of bleeding is almost uniform. Bleeding ends when either the movement of water is blocked by the growth of hydration products or by the solids effectively coming into contact with each other. Bleeding results in a variation in the effective water content throughout the concrete element which produces corresponding changes in the concrete properties.

Wainwright and Ait-Aider [73] investigated the influence of the addition of GGBS (40 and 70%) on the bleed characteristics of concrete. The bleed water was drawn at 10 min intervals during the first 40 min and at 30 min intervals thereafter until cessation of bleeding. Partial replacement of OPC with 40 and 70% of GGBS led to increase in the bleeding of the concretes, and increase were more pronounced at the higher replacement levels.

Olorunsogo [56] examined the effect of particle size distribution (PSD) of GGBS on the bleeding rate and bleeding capacity of mortar mixes containing 30 and 70% slag. Water retained on the surface of the mortar mixes was collected at intervals of 10 min within the first 40 min, and 30-min intervals thereafter until cessation of bleeding. They concluded that (1) on the basis of equal w/c, bleeding rate and capacity increased significantly with increase in slag content except for the 30% slag mortar mix, which had the lowest bled capacity at 0.45 w/c; (2) when PSD of slag was varied by altering slope (n) at constant position parameter (x_0) no specific relationship between bleeding characteristics and PSD was observed. However, the highest value of bleeding rate and capacity were exhibited by mixes in which the slag with intermediate value of slope was included. Also, the slag samples with similar size range distribution (having a constant slope, n) the bleeding rate increased with the increase in $x_{0,}$ except the 30% slag mixes that were made to 0.35 w/c. Changing the w/c from 0.35 to 0.45 resulted in increase of 86, 83, and 71% in bleeding rate of the OPC, 30, and 70% slag mixes, respectively.

Wainwright and Rey [74] studied the influence of GGBS on the bleeding of concrete. GGBS from four different sources (S1–S4) and Portland cement from one source were used. One control mix with cement content of 300 kg/m^3 was used in mix proportion of 1:2.5:3.6, having water/binder ratio of 0.56. Slag replacement levels were 55 and 85% by weight of cement. Bleed tests were performed in accordance with ASTM C232 [4], starting 30, 75 and 120 min after completion of mixing for each of the mixes. Bleeding rate is defined as the volume of water collected per second during the first 40 min of the test. The volume is expressed in ml/cm^2 of exposed surface. Bleed capacity is defined as the fraction of the initial volume of the concrete that has separated out as bleed water during the entire course of the test. Bleed capacity and bleed rate of concrete mixes are given in Table 3.6. They concluded that (1) addition of 55% slag increased the bleed capacity by 30% (compared to the plain Portland cement (OPC mix) but had

Table 3.6 Influence of test start time on bleed capacity and rate [74]

	Bleed capacity (ml/ml)			Bleed rate (ml/cm^2/s) $\times$ 10^{-6}		
	30 min	75 min	120 min	30 min	75 min	120 min
OPC	8	8.1 (1.3%)	7.9 (1.3%)	17.8	10.5 (41%)	9.0 (49%)
55% GGBS	10.7	7.7 (28%)	6.7 (36%)	18.5	11.5 (38%)	9.8 (47%)
85% GGBS	11.1	4.2 (62%)	3.5 (68%)	16.5	14.5 (12%)	10.5 (36%)

Values in bracket represent % change from 30 min reading

little effect on bleed rate; (2) Increasing slag content to 85% had no further significant effect on bleeding; (3) source of slag was also found to have little effect on the bleeding; (4) delaying the start of the bleed tests from 30 to 120 min reduced the bleed capacity of the OPC mix by more than 55% compared with 32% for the slag mixes. The reduction in bleed rate was similar for all mixes at about 45%.

3.3.2 Workability

As per ACI Committee [1], the greater solid volume and higher fineness of slag allow more coarse aggregate to be used without the loss of workability. This often reduces the stickiness of the mix.

Meusel and Rose [52] experimented with highly active slag at contents of 30–50% in concrete. They observed that inclusion of slag improved the workability of concrete mixes, but greater improvement was achieved with higher slag content, and higher fineness of slag did not had significant effect on the workability. Stutterheim [68] concluded that slag concretes have appreciably better workability than Portland cement concretes, allowing for reduction in water quantity.

Wainwright and Rey [74] reported the influence of GGBS additions on the slump of concrete. GGBS from four different sources (S1–S4) and Portland cement from one source were used. One control mix proportion of 1:2.5:3.6 was used having w/b ratio of 0.56. Slag replacement levels were 55 and 85% by weight of cement. Mixture proportions and slump results are given in Table 3.7.

Wan et al. [75] concluded that workability of mortar increased with the increase in the surface area of GGBS. Four samples (A, B, C, and D) of GGBS were used. Fluidity ratio was determined of the samples. The influence of GGBS on the fluidity of mortar is shown in Table 3.8. The fluidity of sample B and D was much better than that of sample A. The fluidity of mortar was associated with the

Table 3.7 Mix proportions and workability [74]

Mix		Mix proportions (kg/m^3)					Water/ binder	Slump (mm)
		OPC	GGBS	Fines	Coarse	Water		
OPC		300	–	750	1,080	168	0.56	15
55% GGBS	Source 1	135	165	750	1,080	168	0.56	30
	Source 2	135	165	750	1,080	168	0.56	20
	Source 3	135	165	750	1,080	168	0.56	25
	Source 4	135	165	750	1,080	168	0.56	45
85% GGBS	Source 1	45	255	750	1,080	168	0.56	30
	Source 2	45	255	750	1,080	168	0.56	40
	Source 3	45	255	750	1,080	168	0.56	25
	Source 4	45	255	750	1,080	168	0.56	20

Table 3.8 Test results of fluidity ratio for different surface area and particle sizes of GGBS [75]

Sample	Surface area (m^2/kg)	Average diameter (μm)	Fluidity ratio (%)
Sample A	510	13.69	1.02
Sample B	685	9.12	1.06
Sample C	515	11.72	1.03
Sample D	512	13.15	1.06

morphological characteristics of the GGBS particles. The smoother the surface of particles, the better was the fluidity of mortar.

3.3.3 Setting Times

The setting time of concrete with GGBS is generally greater than that of similar concrete with Portland cement only. Setting time increased with the increase in GGBS content as shown in Fig. 3.7.

Wainwright and Ait-Aider [73] investigated the effect of GGBS on the setting times and consistency of cements. Cement from three different sources and GGBS from one source was used. Cements were partially replaced with 40 and 70% of GGBS. Table 3.9 gives the results of setting times and consistency of OPC and blended cements with GGBS. They concluded that (1) consistency and setting times were almost similar for all three sources of cements; (2) inclusion of GGBS affected the consistency of cements, and it deceased with the increase in GGBS content; and (3) setting time of cements was increased with the increase in GGBS content.

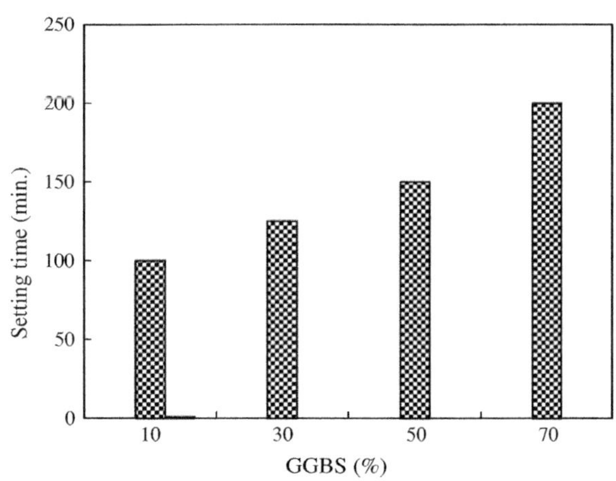

Fig. 3.7 Effect of GGBS on setting time [76]

Table 3.9 Setting times and consistency of cements [73]

Cement type	GGBS (% by wt)	Consistency (%)	Initial setting (min)	Final setting (min)
Source 1	0	31.2	108	182
Source 1	40	29.0	164	224
Source 1	70	26.5	241	318
Source 2	0	30.6	105	171
Source 2	40	28.2	153	219
Source 2	70	26.8	248	322
Source 2	0	30.4	118	169
Source 2	40	30.1	170	213
Source 2	70	27.0	249	310

3.4 Properties of Hardened Concrete Containing GGBS

3.4.1 Water Absorption

Pavía and Condren [60] studied the effect of GGBS on the water absorption of mortar subjected to immersion in silage effluent solution and magnesium sulfate solution. Cement was replaced with 0, 30, and 50% GGBS. The chemical composition of synthetic silage effluent was lactic acid (15 g/kg), acetic acid (5 g/kg), formalin (3 g/kg), KOH (3.67 g/kg), NaOH (0.78 g/kg), Ca $(OH)_2$ (1.39 g/kg), and Mg $(OH)_2$ (0.44 g/kg). Three 28-day cycles of effluent exposure were repeated over a 4-month period. It was observed that with the inclusion of GGBS, final-water absorption of mortar decreased. OPC mortar specimens absorbed 18.65% water whereas it was 12.65 and 8.26% with 30 and 50% GGBS.

Elahi et al. [25] reported the water absorption (sorptivity) results of concrete containing GGBS (50 and 70%) at the age of 44 and 91 days. The sorptivity values of control concrete were 100 $m^3 \times 10^{-7}$/min½ at 44 and 91 days, respectively. The mix with 50% GGBS yielded lower value of sorptivity (77.1 $m^3 \times 10^{-7}$/min½) at 44 days compared to that at 91 days (96.9 $m^3 \times 10^{-7}$/min½). Increasing the GGBS content to 70% significantly increased the sorptivity (120.1 $m^3 \times 10^{-7}$/min½) at 44 days but considerably reduced the sorptivity (103.8 $m^3 \times 10^{-7}$/min½) at 91 days.

3.4.2 Microstructure

GGBS can be effectively used in concrete to reduce the pore size [16]. Higher GGBS replacement percentage has denser structure and prevents concrete from water penetration. The GGBS reacts with water in alkali environment and then with calcium hydroxide to form cement hydration product through pozzolanic reaction to form extra C–S–H gel in the paste and slow down the strength

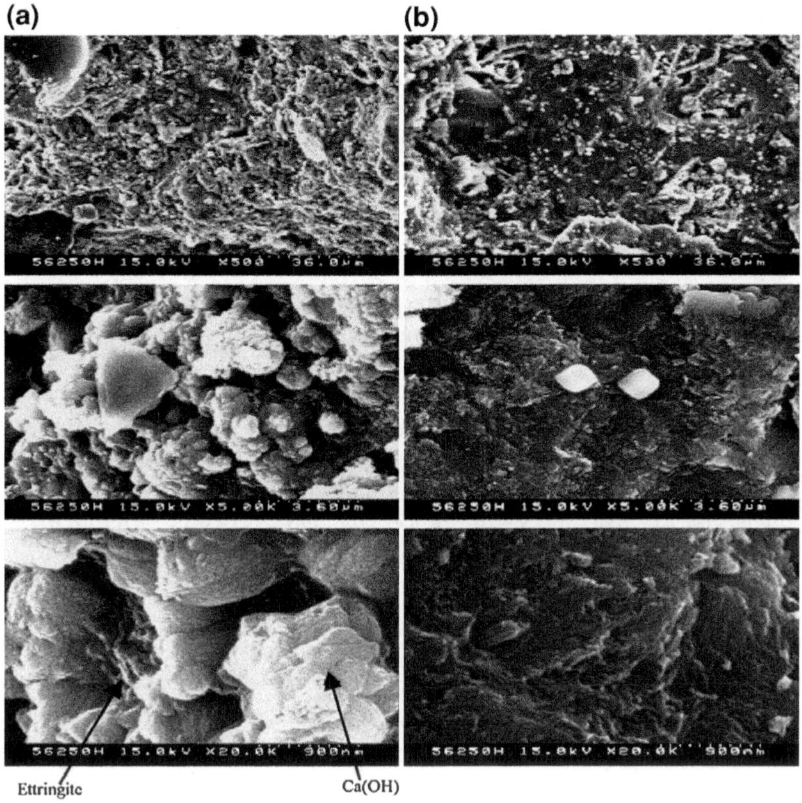

Fig. 3.8 SEM micrograph of: **a** ordinary Portland cement concrete and **b** GGBFS (60%) concrete [22]

development at early age. Denser microstructure or lower porosity results from higher C–S–H content that represents higher GGBS replacement percentage and higher durability of concrete.

SEM micrographs (Fig. 3.8) by Daube and Bakker [22] indicated that the addition of GGBFS modifies the products and the pore structure in a hardened cementitious material. It was found that a great number of calcium hydroxide and large capillary pores (0.05–60 μm) were found in OPC specimens. But few needle shape ettringite existed in GGBFS concrete specimens and the capillary pores were less than (10–50 μm) which could be filled up with pozzolanic reaction product such as low density C–S–H gel. Hydration product; $Ca(OH)_2$ activates the slag hydration to from a mixture of low CaO/SiO_2 (C/S) ratio $CaO–SiO_2–H_2O$ (C–S–H) and AFm (cementitious product from the reaction of reactive alumina and calcium hydroxide) phases. Pozzolanic reaction is also found to increase the C/S ratio to a value of about 1.7 in slag-cement blends due to unstable low calcium C–S–H and $Ca(OH)_2$ mixture. When supplementary cementitious material like

GGBFS are used in concrete, they do not only reduce the porosity but also the pores become finer and the change in mineralogy of the cement hydrates leads to the reduction in mobility of chloride ions.

Luo et al. [47] studied the pore structure of three types of concretes made with OPC, OPC/70% GGBS and OPC/65% GGBS/5% gypsum. The mix ratio for three concretes was 1:1.7:3.29 with water–binder ratio of 0.34. It was reported that coarse pores of GGBS concrete were much less and the pore structure of OPC concrete was improved greatly when 70% GGBS was added, especially for the pore structure for the time of 60 days. But sulfates did not improve the pore structure of GGBS as expected, and it showed that some other causes, except the expanding reaction between sulfates and C_3A, influence the pore structure of concretes.

Li and Zhao [45] investigated the effect of combination of GGBS and fly ash (FA) on the microstructure of concretes. Three types of concretes viz GGFAC (concrete incorporating GGBS and FA), HFAC (high-volume FA concrete), and PCC (control Portland cement concrete) were made. PCC had 500 kg/m^3 of cement, HFAC had 300 kg/m^3 of cement and 200 kg/m^3 FA, GGFAC had 300 kg/m^3 of cement and 125 kg/m^3 FA and 75 kg/m^3 of GGBS. Microstructure of concretes was studied at the age of 7 and 360 days using scanning electron microscope.

The image characteristics (SEM) of concretes at 7 days are shown in Figs. 3.9a, b, 3.10a, b and 3.11a. As can be seen from Fig. 3.9a, there were great deals of needle-shaped ettringite and plated-shaped calcium hydroxide in PCC, and large pores could also be observed. Figure 3.11a shows numerous un-hydrated FA particles in

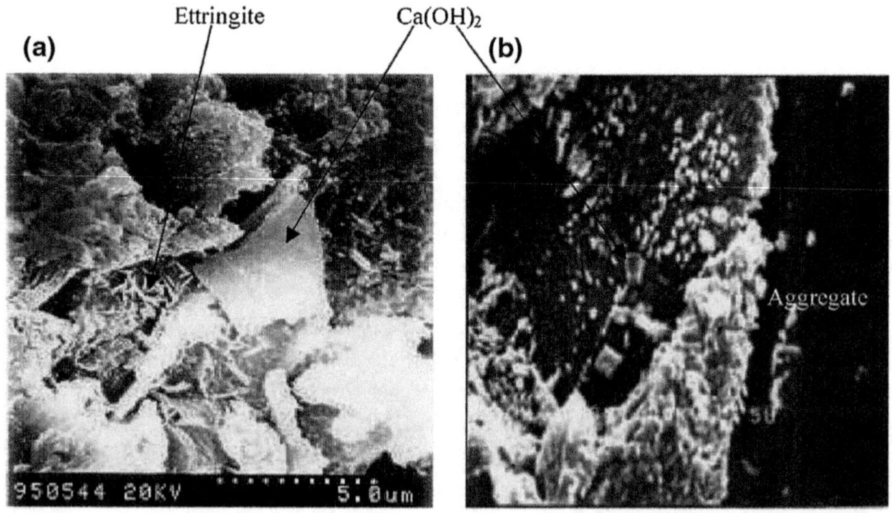

Fig. 3.9 SEM micrograph of PCC; **a** at the age of 7 days, **b** at the age of 360 days [45]

Fig. 3.10 SEM micrograph of GGFAC; **a** at the age of 7 days, **b** at the age of 360 days [45]

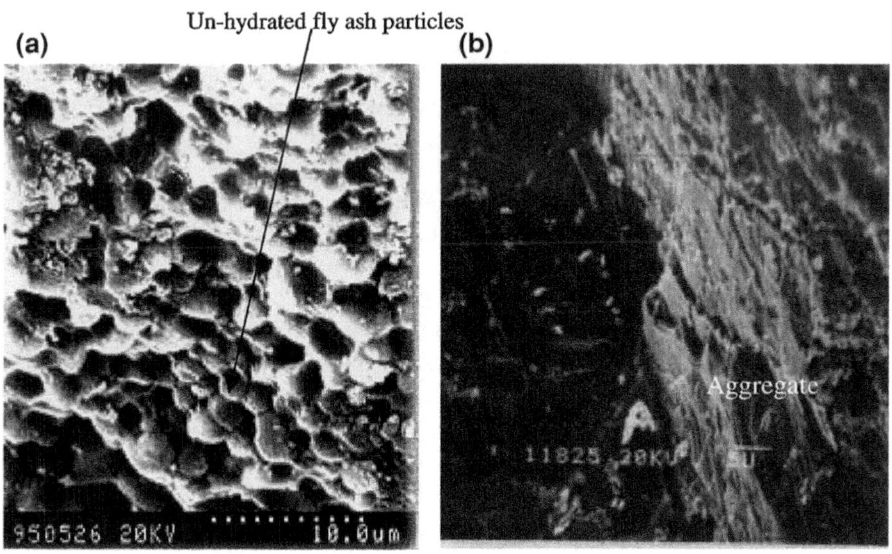

Fig. 3.11 SEM micrograph of HFAC; **a** at the age of 7 days, **b** at the age of 360 days [45]

HFAC. However, SEM observation of GGFAC (Fig. 3.10a) shows that microstructure changed greatly with the incorporation of GGBS, and no FA particles can be observed. The main hydration products were cotton-shaped C–S–H gel and a certain amount of needle-shaped ettringite. The characteristics of concretes at

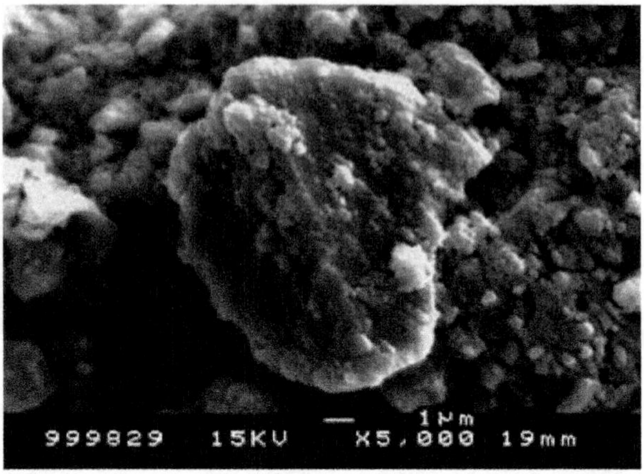

Fig. 3.12 SEM picture of concrete containing GGBS at 7 days of curing [28]

360 days are shown in Figs. 3.9b, 3.10a, b and 3.11a, b. It is seen from Fig. 3.9b that though the microstructure of PCC was very compact; there were great deals of plated-shaped calcium hydroxide in it. The microstructures of HFAC and GGFAC were also very compact, and no needle-shaped ettringite or plated shaped calcium hydroxide was observed.

Gao et al. [28] studied the morphology (Fig. 3.12) of the hydration products in concrete made with (GGBS) using SEM. Specific surface area of cement and GGBS were 361, 600 m^2/kg, respectively. Cement was replaced with 40% GGBS. At curing age of 7 days, the GGBS particles' surface was covered with hydration product

Binici et al. [12] made the micro-structural investigations of blended cements prepared with clinker (4%), and made with basaltic pumice and GGBS. Different amounts of additives (10 and 30%) were incorporated into these blends. Cement paste and mortars were prepared using OPC, two types of grinding systems (intergrinding and separate grinding) at two Blaine values ($2{,}800 \pm 30$ and $4{,}800 \pm 30$ cm^2/g). For micro-structural studies, SEM photos were taken at the end of 7-day testing (Fig. 3.13). A large amount of needle-shaped ettringite and plate-shaped calcium hydroxide can be observed in the blended cement. These figures show the formation of Portlandite (P) and CSH in OPC. The main hydration products are cotton-shaped CSH gel and a certain amount of needle-shaped ettringite.

Bouikni et al. [17] studied the microstructure of concrete made with 50 and 65% slag. A high alkali Portland cement with 1% sodium oxide and slag with 0.53% sodium oxide equivalent were used. Two concrete mixtures having high workability and low water–binder ratios were used. The mix proportions (by mass) were 1:1.75:2.53:0.43 and 1:1.67:2.41:0.40 (cement + slag:sand:aggregate:water), respectively, for the 50 and 65% slag content, respectively. Scanning electron microscopy (SEM) was used to study the internal microstructure. The fractured sections of concrete prisms were examined by SEM at about 6 months. Concrete

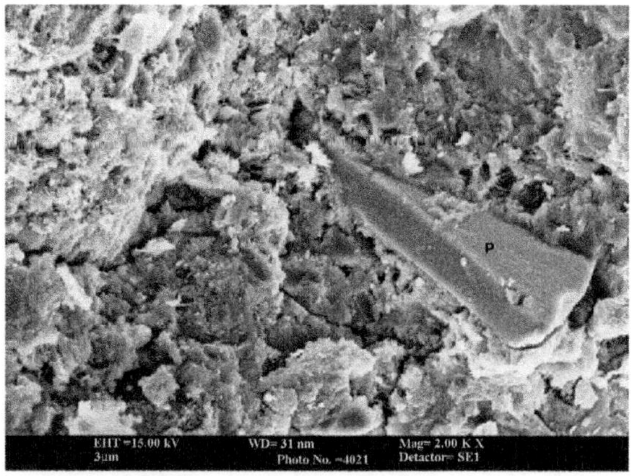

(a)

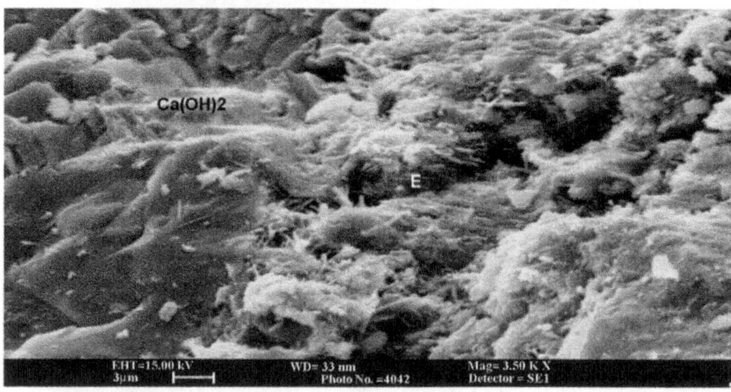

(b)

(c)

Fig. 3.13 SEM images of blended cement samples:**a** OPC sample (clinker 96%, gypsum 4%, fineness 2800 cm^2/g), **b** Blended cement sample—separate grinding (clinker 66%, gypsum 4%, GGBS 15%, blast pumice 15%, fineness 2800 cm^2/g), **c** Blended cement sample (clinker 96%, gypsum 4%, fineness 4800 cm^2/g)[12]

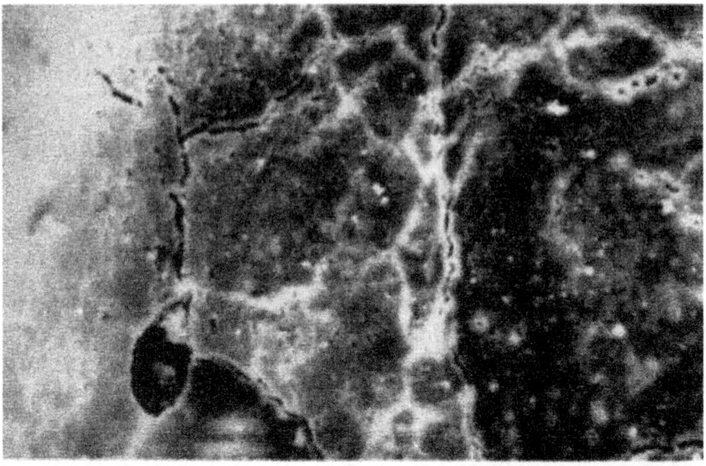

(a) concrete with percent 50 percent (Magnification 320x)

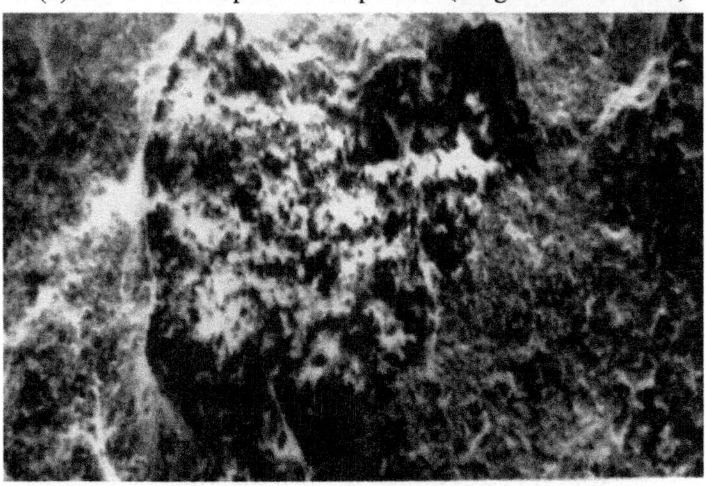

(b) concrete with 65 percent slag(Magnification 320x)

Fig. 3.14 SEM micrographs of **a** concrete with 50% and **b** concrete with 65% slag subjected to 7 days fog + Lab curing [17]

cured continuously in a wet environment, irrespective of the slag replacement level, showed development with time of a compact and dense texture. Concretes exposed to a prolonged drying environment invariably showed the presence of pores and micro cracks. Figure 3.14 shows the microstructure of the two concretes subjected to 7 days initial water curing prior to drying in ambient conditions. These pictures confirm the drying environment to be the major cause of increased porosity in slag concretes as the replacement level and the drying exposure period both increase.

3.4.3 *Compressive Strength*

3.4.3.1 Compressive Strength of Mortar

According to Neville [55], slag cements tend to be finer than Portland cement, but the strength gain in first 28 days is somewhat slower. At later-ages, the strength is similar. The low heat of hydration of slag cements make them vulnerable to frost damage in cold weather concreting.

Hogan and Meusel [35] reported strength of slag-blended cements made with using a slag with a target fineness of 55 cm^2/g. They found that (1) mortar strengths of cements developed more slowly than the controls for the first 3 days, after which the rate of strength development increased sharply. The 7-day and later-age strengths were greater than for the controls; (2) slag cement mortars subjected to elevated temperature-curing were stronger than the controls at all ages (1, 7, and 28 days); (3) strengths of slag-cement mortars subjected to low temperature curing were lower than for the controls at 1, 7, and 28 days. Similar results were also found for concrete; and (4) the optimum dosage of slag for strength development was 40%.

Gee [30] reported that early strength development in slag cement is affected by the chemistry of the clinker, since the manner in which it releases calcium and alkali cations affects the rate of hydration of the slag. Clinker can be formulated with high lime content for use in blends with high slag contents.

Wan et al. [75] reported the effect of GGBS on compressive strength and activity index of mortars up to the age of 28 days. Four samples (A, B, C, and D) of GGBS were used. The weight proportion of reference sample (with out GGBS) was: Portland cement/standard sand/water = 540:1,350:238, whereas for samples containing GGBS, it was: Portland cement/GGBS/standard sand/water = 270: 270:1,350:238. Results of compressive strength and activity index are given in Table 3.10. They concluded that strength of mortar incorporating GGBS is both related to the surface area and PSD of GGBS. When GGBS had the same surface area, the more the mortar contained fine particles (<3 μm) of GGBS, the higher its early strength was. The mortars containing 3–20 μm particles of GGBS, the higher was its long-term strength; and (2) compressive strengths increased with the increase of the surface area of GGBS.

Table 3.10 Results of compressive strength and activity index [75]

Sample	Surface area (m^2/kg)	Average dia (μm)	Compressive strength (MPa)		Activity index (%)	
			7 days	28 days	7 days	28 days
Reference (OPC)	–	–	43.4	54.3	–	–
Sample A	510	13.69	35.6	55.4	82	102
Sample B	685	9.12	42.5	63.0	98	116
Sample C	515	11.72	35.2	58.4	81	108
Sample D	512	13.15	34.9	56.5	80	104

Cakir and Aköz [21] studied the effect of curing conditions on the compressive strength of mortars with and without GGBS. In mortar mixes, cement was replaced with 0, 30, and 60% GGBS by weight. One group of mortar mixes (W) was kept in water at 20°C standard conditions, and the other group of mortar mixes (H) was kept in moisture cabinet at 40°C and 100% relative humidity (RH). They concluded that (1) at the age of 7 days, compressive strength of mortars (W) were 33.7, 31.0, and 26.3 MPa at 0, 30 and 60% GGBFS content whereas compressive strength of mortar mixes (H) were 36.9, 34.1, and 32.6 MPa with 0, 30 and 60% GGBFS; (2) compressive strength of mortars increased with time for both curing conditions. In comparison with 7-day strength, strength of control mortars cured in water (0 W) at 20°C temperature increased by 54 and 109% at 28 and 180 days, respectively. Slag replaced mortars 3 W group's (30% GGBS) compressive strength increased by 60 and 117%, 6 W group's (60% GGBS) strength increased by 72 and 140% at 28 and 180 days; and (3) compressive strength of control specimens, cured, in moisture, increased by 34 and 82% at 28 and 180 days, respectively, in comparison with 7-day strength. Compressive strength of slag replaced mortars 3H group (30% GGBS) increased by 26 and 68%, and 6H group (60%) increased by 27 and 69% at 28 and 180 days.

Pavía and Condren [60] investigated the effect of GGBS on the compressive strength of mortar specimens subjected to immersion in silage effluent solution and magnesium sulfate solution. Cement was replaced with 0, 30, and 50% GGBS. The chemical composition of synthetic silage effluent was lactic acid (15 g/kg), acetic acid (5 g/kg), formalin (3 g/kg), KOH (3.67 g/kg), NaOH (0.78 g/kg), $Ca(OH)_2$ (1.39 g/kg), and Mg $(OH)_2$ (0.44 g/kg). Three 28-day cycles of effluent exposure were repeated over a 4-month period. The initial compressive strength of the samples prior to effluent exposure was 13.83, 19.76, and 24.79 N/mm^2 for the OPC, 30% GGBS and 50% GGBS samples, respectively. They concluded that there was 46.7, 32.74, and 21.82% loss in compressive strength for OPC, 30 and 50% GGBS.

Roy and Idorn [64] also reported similar results. However, it was reported that the benefit in strength of concrete containing 20–60% GGBS did not occur until after 28 days of curing, where similar or higher long term strength was obtained as compared with that of normal PC concrete [2, 39, 53].

3.4.3.2 Compressive Strength of Concrete

Douglas et al. [24] investigated the compressive strength development of alkali activated ground granulated blast-furnace slag concretes. These concretes incorporated sodium silicate as an activator but did not contain any Portland cement. Five alkali activated slag concrete mixtures were made. The first three mixtures were made using a sodium silicate solution having a silicate modulus (Ms = 1.47) and low water-to-binder ratios ranging from 0.34 to 0.39. Two other mixtures incorporated a sodium silicate solution of silicate modulus (Ms = 1.36) and higher water-to-binder ratios of 0.50 and 0.46. One of these mixtures was air-entrained.

Table 3.11 Properties of hardened concretes [24]

Mix no.	w/b	Density of cylinders (kg/m^3)	Compressive strength (MPa)					
			1d	3d	7d	14d	28d	91d
1	0.39	2,373	28.1	38.2	41.6	43.4	46.3	48.5
2	0.34	2,387	37.5	41.8	44.6	48.3	51.0	52.6
3	0.35	2,464	38.9	49.4	54.5	–	59.6	62.8
4	0.50	2,435	23.2	33.7	41.0	43.1	45.4	50.2
5	0.46	2,415	20.4	35.0	41.5	44.3	46.8	51.0

Cylinders of size 102 × 203 mm were cast to determine compressive strengths up to one year. Compressive strength test results are given in Table 3.11. They reported that (1) strength of all the mixtures increased with age; (2) 28-day strength ranged from 46.3 to 59.6 MPa, where as 91-day strength was between 48.5 and 62.8 MPa; (3) ground granulated blast-furnace slag activated with sodium silicate can be used to make slag concretes with satisfactory workability and strength up to 91 days.

Ujhelyi and Ibrahim [72] studied the effect of hot weather on the compressive strength of concrete made with various percentages of GBFS and ground rhyolite tuff as partial replacement of cement. The cement in control mix was 350 kg/m^3 with water-to-cement ratio 0.531. Two series of mixes were made: the materials of the first were stored in laboratory (room temperature 20°C ± 3°C, relative humidity 20–35%), that of the second in room (temperature 42°C ± 1°C, relative humidity 20%). Compressive strength was determined at 7, 28 and 90 days. Based on the investigation, it was concluded that (1) concretes containing slag as a partial replacement of cement (up to 40%) achieved higher compressive strength than that of concretes made with Portland cement alone when casting and curing at 42°C; (2) 7-day compressive strength of concretes made with additives were less than that of concretes made with Portland cement; (3) 28- and 90-day compressive strength of concretes made with slag or pozzolana reached that of concretes made with pure Portland cement; and (4) in-spite of reducing Portland cement content with increasing ratio of additive, compressive strengths remain unchanged after casting and curing at high air temperature comparing with that of Portland cement concretes because the active SiO_2 of additives reacts at high temperature with $Ca(OH)_2$ of cement in the early period.

Wainwright and Rey [74] studied the influence of GGBS on the compressive strength of concrete. GGBS from four different sources (S1–S4) and Portland cement from one source were used. Control mix had 1:2.5:3.6 proportion with water/binder ratio of 0.56. Cement was replaced with 55 and 85% of slag by weight. Compressive strength of all mixes up to 28 days is given in Table 3.12. It can be seen from the table that source of the slag appeared to have little influence on strength development, and strength of OPC concrete and those containing GGBS increased with age.

Babu and Kumar [13] determined the cementitious efficiency of GGBS in concrete at various replacement percentages (10–80%) through the efficiency

Table 3.12 Compressive strength results [74]

Mix code	Compressive strength (MPa)		
	7 days	14 days	28 days
OPC	39.5	45.3	53.1
55% GGBS Source 1	18.5	28.5	42.5
55% GGBS Source 2	26.0	38.5	50.5
55% GGBS Source 3	22.5	36.0	50.5
55% GGBS Source 4	25.0	41.5	51.5
85% GGBS Source 1	17.5	28.0	33.5
85% GGBS Source 2	21.5	33.5	41.0
85% GGBS Source 3	17.5	25.5	29.5
85% GGBS Source 4	19.0	26.5	32.5

concept by establishing variation of strength to w/cm ratio relations of the GGBS concretes from the normal concretes at the age of 28 days. The 28-day compressive strength of concretes containing GGBS up to 30% replacement were all slightly above that of normal concretes, and at all other percentages, the relationships were below that of normal concretes. It was also observed that variations due to different percentages of slag replacement were smaller than the corresponding variations in case of FA. The result showed that the slag concretes based on overall efficiency factor (k), will need an increase of 8.6 for 50% replacement and 19.5 for 65% replacement in the total cementitious materials for achieving strength equivalent to that of normal concrete at 28 days.

Quillin et al. [62] studied the early strength development of 'BRECEM' concretes, made from 50:50 mixtures of calcium aluminate cement (CAC) and GGBS, using 100-mm cubes at w/c ratios of 0.35 and 0.45. Concretes were both air-cured and water-cured. Tests were conducted up to the age of 90 days. Compressive strength results are shown in Fig. 3.15. They concluded that (1) BRECEM concrete with w/c ratio of 0.35 developed strength more rapidly than those made with

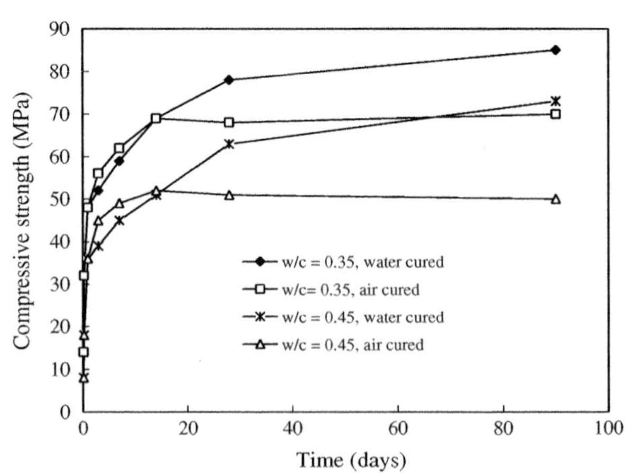

Fig. 3.15 Compressive strength development in BRECEM concretes [62]

w/c ratio of 0.45. The 90-day strength was also higher than those made with w/c ratio of 0.45, although for water stored concretes, strengths were still increasing in both cases; (2) BRECEM concretes stored in air initially developed strength more rapidly than equivalent concretes stored in water. The initial strength development was predominantly due to the hydration of the more reactive CAC component. However, the compressive strength of air-stored concretes rapidly leveled out after 14 days and at later test age's water-stored concretes had a higher compressive strength than air stored ones. The continuing increase in compressive strength with time for water-stored samples after 14 days will have risen due to the ongoing hydration of the GGBS and any remaining CAC.

Jianyong and Yan [41] reported the 28-day compressive strength of three high-performance concrete (HPC) mixtures. Mix proportions and compressive strength results are given in Table 3.13. Concretes B and C acquired much higher compressive strength than Concrete A at each testing age. At the age of 3 days, the compressive strengths of concrete A, B and C were 63.8, 69.3 and 69.3 MPa, respectively. At 28 days, compressive strengths of Concrete B and C increased greatly to 100.4 and 104.0 MPa, respectively, compared with 81.1 MPa of Concrete A. The development of compressive strength reflected the strengthening effect of ultra-fine GGBS and silica fume (SF) on mechanical properties of concrete.

Li and Zhao [45] studied the influence of combination of GGBS and FA on the compressive strength of high-strength concrete. Three types of concretes; GGFAC (concrete incorporating GGBS and FA), HFAC (high-volume FA concrete), and PCC (control Portland cement concrete) were made and their strength was determined up to the age of 360 days. PCC had 500 kg/m^3 of cement content, HFAC had 300 kg/m^3 of cement and 200 kg/m^3 FA, GGFAC had 300 kg/m^3 of cement and 125 kg/m^3 FA and 75 kg/m^3 of GGBS. Table 3.14 gives the strength development of PCC, HFAC (containing 40% of FA) and the concrete (GGFAC)

Table 3.13 Mix proportions (kg/m^3) and compressive strength of HPC [41]

Mix	Mix proportions							Compressive strength (MPa)		
	OPC	GGBS	SF	SP (%)	Fine Agg.	Coarse Agg.	Water	3 days	7 days	28 days
A	600	–	–	1.6	610	1,134	156	63.8	71.2	81.1
B	420	180	–	1.6	610	1,134	156	69.3	83.2	100.4
C	360	180	60	1.6	610	1,134	156	69.3	97.0	104.0

Table 3.14 Compressive strength gain of concretes [45]

Binder combination	Cube compressive strength (MPa)							Strength gain from 28 days to 1 year (%)
	1 day	3 days	7 days	28 days	56 days	112 days	360 days	
PCC	40.5	51.2	66.8	81.1	87.9	91.2	96.3	18.7
HFAC	21.7	32.6	43	65.2	86.7	97.5	107.1	64.3
GGFAC	35.1	49.3	65.4	80.6	89.8	93.7	99.4	23.3

incorporating a combination of 25% FA and 15% GGBS. It can be seen that there is a general trend of increasing strength with age up to 1 year for all concretes. As expected, the behavior of HFAC at early ages is different from that of PCC and GGFAC. Though it had the highest strength at the end test age, its strength was the lowest before 56 days. The strength development of GGFAC is similar to that of PCC, only with slightly lower values before 28 days. This indicates that GGFAC can achieve adequate early compressive strength, while maintaining a high long-term strength.

Cheng et al. [20] reported the effect of GGBS on the compressive strength of concrete. They made three concrete mixes (A, B, and C). Mix A was control mix with mixture proportion 1:1.82:1.97 with water–binder ratio of 0.55. Mixes B and C were made by replacing cement with 40 and 60% of GGBS, respectively. Compressive strength was determined according to ASTM C39 [10]. At 91 days, compressive strength of specimens was 42.4, 45.3 and 48.6 MPa for mix A, B and C, respectively. Compressive strength developments depend upon the GGBS percentage and testing age. The glassy compounds in GGBS react slowly with water and it takes time to obtain hydroxyl ions from the hydration product of Portland cement to breakdown the glassy slag parcels at early age. However, GGBS concrete had higher compressive strength than ordinary Portland cement concrete after GGBS hydration and pozzolanic reaction was almost accomplished. Higher GGBS replacement percentage had higher ultimate strength.

Khatib and Hibbert [42] investigated the influence of GGBS on the compressive strength of concrete. Portland cement (PC) was partially replaced with 0–80% GGBS. First mix was control (M1) having a proportion of 1 (PC): 2 (fine aggregate): 4 (coarse aggregate). In the next three mixes, PC was partially replaced with 40, 60 and 80% GGBS (by mass).

The compressive strength development for concretes containing 0, 40, 60 and 80% GGBS (mixes M1–M4) is shown in Fig. 3.16. There was a systematic decrease in compressive strength with the increase in GGBS content during the

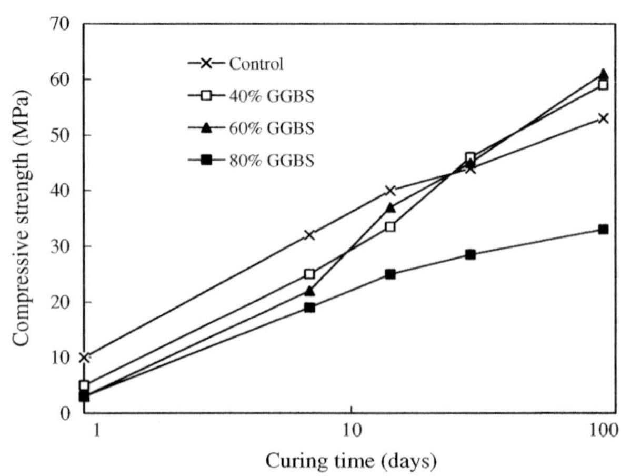

Fig. 3.16 Effect of GGBS on compressive strength development [42]

early stages of hydration. Beyond 28 days and up to at least 90 days, the presence of GGBS was highly beneficial at 40 and 60% replacement with a strength exceeding that of the control. A noticeable strength reduction at all ages is observed at 80% GGBS.

Atis and Bilim [11] investigated the compressive strength of ground granulated blast-furnace slag concrete under dry and wet curing conditions. Total of 45 concretes, including control normal Portland cement (NPC) concrete and GGBS concrete, were produced with three different water–cement ratios (0.3, 0.4, 0.5), three different cement dosages (350, 400 and 450 kg/m^3) and four partial GGBS replacement ratios (20, 40, 60, and 80%). Twelve cubic samples produced from fresh concrete were de-moulded after a day, then, six cubic samples were cured at 22 ± 2°C with 65% relative humidity (RH), and the remaining six cubic samples were cured at 22 ± 2°C with 100%RH until the samples were used for compressive strength measurement at 28 days and 3 months. Average 28-day compressive strength of control and GGBS concrete are given in Table 3.15 for dry- and wet-curing conditions. It can be seen from this table that, in general, wet-cured compressive strength of GGBFS was higher than that of control NPC concrete for 20 and 40% replacement ratios at 28 days. Compressive strength of GGBS was found to be equivalent to that of control NPC concrete for 60% replacement ratio. However, compressive strength of GGBS was found to be satisfactory when compared to control NPC concrete for 80% replacement ratio. It is also evident that, for dry curing conditions, compressive strength of GGBS concrete is found to be equivalent to that of control NPC concrete for 20 and 40%

Table 3.15 Compressive strength of concrete at 28 days [11]

Mixture	Compressive strength (MPa)					
	Wet cured			Dry cured		
Name	0.3	0.4	0.5	0.3	0.4	0.5
CP-350-00	75.8	63.9	53.6	72.4	63.9	52.3
GS-350-20	81.4	65.8	57.0	73.3	60.4	50.6
GS-350-40	81.0	67.2	55.	72.0	57.6	46.5
GS-350-60	73.3	61.8	45.1	57.2	52.6	40.4
GS-350-80	62.7	50.4	29.9	50.2	42.7	26.9
CP-400-00	80.7	63.9	51.4	73.1	65.6	36.9
GS-400-20	81.4	66.0	52.6	69.6	63.1	37.5
GS-400-40	82.0	66.9	51.6	66.5	61.6	35.4
GS-400-60	77.8	61.1	40.1	68.2	59.0	30.5
GS-400-80	67.7	53.1	25.3	54.4	47.2	19.7
CP-450-00	80.3	64.3	48.7	75.0	68.2	41.7
GS-450-20	81.8	73.5	50.4	73.2	69.1	36.3
GS-450-40	83.8	66.4	49.3	76.4	61.7	35.1
GS-450-60	80.6	61.8	39.5	58.2	54.2	28.3
GS-450-80	66.3	46.8	27.7	56.0	42.6	17.6

replacement ratio at 28 days. Compressive strength of GGBFS is found to be satisfactory when compared to control NPC concrete for 60% replacement ratio. However, concrete containing 80% GGBS developed lower strength than that of control NPC concrete.

Hui-sheng et al. [38] studied the influence of GGBS on the compressive strength of HPC up to the age of 180 days. Cement, fine aggregate, and coarse aggregates contents were 550, 687, and 1,030 kg/m^3, respectively. GGBS contents were 0, 15, 30, 45 and 60% by weight of cement, with water–binder ratio of 0.30 and 0.35. It was observed that (1) HPC showed significantly higher compressive strength at lower w/b than that at higher w/b. As expected, prolonging curing period evidently benefits compressive strength development, which is attributed to the fact that the pozzolanic reaction between GGBS and cementitious system is slow; hence, the beneficial effects were exhibited at a later curing age; and (2) for HPC with GGBS at w/b of 0.30, compressive strength reached highest value at optimum replacement of 15%. However, at higher w/b 0.35, compressive strength evidently decreased with the increase of replacement at each curing age.

3.4.4 Tensile and Flexural Strength

Ujhelyi and Ibrahim [72] investigated the effect of hot weather on the flexural strength of concrete containing various percentages of GBFS and ground rhyolite tuff as partial replacement of cement. The cement content in control mix was 350 kg/m^3 with water-to-cement ratio 0.531. Two series of mixes were made: the materials of the first were stored in laboratory room (temperature: +20°C ± 3°C, relative humidity: 20–35%), that of the second in special room (temperature 42°C ± 1°C, relative humidity 20%). Following conclusions were drawn: (1) Concretes containing slag as a partial replacement of cement (up to 40%) had higher flexural strengths casting and curing at +42°C than that of concretes made with Portland cement alone; (2) flexural strength of concretes made with slag was equal (or higher) comparing to Portland cement concrete cast and cured in high temperature; and (3) flexural strengths remain unchanged despite reduction in cement content and with increasing ratio of additive, curing at high air temperature comparing with that of Portland cement concretes because the active SiO_2 of additives reacts at high temperature with $Ca(OH)_2$ of cement in the early period.

Jianyong and Yan [41] reported the 28-day compressive strength of three HPC mixtures. Concrete A was made of OPC (600 kg/m^3). Concrete B had (420 kg/m^3 of OPC and 180 kg/m^3 GGBS), and concrete C had (360 kg/m^3 of OPC and 180 kg/m^3 GGBS, 60 kg/m^3 of SF). Splitting tensile strength results are given in Table 3.16. It is evident that Concretes B and C acquired more splitting tensile strength than Concrete A at each age. The development of splitting tensile strength showed the same tendency as that of compressive strength. Such a tendency reflects the strengthening effect of ultra-fine GGBS and SF on mechanical properties of concrete.

Table 3.16 Splitting tensile strength of HPC [41]

Mix	Cementitious materials (kg/m^3)			Splitting tensile strength (MPa)		
	OPC	GGBS	SF	3 days	7 days	28 days
A	600	–	–	3.8	4.54	5.54
B	420	180	–	4.06	5.03	5.91
C	360	180	60	5.2	5.44	6.14

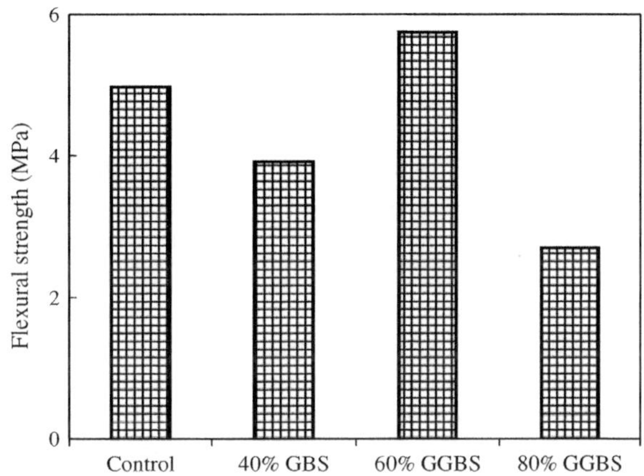

Fig. 3.17 Effect of GGBS on flexural strength [42]

Khatib and Hibbert [42] studied the effect of GGBS on the flexural strength of concrete. Portland cement (PC) was partially replaced with 0–80% GGBS. Control mix had a proportion of 1 (PC): 2 (fine aggregate): 4 (coarse aggregate). In the other three mixes, cement was partially replaced with 40, 60 and 80% GGBS (by mass). Figure 3.17 shows 90-day flexural strength values for mixes containing 0, 40, 60, and 80% GGBS. The flexural strength of concrete containing 60% GGBS was noticeably higher than the control, whereas a slight decrease at 40% and marked decrease at 80% replacement were observed.

Guo et al. [29] investigated the flexural fatigue performance of concretes with 50 and 80% proportions of ground granulated blast-furnace slag by mass of total cementitious materials. Control concrete had 460 kg/m^3 of cement, aggregate-binder ratios and water–binder ratios were 2.4 and 0.35, respectively. The flexure fatigue tests were carried at cyclic loading frequency of 10 Hz. Six nominal stress levels (0.90, 0.85, 0.80, 0.75, 0.70 and 0.65) were chosen. Flexural strength of concrete was 7.65, 7.14 and 5.87 MPa with 0, 50, and 80% GGBS content. The flexural fatigue life of concretes is given in Table 3.17. It is clear that the fatigue life of 50% GGBS concrete is the longest and that of 0% GGBS concrete is the shortest with stress level of 0.80 or more. However, when stress level was lower than 0.80, the fatigue life of 80% GGBS concrete was the longest among three

Table 3.17 Average cycles of flexural fatigue life [29]

Stress levels	Average cycles		
	0% GGBS concrete	50% GGBS concrete	80% GGBS concrete
0.90	68	114	78
0.85	1,015	1,355	620
0.80	4,212	9,506	5,904
0.75	50,433	43,679	59,971
0.70	413,988	469,963	565,779
0.65	2,000,801	$>2 \times 10^6$	$>2 \times 10^6$

mixes. When stress level is 0.65, 0% GGBS concrete was at about 2×10^6 circles, whereas the fatigue life of concretes with GGBS was more than 2×10^6 circles. This is because of the bond strength of interface transition zone (ITZ) between coarse aggregate and matrix is weakened by incorporation of GGBS with a specific surface area of 372 m^2/kg. The potential reason is that the grain size of GGBS (372 m^2/kg) was similar to those of Portland cement (309 m^2/kg) and the elastic modulus of GGBS grain was lower than those of cement grain. Moreover, the effect of GGBS on bond strength of ITZ in concrete is more negative as increasing the mass fraction of GGBS in matrix.

Cakir and Aköz [21] investigated the influence of curing conditions on the flexural strength of mortars with and without GGBS. In mortar mixes, cement was replaced with 0, 30, and 60% GGBS by weight. One group of mortar mixes (W) was kept in water at 20°C standard conditions, and the other group of mortar mixes (H) was kept in moisture cabinet at 40°C and 100% relative humidity (RH). They reported that (1) at the age of 7 days, flexural strength of mortars (W) were 6.8, 6.5, and 6.0 MPa at 0, 30 and 60% GGBS whereas flexural strength of mortar mixes (H) were 7.0, 6.3, and 6.1 MPa with 0, 30 and 60% GGBS; (2) Flexural strength of mortars increased with time for both curing conditions. In comparison with 7-day strength, strength of control mortars cured in water (0 W) at 20°C temperature increased by 7 and 21% at 28 and 180 days, respectively. Slag replaced mortars 3W group's (30% GGBS) flexural strength increased by 8 and 23%, 6 W group's (60% GGBS) strength increased by 13 and 25% at 28 and 180 days; (3) flexural strength of control specimens which cured in moisture cabinet increased by 3 and 7% at 28 and 180 days, respectively, in comparison with 7-day strength. Flexural strength of slag replaced mortars 3H group (30% GGBS) increased by 8 and 13%, and 6H group (60%) increased by 7 and 15% at 28 and 180 days; and (4) slag replaced mortars' flexural strengths were negatively affected at elevated temperature and lower humidity condition. The flexural strength of mortars produced with Portland cement and cured in water at 40°C was lower than the flexural strength of mortars cured in water at 20°C.

3.5 Durability Properties of Concrete Containing GGBS

3.5.1 Creep and Shrinkage

The creep of concrete is the deformation of hardened concrete caused by a long-lasting constant load applied on it. Drying shrinkage of concrete is the shrinkage caused by evaporation of internal water in hardened concrete. Creep and drying shrinkage are very important time-dependent properties of HPC.

Jianyong and Yan [41] studied the creep and drying shrinkage of three HPC mixtures. Concrete A was made of OPC (600 kg/m^3). Concrete B had (420 kg/m^3 of OPC and 180 kg/m^3 GGBS), and concrete C had (360 kg/m^3 of OPC and 180 kg/m^3 GGBS, 60 kg/m^3 of SF). Specimens for creep and shrinkage were of size 100 × 100 × 300 and 100 × 100 × 500 mm, respectively. The creep and shrinkage tests were conducted up to the age of 180 and 210 days, respectively. Creep and shrinkage results are shown in Figs. 3.18 and 3.19, respectively.

It is evident that up to 180-day of testing, Concretes B and C always obtained much smaller creep than Portland cement concrete A. When ultra-fine GGBS and SF were used at the same time (Concrete C), the creep value was the lowest, which is in accordance with the development trend of mechanical strengths. At age of 180 days, creep for Concretes A, B and C were 1,293 × 10^{-6}, 623 × 10^{-6} and 450 × 10^{-6}, respectively. According to creep rate, three groups of concrete were subject to a faster development of creep at early ages than at late ages, and the age of 60 days was the turning point. For the individual concrete, the value of creep rate was different. Before 60 days, Concrete A was much greater than B and C, and C was the lowest. After 60 days, the creep rates of A, B and C became similar while Concretes B and C can be thought to have equal creep rate.

Figure 3.19 shows the drying shrinkage of concretes. It can be seen that at early ages, the difference between drying shrinkage of Concrete A, B and C was small. After 28 days, Concrete A obtained substantially greater drying shrinkage than

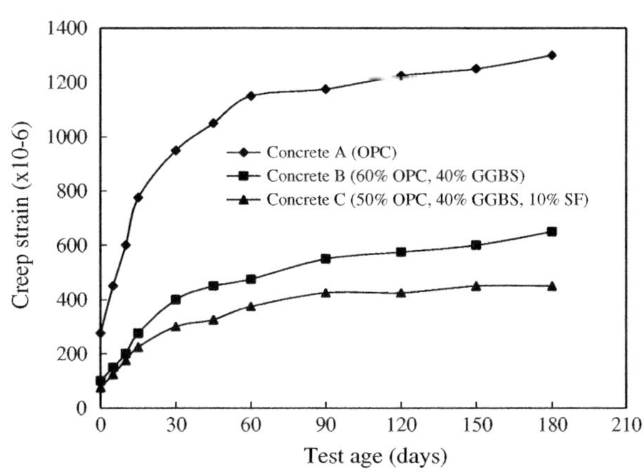

Fig. 3.18 Development of creep of concrete [41]

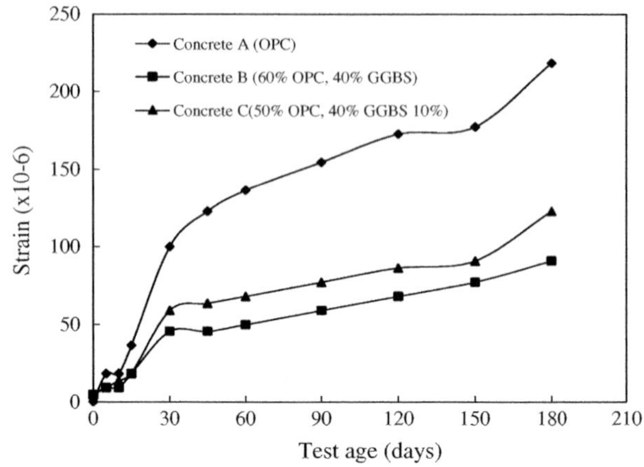

Fig. 3.19 Test results of drying shrinkage of concrete [41]

Concretes B and C, while the latter two always had similar shrinkage amount. At age of 180 days, the amount of shrinkage for Concretes A, B and C was 220×10^{-6}, 96×10^{-6} and 127×10^{-6}, respectively.

3.5.2 Chloride Binding Capacity/Resistance

Dhir et al. [23] determined the chloride binding capacity of paste samples containing Portland cement (PC), and blends of PC and GGBS. GGBS was added as 0, 33.3, 50 and 66.7% of the total binder. The water–binder ratio was kept constant at 0.55. They concluded that (1) with the increase in GGBS content, the chloride binding capacity increased for all chloride concentrations. For a GGBS replacement level of 66.7%, the chloride binding capacity was around five times that of the PC control for the case of 5 mol/l exposure concentrations; and (2) as the chloride concentration increased, the binding capacity increased for all GGBS contents. The chloride binding capacity appeared to be directly proportional to the exposure concentration. This could be attributed to the sensitivity of the chloride binding capacity to the ratio of Cl^-/OH^- ions in the cement pore solutions (in this case, the chloride solutions to which the pastes were exposed). As the Cl^-/OH^- ratio increased chloride binding capacity increased.

Dhir et al. [23] determined the strength and intrinsic permeability of concrete mixes. Four concrete mixes were made. All the mixes had the same total binder content of 300 kg/m^3 with GGBS replacement levels of 0, 33.3, 50 and 66.7%. An air permeability test was conducted at ages of 28 and 90 days, whilst chloride diffusion testing was carried out at 28 days. Strength and intrinsic permeability results of PC and GGBS concrete are given in Table 3.18. It is evident from this table that (1) 280-day cube strength decreased with the increase in GGBS content; (2) however, at 90 days, strength of GGBS and PC concrete mixes were much closer; (3) at 28 days, despite the lower compressive strengths of the GGBS

Table 3.18 PC and GGBS concrete details [23]

GGBS (%)	Mix details			Cube strength (N/mm^2)		Intrinsic permeability (m^2) × 10^{-17}		Coefficient of chloride diffusion (cm^2/s) × 10^{-9} at 28-day
	PC (C) kg/m^3	GGBS (G) kg/m^3	W/ (C + G)	28 days	90 days	28 days	90 days	
0	300	–	0.55	45.0	51.0	2.08	1.85	10.90
33.3	200	100	0.55	41.0	51.5	1.97	1.61	0.873
50.0	150	150	0.55	34.0	47.0	2.06	1.72	0.329
66.7	100	200	0.55	32.0	45.0	2.31	1.93	0.100

concrete, compared to the control, the intrinsic permeabilities were similar. At 90 days, the intrinsic permeability of the GGBS became much better; (4) GGBS enhanced the protection against chloride ingress and with the increase in GGBS replacement level, the coefficient of chloride diffusion steeply decreased.

Xu [77] investigated the chloride binding properties and the concentrations of chloride and hydroxyl ions in the pore solutions of Ordinary Portland cement (OPC) and blended (35% OPC, 65% GGBS) cements, containing 2.0 to 9.0% sulfates derived from sodium sulfate and calcium sulfate. Chlorides derived from sodium and calcium chlorides were introduced at the time of mixing. The results indicated that calcium sulfate had a different effect on chloride binding and the pore solution chemistry than sodium sulfate. The slag cement had higher chloride binding capacities as a result of simple replacement for OPC, but at the same sulfate contents, the slag cement did not give the expected higher binding capacities, suggesting that the difference in sulfate content between the two cements may be the main reason for their different chloride binding behavior.

Huang and Yeih [37] studied the diffusivity of chloride ions through high strength concrete containing slag. Numerical methods were used to compute the diffusion coefficients of chloride ions in concrete, and were found to be $2.53 \times 10^{-14} m^2/s$ to $9.84 \times 10^{-4} m^2/s$ for compressive strengths between 62.5 and 91.1 MPa.

Luo et al. [47] measured the chloride diffusion coefficients and chloride-binding capacity of the three kinds of concretes made with OPC, OPC/70% GGBS and OPC/65% GGBS/5% gypsum. The mix ratio for the three kinds of concretes was 1:1.7:3.29 with water–binder ratio of 0.34. Both the total chloride diffusion coefficient and the free chloride diffusion coefficient were determined. The total chloride diffusion coefficient is the coefficient for all the diffused chlorides including the bound chloride by the concrete, and the free chloride diffusion coefficient is only the unbound chloride diffused to the concrete. Results of chloride diffusion coefficients are given in Table 3.19. They concluded that (1) GGBS deceased the chloride diffusion coefficient greatly; (2) both the total chloride diffusion coefficient and the free chloride diffusion coefficient decreased substantially with 70% GGBS; (3) but chloride diffusion coefficient increased greatly with 5% gypsum; (4) GGBS increased the chloride-binding capability

Table 3.19 Chloride diffusion coefficient of concrete (10^{-8} cm^2/s) [47]

	OPC concrete	GGBS concrete	Concrete with gypsum
Total chloride	3.02	1.95	2.44
Free chloride	2.65	1.58	2.14

greatly, especially the chemical chloride-binding capability, but sulfates and alkalinity decreased the chloride-binding capability greatly.

Cheng et al. [20] studied the influence of GGBS on the rapid chloride permeability (RCPT) and water permeability of concrete. Three were three concrete mixes A, B, and C. Mix A was control one with mixture proportion 1:1.82:1.97 and water–binder ratio of 0.55. Mixes B and C were made by replacing cement with 40 and 60% of GGBS, respectively. ASTM C1202 [7] procedure was followed for carrying out RCPT test. For each mix, cylindrical specimens (100 × 200 mm) were cast and moist-cured for 91 days. 50-mm-thick samples were cut from the middle portion of each cylinder. For measuring water permeability 100 × 200 mm cylinders were cast. Based on the results obtained they reported that (1) water permeability was 2.56×10^{-13}, 1.52×10^{-13}, and 1.32×10^{-13} m/s for mixes A, B, and C, respectively. It appeared that higher GGBS percentage made structure denser and prevented concrete from water penetration; (2) RCPT results indicated the highest total charge-passed (10,271 C) obtained in mix A specimen and the lowest total charge passed (1,864 C) in mix C specimen, which represented highest chloride-ion penetration resistance. The reaction products were very efficient to fill up the large capillary pores and refine the pore system in concrete thereby enhancing its permeability resistance.

Hootan and Titherington [36] investigated the strength and chloride penetration resistance of HPC after curing either at 23°C or accelerated by heating to 65°C. Six air-entrained concrete mixtures with different cement replacement levels of slag (25–27%) and SF (4–8%) (by mass) were made. The water to cementitious materials ratio (w/cm) was kept constant, at 0.30. The total cementing materials content was 460 kg/m^3 for all mixtures. Chloride penetration resistance of HPC was measured as per ASTM C 1202 [7]. They concluded that (1) concretes containing SF or ternary blends of SF and GGBS exhibited improved chloride penetration resistance compared to those of plain Portland cement concretes; (2) all concretes recorded are below 4,000 C, which was also possible because of water/cement ratio, 0.30. The 100% OPC mixture, whether accelerated or ambient cured, was in the moderate permeability category (between 2,000 and 4,000 C). All mixtures containing supplementary cementing materials, for both accelerated and ambient curing conditions, were in the very low permeability category of <1,000 C; and (4) chloride penetration resistance of Portland cement concrete was adversely affected by accelerated curing.

Roy et al. [65] studied the characteristics and performance of alkali-activated cementitious materials (AAC) whose properties equal or exceed those of normal Portland cement-based materials. The materials activated included: GGBS, FA, SF, and other pozzolans whether alone or when combined with Portland cement.

The activators used were NaOH, KOH, Na_2SO_4, Na_2CO_3, $CaSO_4$, and soluble silicates of sodium and potassium. Steady state chloride diffusion studies were conducted of pastes of Type I Portland cement, and its blends with different proportions of ground granulated blast-furnace slag. Very substantial reductions in diffusion rates have been found with the increased proportion of slag. In addition, alkali activation has been shown to reduce the diffusion rate by at least a factor or two.

Hui-sheng et al. [38] investigated the effect of GGBFS on the nitrogen gas permeability of HPC up to the age of 180 days. Control mixture contained cement (550 kg/m^3), fine aggregate (687 kg/m^3), and coarse aggregates (1,030 kg/m^3). Cement was replaced with 0, 15, 30, 45 and 60% GGBFS by weight of cement, with water–binder ratio of 0.30 and 0.35. Gas permeability of control HPC mix with water–binder ratio of 0.30 and 0.35 ranged between 29 and 31 $\times$ 10^{-17} m^2. Influencing trends of GGBFS on gas permeability of HPC differed little for different w/b ratios. At w/b of 0.30, gas permeability coefficient slightly increased with the increase in GGBFS replacement up to 60%. At w/b of 0.35, gas permeability coefficient at an optimum replacement of 30% is slightly lower than that of the control HPC.

Elahi et al. [25] reported the air-permeability of concrete containing GGBS (50 and 70%) at the age of 44 and 91 days. The air-permeability indices of control concrete were 100 ln (pressure)/min at 44 and 91 days. Binary mixes, obtained after inclusion of GGBS, resulted in an increase in air permeability at both 44 and 91 days. However, the increase in air permeability was more pronounced at 91 days. The mix with 50% GGBS yielded air-permeability indices of 116 ln (pressure)/min at 44 days and 421.7 ln (pressure)/min at 91 days. Increasing the GGBS content to 70% significantly increased the air-permeability indices of 268.1 ln (pressure)/min at 44 days and 673.9 ln (pressure)/min at 91 days.

3.5.3 *Sulfate Resistance*

Portland cement mortar/concrete is attacked by the solutions containing sulfate (sodium or magnesium sulfate) in natural or polluted ground water. Sulfate attack can lead to expansion, cracking, strength loss, and disintegration. The constituents of hydrated cement paste taking part in the expansive reactions are monosulfoaluminate hydrate, calcium aluminate hydrate, and calcium aluminate hydroxide.

In Portland cement mortar and concrete, sulfate attack can be minimized by reducing the presence of monosulfoaluminate and calcium aluminate hydrates. The addition of supplementary cementing materials (GGBS) to Portland cement minimizes sulfate attack in three ways: (1) GGBS does not contain any C_3A, and its inclusion in concrete reduces the overall proportion of C_3A in the mix; (2) GGBS reacts with $Ca(OH)_2$ to substantially reduce its presence in the concrete, leaving significantly less $Ca(OH)_2$ to react to form ettringite; and (3) greatly reduced permeability of GGBS concrete.

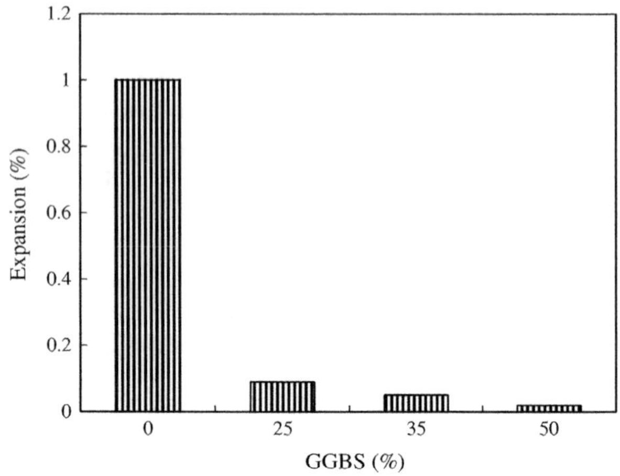

Fig. 3.20 Effect of GGBS on sulfate resistance [76]

The rate and extent of sulfate attack depends upon the ease with which sulfate ions are able to penetrate the concrete and upon the chemical resistance of the cement paste. The effect of GGBS in limiting the impact of sulfate attack, as measured from the expansion it produces in concrete, is shown in Fig. 3.20.

Fearson [27] observed that by increasing the slag contents in Portland cement-slag mortars increased the sulfate resistance substantially. The sulfate resistance of blended cements made with slag was less influenced by the water-to-cement ratio than by the amount of cement replaced by slag.

Mangat and El-Khatib [49] investigated the effect of initial curing conditions on the sulfate resistance of concrete made with ordinary Portland cement and using pulverized fuel ash (PFA), SF and GGBS as partial replacement of cement. Five different concrete mixes were used. First was control mix having proportions 1:2:3.4 with water/cement ratio of 0.45. The cement content was 350 kg/m^3. In the remaining mixes, cement was partially replaced by 22% PFA, 9% SF, 40 and 80% GGBS by weight, respectively. Three different initial curing conditions were adopted, namely: AI (air 45°C, 25% RH); CH (wet/air 45°C, 25% RH); AL (air 20°C, 55% RH). After the initial curing period of 28 days under the three curing regimes CH, AI and AL, specimens were immersed in fresh water for another 28 days. The temperature of the fresh water was 20°C for specimens cured initially at 20°C and 45°C for specimens cured initially at 45°C. After soaking the specimens for 28 days, extensometer measurements were taken for the specimens across each face. These initial measurements were considered as datum for sulfate attack studies. After that, specimens were immersed in a sulfate solution (7% Na_2SO_4 + 3% $MgSO_4$ by weight). The temperature of the sulfate solution was 20°C for specimens cured initially at 20°C and 45°C for specimens cured initially at 45°C. Expansion measurements across the demec points to measure the change in length were taken after 47, 83, 152, 207, 337, 502 and 660 days of immersion in the sulfate solution. Based on the results, following conclusions were drawn (1) initial moist curing of concrete (at 45°C) followed by dry curing at 45°C, 25%

RH resulted in lower sulfate resistance than initial air curing at low relative humidity (25 and 55% RH); (2) sulfate resistance of concrete increased with the replacement of cement with 22% PFA, 9% SF, 80% GGBF slag. The sulfate resistance also increased due to drying out of concrete during early curing at low relative humidity and due to carbonation; (3) the use of PFA (22%), SF (9%) and GGBS (80%), for partial replacement of OPC, resulted in higher sulfate resistance than for plain concrete under curing conditions.

Gollop and Taylor [31] examined the cubes of a blended cement paste stored for 6 months in solutions of Na_2SO_4 or $MgSO_4$ by scanning electron microscopy. The blend of cement contained 31% of PC and 69% of blast furnace slag (GGBS). It was observed that (1) tendency of the materials made with slag cements to soften and disintegrate as a result of sulfate attack rather than to expand; (2) when a blend containing 69% of blast furnace slag was stored in 0.25 mol l^{-1} Na_2SO_4 solution, the principal chemical reaction resulted in partial decalcification of the C–S–H and replacement of an AFm phase, loosely described as mono-sulfate, by ettringite. The results suggested that neither Al^{+3} substituted in the C–S–H, nor that present in hydrotalcite, was available for reaction with sulfate; (3) storage in 0.25 mol l^{-1} $MgSO_4$, solution produced the same reaction, but the dominant effects in this case were decomposition of the C–S–H and formation of brucite and gypsum in the surface regions below the cube faces and of magnesium silicate hydrate and gypsum at the cube edges; (4) SO_4^{2-} ions penetrated to a depth of about 1 mm below the cube faces and probably more deeply at the edges. These results were similar to those observed with a sulfate-resisting PC under similar conditions; (5) the micro-structural features of the deposits of gypsum, and of the brucite formed on reaction with $MgSO_4$ differed from those observed with the plain PC. The quantity of gypsum formed in the paste stored in Na_2SO_4 solution was probably less for the slag blend; (6) damage from sulfate attack was attributed partly to decalcification, which weakened the C–S–H matrix, and partly to ettringite formation, which caused expansion and cracking; (7) In typical cases of blends high in slag, at least as much sulfur entered the hydration products in sulfide as in sulfate.

El-Darwish et al. [26] studied the replacement of OPC by ground steel slag up to 30% volume replacement in concrete. Concrete specimens exposed to sodium and magnesium sulfate solutions containing 3.5% SO_3 and seawater were monitored for expansion, strength loss and stiffness up to one year. Samples in seawater and sodium sulfate solutions performed satisfactorily. However, C–S–H gel transformed into M–S–H in magnesium sulfate solution

Cao et al. [18] investigated the sulfate resistance of Portland cements and blended cements. Four Portland cements of different characteristics and blended cements containing FA, GGBS and SF were used. Mortar cubes were made with water-to-binder ratio (w/b) of 0.6, sand-to-binder ratio of 2.75 and cured for 7 days prior to immersion in sulfate solutions. Three sulfate solutions were (1) 5% Na_2SO_4 solution having pH 12; (2) 5% Na_2SO_4 and maintained at nominal pH of 7 and 3 by automatic titration with 10% H_2SO_4 at regular set intervals. ASTM C 1012 [9] procedure was followed to evaluate the performances of binders in sulfate solutions maintained at different pH levels ranging from 3 to 12 using expansion of mortar prisms (ASTM C 1012) and strength development of mortar cubes. The results indicated that sulfate resistance of cementitious materials was dependent on its composition and on the pH of the environment. Portland cement with low C_3A and low C_2S performed well in all sulfate solutions. Blended cements containing SF and FA (particularly at 40% replacement) showed a much superior performance than any of the Portland cements used. For slag blended cement, this can be achieved when the replacement percentage was higher than 60%. The selection of the mineral admixture and its replacement dosage became more critical as the pH of the sulfate solution decreased. It was found that 40% FA blend, 5% SF blend or a high slag blend (80%) provided a good overall performance related to resistance to sulfate attack over a wide pH range.

Higgins [33] studied the influence of addition of a small percentage of calcium carbonate or calcium sulfate on the sulfate resistance of concrete containing GGBS. Six concrete mixtures were made having water/cementitious ratio of 0.5 and total cementitious content of 350 kg/m^3. Three test methods were employed to assess sulfate resistance at 20°C (1) 100-m cubes were immersed in magnesium (containing 1.5% SO_3) and sodium sulfate solutions (containing 1.5 and 2.4% SO_3) and monitored for corner-loss and strength-loss, over 6 years; (2) 20 × 20 × 160-mm prisms made of mortar were immersed in magnesium and sodium sulfate solutions and their expansions monitored for up to 6 years.

Compressive strength of the cubes stored in water and in the sulfate solutions are given in Table 3.20. The strength in the sulfate solutions is expressed as a percentage of the strength of equivalent concretes stored in water for the same age. Table 3.21 shows the 'wear ratings' of the cubes, measured annually. 'Wear rating' is taken as the measure of the attack on the corners of a cube. It is the average depth of erosion or damage for one corner (in mm).

In sodium sulfate solution, the PC concrete had almost completely disintegrated by 6 years, while the GGBS concretes showed minor strength- and corner-loss. As would be expected, 70% GGBS concrete was more resistant to attack than 60% GGBS concrete. The lower level of addition of calcium sulfate (to increase the SO_3 level of the cementitious to 2%), significantly increased the degree of attack. However, at the higher level of addition of calcium sulfate (to increase the SO_3 level of the cementitious to 3%) the degree of attack was reduced, as was also the case with the addition of calcium carbonate.

In magnesium sulfate solution, the PC concrete performed somewhat better, whereas the GGBS concretes performed distinctly less in the magnesium than the sodium solution. This is what is normally observed [58] with GGBS concrete generally more susceptible to attack by magnesium sulfate than sodium sulfate. Once again 70% GGBS concretes were more resistant than 60% and the lower level of addition of calcium sulfate gave no advantage. The higher level of addition of calcium sulfate and the addition of calcium carbonate both reduced the degree of attack.

For PC concrete, the expansion was rapid, reaching 1% after about 9 months in all solutions. In sulfate solutions, 60% GGBS specimens disintegrated at a

Table 3.20 Compressive strength in water and in sulfate solutions [33]

	Compressive strength in water (N/mm^2)						Strength in Na_2SO4 (%)			Strength in MgSO4 (%)		
	3 days	7 days	28 days	1 year	2 years	6 years	1 years	2 years	6 years	1 year	2 years	6 years
Portland cement	34	41	53	66	68	69	97	87	0	85	74	28
60% GGBS	17	31	48	65	69	73	106	97	62	95	75	18
70% GGBS	13	23	49	63	66	71	105	89	90	94	77	28
60% GGBS + $CaCO_3$	15	23	45	67	69	76	99	97	84	96	83	41
60% GGBS 2% SO_3	19	29	47	68	67	72	97	99	33	94	81	19
60% GGBS 3% SO_3	18	32	50	62	69	75	102	99	80	94	88	50

Table 3.21 Wear rating in sulfate solutions [33]

	Wear rating in Na_2SO_4 solution						Wear rating in $MgSO_4$ solution					
	1 year	2 years	3 years	4 years	5 years	6 years	1 year	2 years	3 years	4 years	5 years	6 years
Portland cement	1	1	1	3	10	71	1	1	1	0	0	2
60% GGBS	2	2	3	5	9	12	4	6	7	10	15	23
70% GGBS	2	3	3	3	2	3	5	4	6	6	7	6
60% GGBS + $CaCO_3$	0	2	2	2	2	3	1	2	4	4	7	9
60% GGBS 2% SO_3	1	4	12	11	15	22	1	4	8	9	14	20
60% GGBS 3% SO_3	1	2	2	1	1	3	1	2	4	2	2	5

relatively low expansion of less than 0.1%. The 70% GGBS and specimens with additions of calcium sulfate or carbonate, all survived with low expansions, although the lower level of calcium sulfate addition showed signs of more rapid expansion at 72 months. In magnesium sulfate, expansion is more rapid and severe. All the specimens had disintegrated by 48 months. Once again the higher level of addition of calcium sulfate and the addition of calcium carbonate, proved particularly beneficial.

Higgins and Crammond [34] investigated how cement-type, aggregate-type and curing, affect the susceptibility of concrete to the thaumasite form of sulfate attack (TSA). The cements were Portland cement (PC), sulfate-resisting Portland cement (SRPC) and a combination of 70% GGBS with 30% PC. These were combined with various carbonate aggregates or a non-carbonate control. Initial curing was either in water or in air. Concrete cubes were immersed in four strengths of sulfate solution at 5 and 20°C up to 6 years. They concluded that (1) deterioration, consistent with TSA was found to occur on all of the PC and SRPC concretes made with carbonate aggregate and stored in sulfate solutions at 5°C; (2) degree of TSA increased with the sulfate concentration of the test solution and with time; (3) at 5°C, there was no discernible difference between the performance of SRPC concretes containing carbonate aggregates and those made with PC; (4) concretes made with 70% GGBS/30% PC and normal quality carbonate aggregates performed extremely well and showed no evidence of TSA in any of the solutions, with either normal or high-alumina GGBS. However, concretes made with 70% GGBS/30% PC and inferior quality carbonate aggregates did not perform well at either 5°C or 20°C; (5) an initial air-cure proved beneficial against both conventional sulfate attack and TSA. After six years, the air-cured 70% GGBS concretes showed no evidence of attack, even in the strongest sulfate solution; (6) presence of carbonate in the mix substantially improved the resistance of 70% GGBS/30% PC concretes to conventional sulfate attack; (7) overall, the worst performers were concretes made with PC. However, the concretes made with PC, appeared slightly less susceptible to conventional sulfate attack than those made with 70% GGBS/30% PC and it would appear that the low C_3A PC used in the present study, had some sulfate-resisting properties.

3.5.4 Alkali Silica Reaction

Alkali-Silica Reaction (ASR) is a chemical reaction which occurs in concrete between the highly alkaline cement paste and reactive non-crystalline (amorphous) silica, which is found in many common aggregates. The ASR reaction is the same as the Pozzolanic reaction which is a simple acid–base reaction between calcium hydroxide, also known as Portlandite, or ($Ca(OH)_2$), and silicic acid (H_4SiO_4, or $Si(OH)_4$). For the sake of simplicity, this reaction can be schematically represented as following:

$$Ca(OH)_2 + H_4SiO_4 \rightarrow Ca^{2+} + H_2SiO_4^{2-} + 2\ H_2O \rightarrow CaH_2SiO_4 \cdot 2\ H_2O$$

This reaction causes the expansion of the altered aggregate by the formation of a swelling gel of CSH. This gel increases in volume with water and exerts an expansive pressure inside the material, causing spalling and loss of strength of the concrete, finally leading to its failure.

The mechanism of ASR can be detailed in following four steps; (1) alkaline solution attacks the siliceous aggregate to convert it to viscous alkali silicate gel; (2) consumption of alkali by the reaction induces the dissolution of Ca^{2+} ions into the cement pore water. Calcium ions then react with the gel to convert it to hard CSH; (3) penetrated alkaline solution converts the remaining siliceous minerals into bulky alkali silicate gel. The resultant expansive pressure is stored in the aggregate; and (4) accumulated pressure cracks the aggregate and the surrounding cement paste.

GGBS could be very effective in controlling ASR because (1) GGBS reduces the alkalinity of the concrete, and thus the alkali-silica ratio; (2) GGBS reduces mobility of alkalis in the concrete; and (3) GGBS reduces free lime in concrete which is regarded as an important factor for alkali silica reaction.

Hogan and Meusel [35] reported that partial replacement of high-alkali cement with slag dramatically reduces the likelihood of alkali aggregate reaction in concrete. Mehta [51] mentioned that even if the alkali content of the slag is high, its solubility in the high-pH environment of concrete may be low, so that even a high alkali content of a blended cement may not cause any problems. He further observed that depending on the characteristics of the slag combinations of high-alkali Portland cement with 40–65% slag are effective in limiting the alkali-aggregate expansions to acceptable limits.

Kwon [44] studied the effect of ground granulated blast-furnace slag on the alkali-aggregate reaction of high-strength concrete. High-strength concrete with a cement content of 450 kg/m^3, and three types of GGBFS with a fineness of 4,000, 6,000, and 8,000 cm^2/g were used. Percentage replacements of GGBFS were 0, 30, 45 and 60. They observed that (1) the expansion coefficients were 0.20, 0.105, 0.06, and 0.02% with GGBFS replacement ratio of 60, 45, 30, and 0%, indicating the AAR-inhibiting effect of GGBFS; and (2) fineness of GGBFS had a significant effect, and differences were most significant when the replacement ratio was 30%. The AAR-inhibiting effect was highest with a fineness of 8,000 cm^2/g, followed by 6,000 and 4,000 cm^2/g in this order.

Hester et al. [32] made a comparative assessment of the effects of alkali-level of GGBS on potential alkali–silica reactivity of concrete. Expansion tests were conducted on concrete mixes made with Portland cement (NPC), two slags of differing alkali content (<1 and >1%), three aggregates and two alkali loads of 5 and 6 kg Na_2O_{eq}/m^3. Cement was replaced with 50% slag. Expansion at 52 weeks is given in Table 3.22. It was reported that (1) partial replacement of Portland cement with slag (50%) significantly reduced the expansion of the concrete, and slag concretes had very low expansion levels. There was no significant difference

Table 3.22 Average measured expansion at 52 weeks (Hester et al. 2009)

Mix series	Coarse aggregate	Fine aggregate	Binder composition	Alkali load (kg Na_2O_{eq}/m^2)	Average expansion at 52 weeks (%)
CP-11	N-1	H-1	NPC	6	0.048
CP-12	L-1	H-1	NPC	6	0.151
CP-13	R-1	N-2	NPC	6	0.154
CP-14	N-1	H-1	NPC/slag 1	6	0.005
CP-15	L-1	H-1	NPC/slag 1	6	0.006
CP-16	R-1	N-2	NPC/slag 1	6	0.007
CP-17	N-1	H-1	NPC/slag 2	6	0.007
CP-18	L-1	H-1	NPC/slag 2	6	0.001
CP-19	R-1	N-2	NPC/slag 2	6	0.060
CP-21	N-1	H-1	NPC	5	0.005
CP-22	L-1	H-1	NPC	5	0.078
CP-23	R-1	N-2	NPC	5	0.133
CP-24	N-1	H-1	NPC/slag 1	5	0.002
CP-25	L-1	H-1	NPC/slag 1	5	0.000
CP-26	R-1	N-2	NPC/slag 1	5	0.027
CP-27	N-1	H-1	NPC/slag 2	5	0.000
CP-28	L-1	H-1	NPC/slag 2	5	0.004
CP-29	R-1	N-2	NPC/slag 2	5	0.000

N-1: Pure limestone virtually free of clay, some traces of dolomite
N-2: 97% limestone sand, both clean and clay free. Traces of dolomite and quartz
L-1: Argillaceous limestone with moderate amounts of clay, some quartz silt, and organic matter
H-1: Consists of over 40% limestone, 30% chert, and 20% quartz. Also contains about 6% fine greywacke and siltstone. Small amounts of sandstone, quartz and aplite
R-1: Greywacke aggregate, consists of 65% arkosic greywacke/sandstone. Remainder is lithic greywacke. The arkosic sandstone matrix contains feldspar, quartz, clay, clay minerals

in behavior irrespective of aggregate type or alkali load, indicating that the alkali level of the slag was is not a contributory factor at the 50% replacement level.

Bouikni et al. [17] investigated the alkali–silica reaction of concrete made with 50 and 65% slag. A high alkali Portland cement with 1% sodium oxide and slag with 0.53% sodium oxide equivalent were used. Two concrete mixtures having high workability and low water–binder ratios were used. The mix proportions, by mass were 1:1.75:2.53:0.43 and 1:1.67:2.41:0.40 (cement + slag: sand: aggregate: water), respectively, for the 50 and 65% slag content, respectively. Table 3.23 shows that continued exposure to a hot and humid regime resulted in increasing expansion, but the concrete containing 65% slag replacement always exhibited less expansion than concrete incorporating 50% slag at corresponding ages. On removing the specimens from the hot room to dry ambient conditions, the specimens showed recovery and a reduction in expansion. Subsequent exposure to the hot and humid environment, again accelerated expansion, but this time, concretes with 65% slag replacement reached higher expansion than concretes with 50% slag. Further, concretes with 50% slag did not reach the same high expansion of 0.31% which it had reached in its previous exposure to the hot, humid regime.

Table 3.23 Expansion and pulse velocity changes due to alkali–silica reactivity in slag concretes [17]

Exposure	Age (days)	Expansion (%)	
		Slag (50%)	Slag (65%)
Laboratory[a]	1	0.000	0.000
	7	0.014	0.008
	14	0.022	0.015
Fog[b]	28	0.020	0.023
Hot room	38	0.144	0.022
	81.86[c]	0.310	0.216
Laboratory	127	–	0.134
	152	0.137	–
Hot room	200	–	0.328
	257	0.220	–

[a] Represents dry ambient conditions
[b] Represents humid ambient conditions
[c] Represents 86 days

However, continued exposure to the hot, humid environment appeared to stabilize the subsequent progress of expansion in both concrete at about 6 months, and there was little increase in expansion with further exposure.

Puertas et al. [61] evaluated the alkali–silica reaction in waterglass-alkali-activated slag (waterglass-AAS) and ordinary Portland cement (OPC) mortars using three types of (siliceous and calcareous) aggregates. The slag was alkali-activated with a waterglass solution (Na_2O. $nSiO_2$. mH_2O + NaOH) containing 4% Na_2O by slag mass and a SiO_2/Na_2O ratio of 1.08. It was concluded that waterglass-AAS mortars are stronger and more resistant to alkali-aggregate reactions than OPC mortars. When the mortars were made with a reactive siliceous aggregate, expansion was four times greater in OPC than in AAS material. When a reactive calcareous (dolomite) aggregate was used, no expansion was detected in any of the mortars after 14 days, although the characterization results showed that the dolomite had reacted and calcareous-alkali products (brucite) had in fact formed in both mortars. These reactive processes were more intense in OPC than in AAS mortars, probably due to the absence of portlandite in the latter. When the calcareous aggregate was non-reactive, no expansions were observed in any of the mortars, although a substantial rise was recorded in the mechanical strength of AAS mortars exposed to the most aggressive conditions (1 M NaOH and 80°C).

3.5.5 Freezing and Thawing Resistance

Hogan and Meusel [35] tested air-entrained concrete with and without slag for frost resistance. The specimens were moist-cured for 14 days before tested as per ASTM C666 [5] (Procedure A) for 300 cycles. The slag concrete did not perform as well as the Portland cement concrete, but both were of the standard of acceptable durability criteria.

Stark and Ludwig [67] mentioned that freezing and thawing resistance, deicing salt resistance of blast furnace cement concrete was related to the carbonation of the surface which increases the capillary porosity and leads to metastable calcium carbonates that are soluble in NaCl. The degree of hydration at the beginning of freezing determines the pore size distribution and the strength of the structure.

Stark [66] investigated the frost and deicing-salt resistance of slag cement and observed that frost resistance of slag concrete was highly dependent on the degree of hydration. Carbonation was more important than the degree of hydration on resistance to deicing salt.

Nanaumi et al. [54] studied the frost resistance of blast furnace slag concrete (with 30–60% per cement unit addition) and plastic micro-spheres using cellulose ether as anti-washout admixture. At 60% addition of slag and 2% (per cement unit) addition of micro-spheres, scaling resistance was improved.

Knaack and Stark [43] studied the super-sulfated cements (more than 75–85% of highly basic slag and alumina content >15%). They mentioned that in comparison to conventional slag cements, super-sulfated cements are mainly activated by a high amount of gypsum with only a small amount of Portland cement. These cements displayed lower performances in scaling and freeze–thaw due to a lower degree of hydration at 28 days.

Yuksel et al. [79] investigated the usage of GBFS, bottom ash (BA), and combination of both as partial replacement of fine aggregate on the concrete durability. Three series of concretes were produced, i.e. of GBFS, BA and GBFS + BA, which were coded as C, K, and CK, respectively. Five test groups were constituted with the replacement percentages as 10, 20, 30, 40 and 50% in each series. Freezing–thawing resistance was measured by freezing and thawing the specimens in water. Weight-loss results are shown in Fig. 3.21. For three series, resistance to freeze–thaw increased up to 20% replacement. However, it decreased again for the replacement level above 20%. C-series gave the best

Fig. 3.21 Freezing–thawing test results [79]

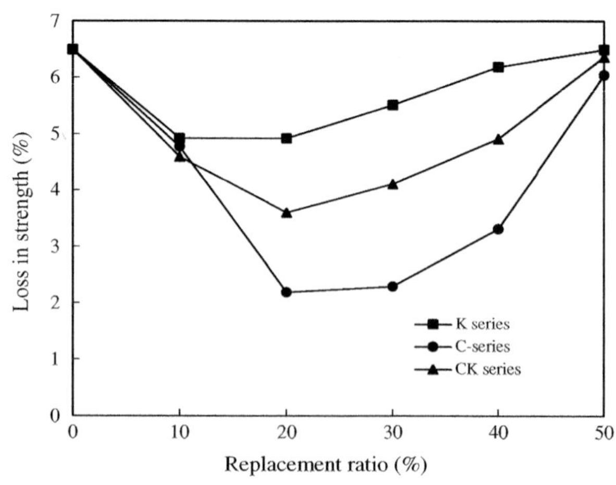

performance among the three series. For all the series, the loss in strength was below the loss in strength value for reference concrete.

3.5.6 Corrosion Resistance

Corrosion can be defined as the degradation of a material due to a reaction with its environment. Degradation implies deterioration of physical properties of the material. This can be a weakening of the material due to a loss of cross-sectional area; it can be the shattering of a metal due to hydrogen embrittlement.

Corrosion of reinforcements is the most common cause of deterioration of concrete structures. Corrosion of steel reinforcement leads to cracking and eventually spalling of concrete. Good quality concrete reduces the deterioration of concrete under adverse environmental conditions.

Li and Roy [46] have indicated that for making good quality durable concrete, use of GGBS is very advantageous. It not only reduces the porosity but also the pores become finer and the change in mineralogy of the cement hydrates leads to a reduction in the mobility of chloride ions.

Mangat and Molloy [50] studied the corrosion of steel reinforcement electrodes embedded in different matrices of concrete, exposed to simulated marine splash zone exposure for about 600 days (1,200 cycles) after initial curing in air for 14 days. The control mix had proportions of 1:2.5:1.2 with a water/cement ratio of 0.58. The cement content was 430 kg/m^3, and GGBS replacement levels were 20, 40 and 60%. The results showed that (1) replacement of cement by up to 40% GGBFS had no significant influence on rebar corrosion. At 60% GGBFS, corrosion rate of reinforcement was significantly reduced; (2) corrosion process of reinforcement in plain concrete, and concrete with low contents of GGBS was similar and was anodically controlled. In GGBFS (60%) concrete, the cathodic polarization increased greatly although control of corrosion remained principally anodic; (3) chloride concentration of the pore fluid gave a more reliable indication of corrosivity of a matrix than Cl^-/OH^- ratio, especially when comparing concretes with different types and contents of replacement materials.

Al-Amoudi et al. [3] studied the long term corrosion resistance of steel rebars in 5% sodium chloride solution. Regression analyses indicated excellent correlation between corrosion resistance and porosity for both plain and blended cement concretes. It was found that the corrosion rate of the steel rebar in GGBFS containing concrete specimen was between one-half and one-twelfth of those in plain concrete specimens.

Torii et al. [71] reported that the resistance to chloride ion penetration of 50% GGBFS concrete was almost the same as that of 10% SF concrete. Tsay [40] evaluated the slag cement subjected to seawater corrosion at various ages and concluded that slag concrete with 20–30% substitution has the best corrosion resistance properties.

Table 3.24 Comparison of weight loss, icorr, half-cell potential and carbonation depth for different slag replaced concrete [59]

Concrete	Weight loss (mg)	Icorr (A/m^2)	Half-cell potential (mV)	Carbonation depth (m)
Reference concrete	0.33	0.1169	−503	2.0
30% slag concrete	0.27	0.0823	−108	2.0
50% slag concrete	0.29	0.1239	−176	1.0
70% slag concrete	0.26	0.0842	−31	1.0

Pal et al. [59] studied the corrosion behavior of embedded reinforcement under different proportions of slag. Corrosion of steel was examined electrochemically and also by an accelerated carbonation test. It can be seen from Table 3.24 that with the increase in slag content, corrosion rate, weight loss, half cell potentials and the carbonation depth were found to be decreasing.

Yeau and Kim [78] examined the influence of GGBS on the corrosion resistance of concrete. Concrete mixtures were made with two types of cements (ASTM Type I or ASTM Type V) and varying percentages (0, 25, 40, and 55%) of GGBS. Concrete mixtures were proportioned to have the 28-day strength of 35 MPa. Concrete specimens of size 150 × 150 × 500 mm with three different covers, 10, 20, and 30 mm. were prepared. Accelerated steel corrosion (ASC) test was conducted to measure half-cell potential and the surface areas of corroded steel bars. Repeated wetting and drying method was used to accelerate steel corrosion. These tests were conducted up to 30 cycles after concrete specimens were cured in water for 28, 56, and 91 days. For each wetting–drying cycle, test specimens were wetted in 3% NaCl solution for 3 days and then dried at 60°C for 4 days. After 30 weeks, the specimens were crushed out and corroded surface area was calculated as a percentage to the total area of steel embedded in concrete. Figure 3.22 shows the results obtained from the measurements of surface area

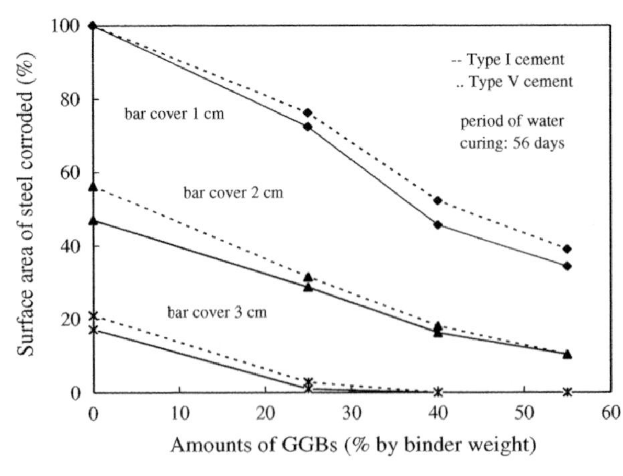

Fig. 3.22 Surface area (%) corroded from steel bars in concrete specimens [78]

corroded from steel bars. They concluded that (1) the corroded surface area (%) depend on the thickness of the concrete cover, and GGBS content; (2) corroded areas of steel embedded in control concrete mixtures were about two and three times larger than those of steel involved in 40% GGBS concrete mixture and 55% GGBS concrete mixture, respectively; and (3) proportion of corroded areas in Type I cement was lower than that of the corroded areas in Type V cement. These results suggested that the resistance to steel corrosion can be superior, since not only Type I cement was used, but also the amount of GGBS was increased.

Cheng et al. [20] studied the influence of GGBS on corrosion behavior of reinforced concrete beams ($150 \times 150 \times 900$ mm) under various loading ratios (37 and 75% of the ultimate load) and exposed to 3.5% NaCl solution. Beams were made of three concrete mixes (A, B, and C). Mix A was control one with mixture proportion 1:1.82:1.97 with water–binder ratio of 0.55. Mixes B and C were made by replacing cement with 40 and 60% of GGBS, respectively. A direct current (1 mA cm^{-2}) was applied to accelerate the corrosion process. Open circuit potential (OCP) and direct current polarization resistance were obtained to evaluate the rebar corrosion. As per ASTM C876 [8], when OCP is higher than −127 mV (SCE), there is only 10% probability that the reinforcement may corrode. If the potential is from −127 to −276 mV (SCE), corrosion probability is uncertain. Corrosion probability may be higher than 90% for OCP between −276 and −427 mV (SCE). Corrosion results indicated that (1) without sustained loading, mixes with 40 and 60% GGBS significantly increased the exposure time span to severe corrosion in comparison with control mix; (2) GGBS replacement for cement refined the pore structure and reduced the corrosion probability of reinforcing steel; (3) although, the pH value was 13.2 for control concrete, 12.8 for 40% GGBS replacement concrete and 12.4 for 60% GGBS replacement concrete, respectively. There seemed no prominent negative effect on the corrosion behavior due to decreasing alkalinity in pore solution; (4) with 37% loading ratio, the OCP of control specimen dropped from −75 to −427 mV (SCE) after 4-day accelerated exposure. Specimens containing 40 and 60% GGBS reached severe corrosion condition after 6–8-day exposures, which revealed that sustained loading or cracks plays paramount role in the corrosion process of reinforcing steel in concrete; (5) for the control specimen with 75% sustained loading, the potential dropped below −427 mV (SCE) even after 2-day accelerated corrosion process and GGBS concretes having cracks wider than 0.02 mm had no better corrosion resistance; (6) at higher sustained loading ratio, OCP did not significantly vary among the three mixes, and the corrosion process was controlled by the pre-determined cracks.

3.5.7 Carbonation

Carbonation occurs in concrete because the calcium bearing phases present are attacked by carbon dioxide of the air and converted to calcium carbonate. Cement paste contains 25–50 wt% calcium hydroxide ($Ca(OH)_2$), which means that the pH of the fresh cement paste is at least 12.5. The pH of a fully carbonated paste is about 7.

The concrete will carbonate if CO_2 from air or from water enters the concrete according to

$$Ca(OH)_2 + CO_2 \rightarrow CaCO_3 + H_2O$$

When $Ca(OH)_2$ is removed from the paste, hydrated CSH will liberate CaO which will also carbonate. The rate of carbonation depends on porosity and moisture content of the concrete.

The carbonation process requires the presence of water because CO_2 dissolves in water forming H_2CO_3. If the concrete is too dry (RH $< 40\%$) CO_2 cannot dissolve and no carbonation occurs. If on the other hand it is too wet (RH $> 90\%$) CO_2 cannot enter the concrete and the concrete will not carbonate. Optimal conditions for carbonation occur at a RH of 50% (range 40–90%).

According to [55], the cement type affects the rate of carbonation with sulfate-resisting cement having a 50% higher rate than Portland cement. Slag cements can have carbonation rates up by 200% more than Portland cement.

Osborne investigated structures made with Portland cement with and without 50 or 70% slag. Total cement contents were in the range 360–380 kg/m^3. They concluded that approximately after 5 years of exposure, carbonation depths were greatest for the 70% slag concretes. Portland cement concrete showed little or no sign of carbonation after 7–17 years exposure. Malhotra et al. [48] reported that after 10 years of outdoor exposure, the carbonation depths of concrete (water to total cementitious material ratios of 0.27–0.29), which included plain concrete and concrete with 28–35% slag, were small (in the range of less than 1–5 mm).

Sulapha et al. [69] studied the carbonation of concrete incorporating ground granulated blast-furnace slag. In the concrete mixtures, OPC was partially replaced with GGBS. The replacement percentages of GGBS (having 4,500 cm^2/g fineness) were 30, 50, and 65%; the replacement level of GGBS (with 6,000 or 8,000 cm^2/g fineness) was 65%. Prisms of size $100 \times 100 \times 400$ mm were made. The carbonation depth of the specimens was monitored every 2 weeks in the first 2 months, and thereafter once a month up to 12 months. After a certain exposure period, the specimens were split and freshly broken surfaces were sprayed with a standard solution of 1% phenolphthalein in 70% ethyl alcohol. In the noncarbonated region with pH values above 9.2, the indicator turned purple-red; in the carbonated portion with pH less than 9.2, the solution remained colorless. The distance between the color change boundary and the concrete surface was measured as the carbonation depth. Based on the results, they concluded that (1) lower w/b (0.3–0.6) and a prolonged curing age in water (from 1 to 28 days) generally led to a slower rate of carbonation, possibly due to pore structure densification; (2) carbonation rate of low fineness GGBS (4,500 cm^2/g) increased with an increase in OPC replacement level; (2) concrete containing low fineness GGBS (4,500 cm^2/g) exhibited higher rates of carbonation than plain OPC concrete. This could be due to the reduction in $Ca(OH)_2$ content, which seemed to have more influence over pore refinement, and hence led to faster rates of carbonation; and (3) concrete incorporating GGBS of higher fineness (6,000 and 8,000 cm^2/g), the carbonation rates were lower than that of plain OPC concrete. The pore modification, being more dominant than the change in $Ca(OH)_2$ content, appeared to control the carbonation rate.

Bouikni et al. [17] studied the carbonation of concrete made with 50 and 65% slag. A high alkali Portland cement with 1% sodium oxide and slag with 0.53% sodium oxide equivalent were used. Two concrete mixtures having high workability and low water–binder ratios were used. The mix proportions, by mass were 1:1.75:2.53:0.43 and 1:1.67:2.41:0.40 (cement + slag:sand:aggregate:water), respectively, for the 50 and 65% slag content, respectively. Carbonation tests were performed on the broken faces after flexural strength tests, by spraying with 2% phenolphthalein in 70% ethyl alcohol, as recommended by RILEM [63]. The carbonation depth measurements were taken at 3 and 6 months, and the results of these measurements for the 50 and 65% slag concretes subjected to the three curing regimes is given in Table 3.25. The results demonstrated that the carbonation depths, were at both ages, higher for the concrete with 65% slag replacement than for the concrete with 50% slag level, and that, of all the curing regimes, concrete exposed to an internal environment without any water curing permitted the highest pen-carbonation measurement. Compressive strength values showed small increase beyond the 28 days, and carbonation did not generally appear to reverse this trend with increased carbonation penetration and ageing.

Hui-sheng et al. [38] investigated the effect of GGBS on the carbonation of HPC up to the age of 56 days. Cement, fine aggregate, and coarse aggregates contents were 550, 687, and 1,030 kg/m^3, respectively. GGBS levels were 0, 15, 30, 45 and 60% by weight of cement, with water–binder ratio of 0.30. Carbonation depth of HPC with GGBFS is shown in Fig. 3.23. It can be seen that the incorporation of GGBS decreased the carbonation resistance of HPC, but at significantly different levels. For HPC with GGBS at w/b of 0.30, carbonation depth significantly increased with the increase in GGBS content.

Table 3.25 Effect of age and curing regime on carbonation depth [17]

Curing regime	Age (months)	50% slag		65% Sslag	
		Compresive strength[a] (MPa)	Carbonation depth (mm)	Compresive strength[a](MPa)	Carbonation depth (mm)
Lab	3	45.8	4.6	36.1	5.1
	6	51.3	9.8	33.9	11.6
7 Fog + lab	3	65.8	3.5	57.8	4.5
	6	71.3	5.1	59.7	7.8
Fog	3	70.9	0.0	64.5	0.0
	6	71.3	0.0	67.6	0.0

[a] Modified cube strength obtained from fractured flexural test specimens

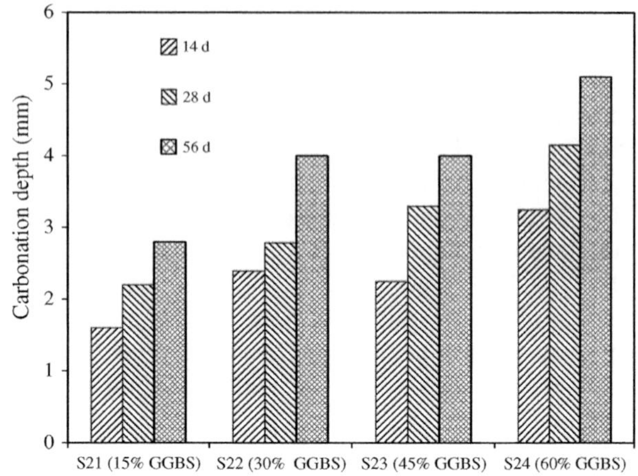

Fig. 3.23 Influence of GGBS on carbonation depth of HPC at w/b ratio of 0.30 [38]

References

1. ACI Committee 226: Ground granulated blast furnace slag as a cementitious constituent in concrete. ACI Mater. J. **84**(4), 327–342 (1987)
2. Aldea, C., Young, F., Wang, K., Shah, S.P.: Effect of curing conditions on properties of concrete using slag replacement. Cem. Concr. Res. **30**(3), 465–472 (2000)
3. Al-Amoudi, O.S.B., Rasheeduzzafar, M.M., Al-Mana, A.I.: Prediction of long-term corrosion resistance of plain and blended cement concretes. ACI Mater. J. **90**(6), 564–570 (1993)
4. ASTM C 232: Standard test methods for bleeding of concrete. ASTM Standards. 4.02:13942, Annual Book of ASTM Standards, American Society for Testing and Materials, West Conshohocken (1992)
5. ASTM C666-Part A: Standard test method for resistance of concrete to rapid freezing and thawing. Annual Book of ASTM Standards, vol. 03.02, 19428–2959, American Society for Testing and Materials, West Conshohocken (1997)
6. ASTM C 989: Standard specification for ground granulated blast-furnace slag for use in concrete and mortars. Annual Book of ASTM Standards, American Society for Testing and Materials, West Conshohocken (1999)
7. ASTM C1202: Standard test method for electrical indication of concrete's ability to resist chloride ion penetration. Annual Book of ASTM Standards, vol. 4.02, American Society for Testing and Materials, West Conshohocken (1997)
8. ASTM C 876: Standard test method for half-cell potentials of uncoated reinforcing steel in concrete, Annual Book of ASTM Standards. American Society for Testing and Materials, West Conshohocken (1991)
9. ASTM C 1012-95a: Standard test method for length change of hydraulic cement mortars exposed to a sulfate solution. Annual book of ASTM Standards, Section 4—Construction, Cement; Lime; Gypsum. vol. 04.01, pp. 450–456. American Society for Testing and Materials, West Conshohocken (1995)
10. ASTM C39/C39 M: Standard test method for compressive strength of cylindrical concrete specimens, Annual Book of ASTM Standards, vol. 4.02, American Society for Testing and Materials, West Conshohocken (2004)
11. Atis, C.D., Bilim, C.: Wet and dry cured compressive strength of concrete containing ground granulated blast-furnace slag. Build. Environ. **42**(8), 3060–3065 (2007)
12. Binici, H., Cagatay, I.H., Shah, T., Kapur, S.: Mineralogy of plain Portland and blended cement pastes. Build. Environ. **43**(7), 1318–1325 (2008)

13. Babu, K.G., Kumar, V.S.R.: Efficiency of GGBS in concrete. Cem. Concr. Res. **30**(7), 1031–1036 (2000)
14. Ballim, Y., Graham, P.C.: A maturity approach to the rate of heat evolution in concrete. Mag. Concr. Res. **55**(3), 249–256 (2003)
15. Ballim, Y., Graham, P.C.: The effects of supplementary cementing materials in modifying the heat of hydration of concrete. Mater. Struct. **42**, 803–811 (2009)
16. Basheer, P.A.M., Gilleece, P.R.V., Long, A.E., McCarter, W.J.: Monitoring electrical resistance of concretes containing alternative cementitious materials to assess their resistance to chloride penetration. Cem. Concr. Compos. **24**(5), 437–449 (2002)
17. Bouikni, A., Swamy, R.N., Bali, A.: Durability properties of concrete containing 50% and 65% slag. Construct. Build. Mater. **23**(8), 2836–2845 (2009)
18. Cao, H.T., Bucea, L., Ray, A., Yozghatlian, S.: The effect of cement composition and pH of environment on sulfate resistance of Portland cements and blended cements. Cem. Concr. Compos. **19**(2), 161–171 (1997)
19. Cheron, M., Lardinois, C.: In: Proceedings of the 5th International Congress on the Chemistry of Cement, Tokyo, vol. IV, pp. 277–285 (1968)
20. Cheng, A., Huang, R., Wu, J.K., Chen, C.H.: Influence of GGBS on durability and corrosion behavior of reinforced concrete. Mater. Chem. Phys. **93**, 404–411 (2005)
21. Cakir, Ö., Aköz, F.: Effect of curing conditions on the mortars with and without GGBFS. Construct. Build. Mater. **22**(3), 308–314 (2008)
22. Daube, J., Bakker, R.: Portland blast-furnace slag cement: a review. Blended Cement, ASTM-STP 897 pp. 5–14 (1986)
23. Dhir, R.K., El-Mohr, M.A.K., Dyer, T.D.: Chloride binding in GGBS concrete. Cem. Concr. Res. **26**(12), 1767–1773 (1996)
24. Douglas, E., Bilodeau, A., Brandstetr, J.: Alkali activated ground granulated blast-furnace slag concrete: preliminary investigation. Cem. Concr. Res. **21**(1), 101–108 (1991)
25. Elahi, A., Basheer, P.A.M., Nanukuttan, S.V., Khan, O.U.Z.: Mechanical and durability properties of high performance concretes containing supplementary cementitious materials. Constr. Build. Mater. **24**(3), 292–299 (2010)
26. El-Darwish, I., Kurdi, A., Mahmoud, H., El-Kair, H.A.: Mechanical properties and durability of Portland cement concrete incorporating ground steel making slag. AEJ Alex. Eng. J. **36**, 1–14 (1997)
27. Fearson, J.P.H.: Sulfate resistance of combination of Portland cement and granulated blast furnace slag. vol. **2**, pp. 1495–1524. ACI Special Publication SP-91, Dr. Farmington Hills (1986)
28. Gao, J.M., Qian, C.X., Liu, H.F., Wang, B., Li, L.: ITZ microstructure of concrete containing GGBS. Cem. Concr. Res. **35**(7), 1299–1304 (2005)
29. Guo, L.P., Sun, W., Zheng, K.R., Chen, H.J., Liu, B.: Study on the flexural fatigue performance and fractal mechanism of concrete with high proportions of ground granulated blast-furnace slag. Cem. Concr. Res. **37**(2), 242–250 (2007)
30. Gee, K.H.. The potential for slag in blended cements. In: Proceedings of the 14th International Cement Seminar, Rock Products, pp. 51–53 (1979)
31. Gollop, R.S., Taylor, H.F.W.: Micro structural and micro analytical studies of sulfate attack IV. Reactions of a slag cement paste with sodium and magnesium sulfate solutions. Cem. Concr. Res. **26**(7), 1013–1028 (1996)
32. Hester, D., McNally, C., Richardson, M.: A study of the influence of slag alkali level on the alkali–silica reactivity of slag concrete. Construct. Build. Mater. **19**(9), 661–665 (2005)
33. Higgins, D.D.: Increased sulfate resistance of GGBS concrete in the presence of carbonate. Cem. Concr. Compos. **25**(8), 913–919 (2003)
34. Higgins, D.D., Crammond, N.J.: Resistance of concrete containing GGBS to the thaumasite form of sulfate attack. Cem. Concr. Compos. **25**(8), 921–929 (2003)
35. Hogan, F.J., Meusel, J.W.: Evaluation for durability and strength development of a ground granulated blast furnace slag. Cem. Concr. Aggreg. **3**(1), 40–52 (1981)

36. Hooton, R.D., Titherington, M.P.: Chloride resistance of high-performance concretes subjected to accelerated curing. Cem. Concr. Res. **34**(9), 1561–1567 (2004)
37. Huang, R., Yeih, W.C.: Assessment of chloride diffusion in high strength concrete using the accelerated ionic migration test. J. Chin. Inst. Eng. **20**, 39–45 (1997)
38. Hui-sheng, S., Bi-wan, X., Xiao-chen, Z.: Influence of mineral admixtures on compressive strength, gas permeability and carbonation of high performance concrete. Construct. Build. Mater. **23**(5), 1980–1985 (2009)
39. Hwang, C.L., Lin, C.Y.: Strength development of blended blast furnace slag cement mortars. ACI SP **91**, 1323–1340 (1986)
40. Jau, W.C., Tsay, D.S.: A study of the basic engineering properties of slag cement concrete and its resistance to sea water corrosion. Cem. Concr. Res. **28**(10), 1363–1371 (1998)
41. Jianyong, L., Yan, Y.: A study on creep and drying shrinkage of high performance concrete. Cem. Concr. Res. **31**(8), 1203–1206 (2001)
42. Khatib, J.M., Hibbert, J.J.: Selected engineering properties of concrete incorporating slag and metakaolin. Construct. Build. Mater. **19**(6), 460–472 (2005)
43. Knaack, U., Stark, J.: Frost and frost-deicing salt resistance of super-sulfated cement concrete, frost resistance of concrete. In: Proceedings of the International RILEM Workshop on Resistance of Concrete to Freezing and Thawing with or Without De-icing Chemicals. pp. 139–146. Essen, Germany (1997)
44. Kwon, Y.: A study on the alkali-aggregate reaction in high-strength concrete with particular respect to the ground granulated blast-furnace slag effect. Cem. Concr. Res. **35**(7), 1305–1313 (2005)
45. Li, G., Zhao, X.: Properties of concrete incorporating fly ash and ground granulated blast-furnace slag. Cem. Concr. Compos. **25**(3), 293–299 (2003)
46. Li, S., Roy, D.M.: Investigation of relations between porosity, pore structure and chloride diffusion of fly ash blended cement pastes. Cem. Concr. Res. **16**(5), 749–759 (1986)
47. Luo, R., Cai, Y., Wang, C., Huang, X.: Study of chloride binding and diffusion in GGBS concrete. Cem. Concr. Res. **33**(1), 1–7 (2003)
48. Malhotra, V.M., Zhang, M.H., Read, P.H., Ryell, J.: Long-term mechanical properties and durability characteristics of high strength/high-performance concrete incorporating supplementary cementing materials under outdoor exposure conditions. ACI Mater. J. **97**(5), 518–525 (2000)
49. Mangat, P.S., El-Khatib, J.M.: Influence of initial curing on sulfate resistance of blended cement concrete. Cem. Concr. Res. **22**(6), 1089–1100 (1992)
50. Mangat, P.S., Molloy, B.T.: Influence of PFA, slag and microsilica on chloride induced corrosion of reinforcement in concrete. Cem. Concr. Res. **21**(5), 819–834 (1991)
51. Mehta, P.K.: Concrete:Structure, Properties and Materials. Prentice Hall, Englewood Cliffs (1986)
52. Meusel, J.W., Rose, J,H.: Production of granulated blast furnace slag at sparrows point and the workability and strength potential of concrete incorporating the slag. vol. 79, pp. 867–890. ACI Special Publication SP, Dr. Farmington Hills (1983)
53. Miura, T., Iwaki, I.: Strength development of concrete incorporating high levels of ground granulated blast-furnace slag at low temperatures. ACI Mater. J. **97**(1), 66–70 (2000)
54. Nanaumi, T., Ayuta, K., Yamakawa, T.: Frost resistance of antiwashout underwater concrete containing expanded plastic micro-spheres and blast-furnace slag. Semento Konkurito Ronbunshu **51**, 90–95 (Chem. Abstr. 128, 208152) (1997)
55. Neville, A.M.: Properties of Concrete, 3rd edn. Pitman, London (1981)
56. Olorunsogo, F.T.: Particle size distribution of GGBS and bleeding characteristics of slag cement mortars. Cem. Concr. Res. **28**(6), 907–919 (1998)
57. Oner, A., Akyuz, S.: An experimental study on optimum usage of GGBS for the compressive strength of concrete. Cem. Concr. Compos. **29**(6), 505–514 (2007)
58. Osborne, G.J.: The effectiveness of a carbonated outer layer to concrete in the prevention of sulfate attack. In: Proceedings of International Conference on the Protection of Concrete, University of Dundee, UK (1990)

59. Pal, S.C., Mukherjee, A., Pathak, S.R.: Corrosion behavior of reinforcement in slag concrete. ACI Mater. J. **99**(6), 1–7 (2002)
60. Pavía, S., Condren, E.: Study of the durability of OPC versus GGBS concrete on exposure to silage effluent. ASCE J. Mater. Civ. Eng. **20**(4), 313–320 (2008)
61. Puertas, F., Palacios, M., Gil-Maroto, A., Vázquez, T.: Alkali-aggregate behaviour of alkali-activated slag mortars: effect of aggregate type. Cem. Concr. Compos. **31**(5), 277–284 (2009)
62. Quillin, K., Osborne, G., Majumdar, A., Singh, B.: Effects of w/c ratio and curing conditions on strength development in BRECEM concretes. Cem. Concr. Res. **31**(4), 627–632 (2001)
63. RILEM draft recommendation: Measurement of hardened concrete carbonation depth. Mater. Struct. **3**(102), 435–90 (1984)
64. Roy, D.M., Idorn, G.M.: Hydration, structure, and properties of blast furnace slag cement, mortars, and concrete. Am. Concr. Inst. J. **79**(6), 444–457 (1982)
65. Roy, D.M., Jiang, W., Silsbee, M.R.: Chloride diffusion in ordinary, blended, and alkali-activated cement pastes and its relation to other properties. Cem. Concr. Res. **30**(12), 1879–1884 (2000)
66. Stark, J., Frost resistance with and without deicing salt—a purely physical problem, frost resistance of concrete. In: Proceedings of the International RILEM Workshop on Resistance of Concrete to Freezing and Thawing with or Without De-Icing Chemicals. pp. 83–99. Essen, Germany (1997)
67. Stark, J., Ludwig, H.M.: Freeze–thaw and freeze-deicing salt resistance of concretes containing cement rich in granulated blast furnace slag. ACI Mater. J. **94**(1), 47–55 (1997)
68. Stutterheim, N.: Portland blast furnace cements—a case for separate grinding of slag. In: Fifth International Symposium on the Chemistry of Cement, Tokyo, pp. 270–276 (1968) (Supplementary Paper IV-113)
69. Sulapha, P., Wong, S.F., Wee, T.H., Swaddiwudhipong, S.: Carbonation of concrete containing mineral admixtures. J. Mater. Civ. Eng. **15**(2), 134–143 (2003)
70. Tasong, W.A., Wild, S., Tilley, R.J.D.: Mechanism by which ground granulated blast furnace slag prevents sulfate attack of lime stabilized kaolinite. Cem. Concr. Res. **29**(7), 975–982 (1999)
71. Torii, K., Sasatani, T., Kawamura, M.: Effects of fly ash, blast furnace slag, and silica fume on resistance of mortar to calcium chloride attack. In: Proceedings of Fifth International Conference on Fly Ash, Silica Fume, Slag, and Natural Pozzolans in Concrete, SP-153 vol. 2, pp. 931–949, American Concrete Institute (1995)
72. Ujhelyi, J.E., Ibrahim, A.J.: Hot weather concreting with hydraulic additives. Cem. Concr. Res. **21**(2–3), 345–354 (1991)
73. Wainwright, P.J., Ait-Aider, H.: The influence of cement source and slag additions on the bleeding of concrete. Cem. Concr. Res. **25**(7), 1445–1456 (1995)
74. Wainwright, P.J., Rey, N.: The influence of ground granulated blast furnace slag (GGBS) additions and time delay on the bleeding of concrete. Cem. Concr. Compos. **22**(4), 253–257 (2000)
75. Wan, H., Shui, Z., Lin, Z.: Analysis of geometric characteristics of GGBS particles and their influences on cement properties. Cem. Concr. Res. **34**(1), 133–137 (2004)
76. http://www.ecocem.ie Concrete mix design using GGBS. Ecocem Ireland Ltd. Dublin
77. Xu, Y.: The influence of sulfates on chloride binding and pore solution chemistry. Cem. Concr. Res. **27**(12), 1841–1850 (1997)
78. Yeau, K.Y., Kim, E.K.: An experimental study on corrosion resistance of concrete with ground granulate blast-furnace slag. Cem. Concr. Res. **35**(7), 1391–1399 (2005)
79. Yuksel, I., Bilir, T., Ozkan, O.: Durability of concrete incorporating non-ground blast furnace slag and bottom ash as fine aggregate. Build. Environ. **42**(7), 2651–2659 (2007)
80. Zhang, Y.J., Zhao, Y.L., Li, H.H., Xu, D.L.: Structure characterization of hydration products generated by alkaline activation of granulated blast furnace slag. J. Mater. Sci. **43**(22), 7141–7147 (2008)

Chapter 4
Metakaolin

4.1 Introduction

Metakaolin (MK) is a pozzolanic material. It is a dehydroxylated form of the clay mineral kaolinite. It is obtained by calcination of kaolinitic clay at a temperature between 500°C and 800°C. Between 100 and 200°C, clay minerals lose most of their adsorbed water. Between 500 and 800°C kaolinite becomes calcined by losing water through dehydroxilization. The raw material input in the manufacture of metakaolin ($Al_2Si_2O_7$) is kaolin clay. Kaolin is a fine, white, clay mineral that has been traditionally used in the manufacture of porcelain. Kaolinite is the mineralogical term that is applicable to kaolin clays. Kaolinite is defined as a common mineral, hydrated aluminum disilicate, the most common constituent of kaolin.

The dehydroxilization of kaolin to metakaolin is an endothermic process due to the large amount of energy required to remove the chemically bonded hydroxyl ions. Above this temperature range, kaolinite becomes metakaolin, with a two-dimensional order in crystal structure. In order to produce a pozzolan (supplementary cementing material) nearly complete dehydroxilization must be reached without overheating, i.e., thoroughly roasted but not burnt. This produces an amorphous, highly pozzolanic state, whereas overheating can cause sintering, to form the dead burnt, non-reactive refractory, called mullite.

Metakaolin reacts with $Ca(OH)_2$, produces calcium silicate hydrate (CSH) gel at ambient temperature. Metakaolin also contains alumina that reacts with CH to produce additional alumina-containing phases, including C_4AH_{13}, C_2ASH_8, and C_3AH_6 [19, 71].

4.1.1 Uses of Metakaolin

Metakaolin finds its usage in many aspects of concrete:

R. Siddique and M. Iqbal Khan, *Supplementary Cementing Materials*, Engineering Materials, DOI: 10.1007/978-3-642-17866-5_4,

- High performance, high strength and lightweight concrete
- Precast concrete for architectural, civil, industrial, and structural purposes
- Fibre cement and ferrocement products
- Glass fibre reinforced concrete
- Mortars, stuccos, repair material, pool plasters
- Improved finishability, color and appearance

4.1.2 Advantages of Using Metakaolin

- Enhanced workability
- Increased compressive strength
- Increased tensile and flexural strengths
- Increased durability
- Reduced permeability
- Increased resistance to chemical attack
- Reduction in alkali–silica reactivity (ASR)
- Reduced shrinkage due to particle packing
- Reduced potential for efflorescence

4.2 Properties of Metakaolin

4.2.1 Physical Properties

Metakaolin particles are extremely small with an average particle size of 3 μm. Its colour is off-white (Fig. 4.1). Some physical properties of metakaolin are given in Table 4.1. SEM picture of typical metakaolin sample is shown in Fig. 4.2 [21].

Fig. 4.1 Metakaolin

Table 4.1 Physical properties of metakaolin

Property	Poon et al. [49]	Al-Akhras et al. [1]	Tafraoui et al. [65]
Specific gravity	2.62	2.5	2.5
Average particle size (μm)		1.0	12
Fineness (m^2/kg)	12,680	12,000	15,000–30,000
Physical form			
Colour		White	

Fig. 4.2 SEM of metakaolin sample [21]

Janotka et al. [31] determined particle size distribution of Slovak poor metakaolin sands with different metakaolin content [36.0% (MK-1), 31.5% (MK-2) and 40.0% (MK-3]. The percentages of metakaolin sands were 10, 20 and 40%. Figure 4.3 shows the particle size distribution of metakaolin sands.

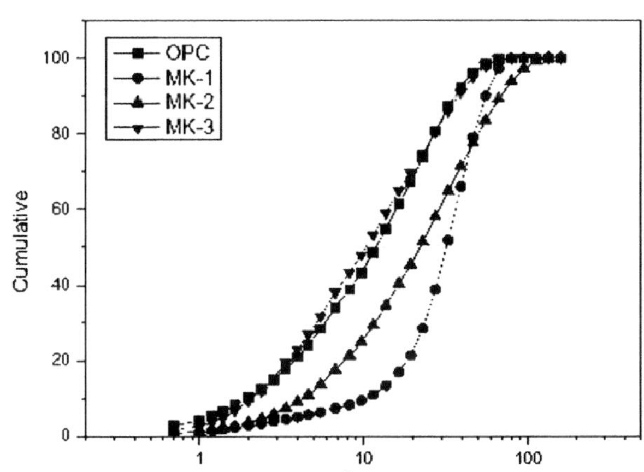

Fig. 4.3 Particle size distribution of Portland cement and metakaolin sands [31]

Table 4.2 Typical chemical composition of metakaolin

Ingredients (%)	Ambroise et al. [2]	Wild and Khatib [68]	Tafraoui et al. [65]
SiO_2	51.52	52.1	58.10
Al_2O_3	40.18	41.0	35.14
Fe_2O_3	1.23	4.32	1.21
CaO	2.00	0.07	1.15
MgO	0.12	0.19	0.20
K_2O	0.53	0.63	1.05
SO_3	–	–	0.03
TiO_2	2.27	0.81	–
Na_2O	0.08	0.26	0.07
L.O.I	2.01	0.60	1.85

4.2.2 Chemical Composition

Major constituents of metakaolin are silica oxide (SiO_2) and alumina oxide (Al_2O_3). Other components include ferric oxide, calcium oxide, magnesium oxide, potassium oxide, etc. The typical chemical composition of metakaolin is given in Table 4.2.

Metakaolin should meet the requirements of ASTM C618 [7], Standard Specification for Coal Fly Ash and Raw or Calcined Natural Pozzolan for use as a mineral admixture in Concrete, Class N, with the following modifications as given in Table 4.3.

4.2.3 Mineralogical Composition

Badogiannis et al. [8] reported mineralogical analysis of metakaolin produced from Poor Greek kaolin and commercial metakaolin (MKC). The semi-quantitative mineralogical estimation of MK and MKC are given in Table 4.4. The estimation is based on the characteristic X-ray diffraction peaks of each mineral, in combination with the bulk chemical analysis of the samples.

Table 4.3 Requirements of metakaolin (ASTM C 618)

Modified specification requirements	
Item	**Limit**
Silicon dioxide (SiO_2) plus aluminum oxide (Al_2O_3) plus iron oxide (Fe_2O_3)	Min 85%
Available alkalies	Max 1.0%
Loss on ignition	Max 3.0%
Fineness: amount retained when wet-sieved on 45-μm sieve	Max 1.0%
Strength activity index at 7 days (% of control)	85
Increase of drying shrinkage of mortar bars at 28 days	Max 0.03%

Table 4.4 Mineralogical analysis of metakaolins [8]

	Kaolinite	Alunite	Quartz	Illite
MKC	96	–	–	3
MK	56	5	41	–

Table 4.5 Mineralogical phases of metakaolin sands [31]

Metakaolin sand	Metakaolin (wt.%)	Quartz (wt. %)	Illite (wt. %)	Albite feldspar (wt. %)	Muscovite (wt. %)
MK-1	36.00	21.85	6.93	4.23	31.00
MK-2	31.50	31.10	9.69	4.62	23.01
MK-3	40.00	18.11	6.18	3.82	31.90

Janotka et al. [31] reported the mineralogical phases (Table 4.5) of Slovak poor metakaolin sands with different metakaolin content. Contents of metakaolin varied between 31.50 and 40 wt. %, the rest was mainly sand.

4.3 Hydration Reaction

In Portland cement concrete, MK reacts at normal temperatures with calcium hydroxide in cement paste to form mainly calcium silicate hydrates (C–S–H), C_2ASH_8 (gehlenite hydrate), and C_4AH_{13} (tetracalcium aluminate hydrate). The formation of secondary C–S–H by this reaction reduces total porosity and refines the pore structure, improving the strength and impermeability of the cementitious matrix.

The optimum replacement percentage of cement with MK is associated with the changes in the nature and proportion of the different reaction products, temperature and reaction time. Hydration reaction depends upon the level of reactivity of MK in terms of the processing conditions and purity of feed clay. The feed clay (kaolin) should be either naturally pure or refined by standard mineral processing techniques; otherwise the impurities would act as diluents [41]. Reactivity level of MK can be determined by Chapelle test [4, 41]. Table 4.6 presents the comparison of reactivity level of MK with silica fume and fly ash. It is expressed as consumption rate of calcium hydrate per gram of pozzolans. The amount of CH in hardened concrete can be determined by thermogravimetric analysis (TG) and differential thermal analysis (DTA).

Kostuch et al. [41] observed that (1) CH was significantly reduced with age for all replacement levels (0, 10, and 20%); and (2) 20% by MK was required to fully

Table 4.6 Pozzolanic activity of various pozzolans [4]

Pozzolan	SF	FA	MK
Reactivity (mg) Ca $(OH)_2$/g pozzolan	427	875	1,050

remove all the CH in concrete at 28 days where as Oriol and Pera [48] reported that between 30 and 40% MK is required to remove all the CH in MK-PC paste at a water-binder ratio of 0.5 when cured in lime-saturated water for 28 days.

Ambroise et al. [2] studied the early hydration period of pastes containing metakaolin using isothermal calorimetry and conductivity. Differential thermal analysis, X-ray diffraction, and Fourier transform infrared spectrometry were used to estimate the consumption of calcium hydroxide (CH) and identification of reaction products. The results showed that CH was quickly consumed, the microstructure was rich in CSH and stratlingite (C_2ASH_8), and the pore size distribution displaced toward smaller values. At up to 30% replacement, MK acted as an accelerating agent, the pore size distribution was displaced toward small values, and the CH content was considerably reduced.

Curcio et al. [22] examined the role of metakaolin as pozzolanic micro-filler in high-performance mortars containing 15% metakaolin as replacement of cement. Water/binder ratio was 0.33. Four commercially available MK (M1, M2, M3, and M4) samples were used. Major chemical composition (SiO_2 and Al_2O_3) of M1, M2, M3, and M4 metakaolin were; 51.89 and 43.53%; 51.42 and 44.81%; 55.0 and 40.0%; 53.0 and 44.0%; respectively. Silica fume had 96% SiO_2. Figure 4.4 shows the height of the DTA peak of CH, occurring around 485°C, as a function of the curing time up to the age of 90 days. DTA calcium hydroxide peak heights on mortars and the pozzolanic activity test on mixtures of 80% cement and 20% microfiller showed that three of the four metakaolins had a remarkable pozzolanic activity comparable to that of silica fume, even if slightly lower at longer ages. The amount of CH changes very little after 28 days for the samples with metakaolins, slightly more for that with SF. MK1,MK2, and MK3 have a remarkable pozzolanic activity, comparable to that of SF, even if slightly lower at longer ages.

Klimesch and Ray [40] studied the effect of metakaolin on binding material and physical properties of cement–quartz pastes autoclaved at 180°C. Control cement–quartz mixture was prepared using 61.5% OPC and 38.5% quartz with the bulk C/S molar ratio (0.83) equivalent to that of $C_5S_6H_5$. MK contents were 6, 12, 18.25, 24,

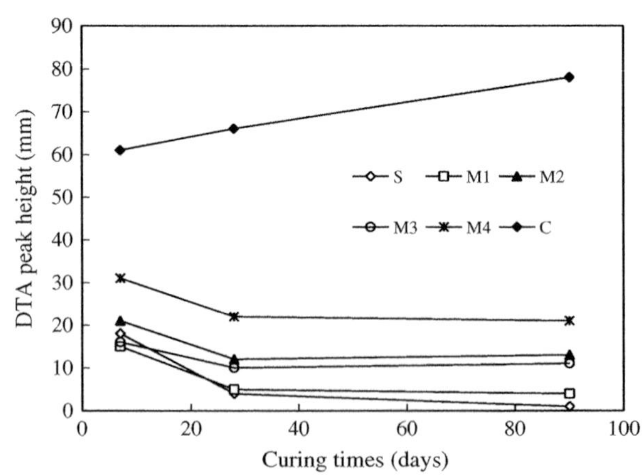

Fig. 4.4 DTA $Ca(OH)_2$ peak height versus curing time [22]

and 30.5% as a quartz replacement in one series and as a cement replacement in another series. Water-to-total solids ratio of 0.46 was maintained in all mixtures. It was concluded that (1) for MK content up to 12%, the amount of tobermorite formation decreased when cement was partially replaced, whereas the opposite effect was observed in the case of partial replacement of quartz; (2) an overall reduction in compressive strength and drying shrinkage occurred with MK additions. In autoclaved systems, MK provided a source of silica that was more reactive than ground quartz; (3) when added as a quartz replacement, the amount of tobermorite decreased while hydrogarnet (C_3ASH_6) increased with the increase in MK content, indicating that hydrogarnet may prevail upon tobermorite by competing with it for Al_2O_3 and possibly SiO_2. However, when MK was used as a cement replacement, an increase in the tobermorite content was observed up to 12% addition followed by a sharp decrease at greater addition; and (4) MK addition, as either quartz or a cement replacement, also indicated that un-reacted MK will be present if the bulk Ca/Al molar ratio is less than 3. The decrease in these physical properties can be attributed mainly to increasing amounts of both hydrogarnet and α-C_2SH, and a general decrease in the C–S–H type binding phases.

Frías and Cabrera [24] studied the hydration rate of cement pastes containing 0, 10, 15, 20 and 25% of MK as partial replacement of cement. The water/binder ratio was 0.55 by weight. To determine the degree of hydration, it is necessary to know the total amount of calcium present in the hydrated MK/OPC systems. The total calcium hydroxide was calculated taking into account the weight loss due to the dehydroxylation of $Ca(OH)_2$ and the decomposition of $CaCO_3$. Figure 4.5 shows the percentage of calcium hydroxide with the curing time up to 360 days. It was observed that (1) calcium hydroxide contents of MK/OPC samples increased with age until 3 ± 7 days. Subsequently, the values start decreasing, more or less depending on MK content; (2) in 10 and 15% of MK mixtures, an inflexion point at 56 and 90 days, respectively was observed. Beyond this point, the calcium

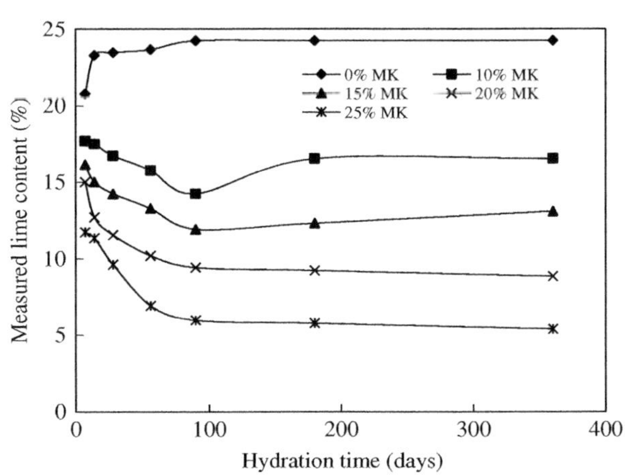

Fig. 4.5 Evolution of calcium hydroxide with time [24]

hydroxide content achieved a progressive increase. This behavior of calcium hydroxide content was due to different hydration mechanisms: the increase in the $Ca(OH)_2$ amounts are due to the OPC hydration, while the decrease in the values is related with the pozzolanic reaction of MK. The inflexion points (for 10 and 15% of MK) represent the end of the pozzolanic reaction, due to the total consumption of MK.

Sha and Pereira [62] investigated the hydration and pozzolanic reactions in cement pastes containing metakaolin (MK) as partial replacement of cement using differential scanning calorimetry (DSC) and thermal analysis. 100-mm cement paste cubes were made at water–solid ratio of 0.45 with 20 or 30% MK. The peak temperature and enthalpy are given in Table 4.7. They observed that pozzolanic effect of metakaolin even at this very early stage of hydration. The amount of calcium hydroxide drops by 25% when MK content increased from 20 to 30%. With this increase in MK content, the amount of OPC dropped by 14%. Also the crystallization peak increased by 24%, much less than expected from the 50% increase of the amount of MK.

Cabrera and Rojas [15] studied the reaction kinetics of a mixture of metakaolin and lime in water at 60°C using thermal analysis. Nature and quantity of the reaction products were determined by differential thermal analysis thermogravimetry and X-ray diffraction Based on the experimentation, they concluded that: (1) microwave drying to stop the hydration process was found to be a very practical and useful method that does not alter the nature of the reaction products. A drying time of 9 min was found to be adequate to drive off the free water from the metakaolin–lime–water mixture; (2) lime was consumed at a very rapid rate in the initial period of reaction (up to 50 h). The reaction mechanism was consistent with diffusion control for the first 120 h; (3) the reaction products detected from 2 h up to 9 days indicated that C_2ASH_8 and C_4AH_{13} phases were meta-stable for up to 9 days at 60°C under the circumstances studied, notwithstanding the fact that hydrogarnet (C_3ASH_6) was also present from 30 h up to 9 days. This might be an indication that hydrogarnet is not formed from a transformation reaction, but results from direct reaction between metakaolin and lime.

Table 4.7 DSC peak temperatures and enthalpy in constant heating runs for samples hydrated for 24 h in air in DSC [62]

Peak	Metakaolin (20%)		Metakaolin (30%)	
	T (°C)	J/g	T (°C)	J/g
C–S–H	118	–	120	–
Fe ettringite	No peak	–	177	0.4
Ettringite	No peak	–	198	0.9
C_4AH_{13}	No peak	–	291	1.2
Solid solutions	No peak	–	393	0.3
CH	455	40	482	30
$CaCO_3$	705	24	737	1.8
Crystallisation	935	42	968	52

Frías and Cabrera [26] examined metakaolin/lime and MK-blended cement pastes samples stored and cured at 20°C and up to 360 days of hydration. The nature of the reaction products and their sequence of appearance were obtained from differential thermal analysis and X-ray diffraction techniques. The results showed a different behavior of the pozzolanic reaction in both pastes. XRD data revealed the absence of crystalline phases in MK-blended cements up to 360 days of curing. MK showed high pozzolanic activity, which provided the quick formation of CSH, C_2ASH_8 (stratlingite), and C_4AH_{13} in MK/lime systems. The CSH was detected at 2 days of hydration time, followed by C_2ASH_8 and C_4AH_{13}. These two phases were detected at the same time.

In an MK/cement system (up to 25% of MK), the development of the hydrated phases presented a different behavior to the MK/lime system. The C_2ASH_8 was detected between 3 and 7 days of curing and it depends on MK contents. Subsequently, the C_4AH_{13} was possibly formed between 180 and 360 days. This phase was only detected with high MK contents, mainly with 25% MK. Below 20% MK, this phase was not identified. In the XRD patterns, for the MK/lime systems, the C_2ASH_8 was only identified as a crystalline hydrated phase, but not the C_4AH_{13}. This fact indicated very low crystallinity of the C_4AH_{13} with respect to the C_2ASH_8 phase. In MK-blended cement, XRD showed that the C_2ASH_8 formed during the pozzolanic reaction is an amorphous compound and on prolonged hydration it was possible to identify a very weak peak with high MK contents. The pozzolanic reaction in a blended cement matrix is more complex that in an MK/lime matrix, because of the presence of different ions (CO_2^{2-}, SO_4^{2-}, alkalis, chlorides), which could alter the kinetics as well as the development and crystallinity of the hydrates phases. It was due to the changes in solubility and evolution of the concentrations of the majority compounds (CaO, Al_2O_3, and SiO_2). These findings indicated that there was no formation of the C_4AH_{13} up to 180 days and the formation of an amorphous C_2ASH_8.

Poon et al. [49] reported the degree of pozzolanic reactions of high-performance cement paste made with 5, 10, and 20% metakaolin. The results (Table 4.8) of the pozzolanic reaction of the pastes, presented as % value of the reacted pozzolan relative to the initial amount of the pozzolan in the paste. Based on the test results, it was concluded that (1) degree of pozzolanic reaction of MK at each age was higher at a replacement level of 5% than at the replacement levels of 10 and 20%. The higher rate of pozzolanic reaction in cement pastes with a lower replacement level can be attributed to the higher concentration of CH available for the pozzolan to react with; (1) at 3 days, the degree of pozzolanic reaction of MK in the pastes at 5 and 10% MK content was 20.6 and 15.3%, respectively. The higher initial

Table 4.8 Degree of reaction of blended cement pastes [49]

Mix	Degree of reaction of pozzolans (%)			
	3 days	7 days	28 days	90 days
5% MK	20.56	29.52	39.44	56.28
10% MK	15.34	26.98	36.28	50.96
20% MK	9.38	18.64	30.82	41.29

reactivity of MK can be attributed to its Al_2O_3 phases [22], which are involved in the formation of gehlenite (C_2ASH_8) and a small amount of crystalline C_4AH_{13} phase [47]; and (3) although the rate of MK reaction became slower after prolonged curing, there was still a considerable increase in the degree of pozzolanic reaction of MK from 28 to 90 days. For cement paste with 10% MK, the degree of pozzolanic reaction of MK was 36.3% at 28 days and 51.0% at 90 days. The reaction of MK was not completed at the age of 90 days and about half of the MK remained unreacted.

Sun et al. [64] studied the hydration process of K-PS geopolymer cement under an 80% RH environment using environmental scanning electron microscope (ESEM). The ESEM micrographs showed that metakaolin particles pack loosely at 10 min after mixing, resulting in the existence of many large voids. As hydration proceeded, a lot of gels were seen and gradually precipitated on the surfaces of these particles. At later stage, these particles were wrapped by thick gel layers and their interspaces were almost completely filled. The corresponding energy dispersive X-ray analysis (EDXA) results illustrated that the molar ratios of K/Al increased while Si/Al decreased with the development of hydration. As a result, the molar ratios of K/Al and Si/Al of hydration products at an age of 4 h amounted to 0.99 and 1.49, respectively, which were close to the theoretical values (K/Al = 1.0, Si/Al = 1.0 for K-PS geopolymer cement paste). In addition, well-developed crystals could not be found at any ages; instead, sponge like amorphous gels were always observed.

Badogiannis et al. [9] investigated the hydration rate of metakaolin pastes. The metakaolinite content in metakaolin (MK4, derived from poor Greek kaolins) was 49%, where as it was 95% in a commercial metakaolin (MKC) of high purity. Cement was replaced with 0, 10, and 20% metakaolin. The weight loss up to 550°C, which corresponds to the total water incorporated in the cement paste (total combined water), was determined. The $Ca(OH)_2$ content, which for PC is directly related to the hydration of silicate compounds, was also measured. When pozzolanic materials were combined with PC, the $Ca(OH)_2$ content (when compared to that of PC alone) also provided an indication of the pozzolanic reaction. The weight loss in the range 600–700°C, if any, corresponds to the decomposition of $CaCO_3$ and it has to be converted to the equivalent $Ca(OH)_2$. The water combined in the hydration products (other than calcium hydroxide) corresponds to the weight loss up to 300°C. Any changes of this value indicated that the quantity, type and relative proportions of the hydration products are changing.

Figure 4.6 shows the calcium hydroxide content, total combined water and water in the hydration products at the age of 7 days. There was significant decrease in calcium hydroxide content in samples containing metakaolin in comparison to PC because of pozzolanic reaction. Also the pozzolanic reaction was more rapid in samples containing the commercial metakaolin MKC. This phenomenon is related to the higher fineness of MKC as well as to its higher metakaolinite content (95%). In contrast, pastes containing MK4 exhibited higher values of total combined water and water in the hydration products than the MKC pastes, probably due to the different hydration products.

Fig. 4.6 Calcium hydroxide, total combined water and water in the hydration products for a hydration age of 7 days [9]

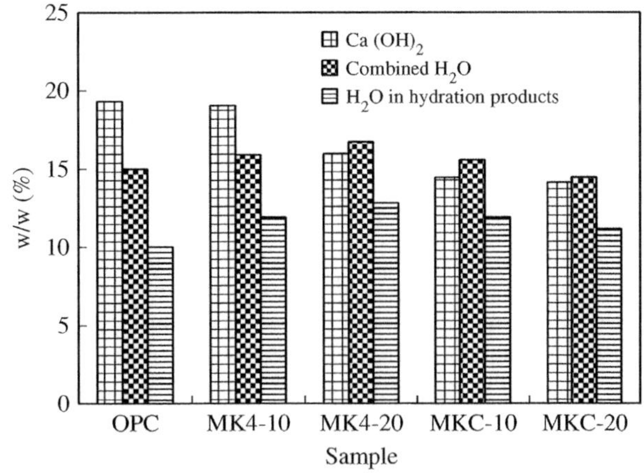

Figure 4.7 shows the calcium hydroxide content, total combined water and water in the hydration products for the sample MK4-20 in relation to the hydration age. A steep decrease of $Ca(OH)_2$ content was observed between 7 and 28 days, due to the acceleration of the pozzolanic reaction. At the same period, the change of the water combined in the hydration products (other than calcium hydroxide) could be attributed to the change of the quantity, type and relative proportions of the hydration products.

Wong and Razak [70] suggested an approach for the evaluation of efficiency factor (k) of a pozzolanic material. The cementing efficiency factor (k) of a pozzolan is defined as the number of parts of cement in a concrete mixture that could be replaced by one part of pozzolan without changing the property. The method, developed following Abram's strength—w/c ratio rule, calculates efficiency in terms of relative strength and cementitious materials content. The advantage of this method is that only two mixtures are required to determine the k-factor of a

Fig. 4.7 Calcium hydroxide, total compound water and water in the hydration products in relation to the hydration age for the sample MK4-20, 20% MK content [9]

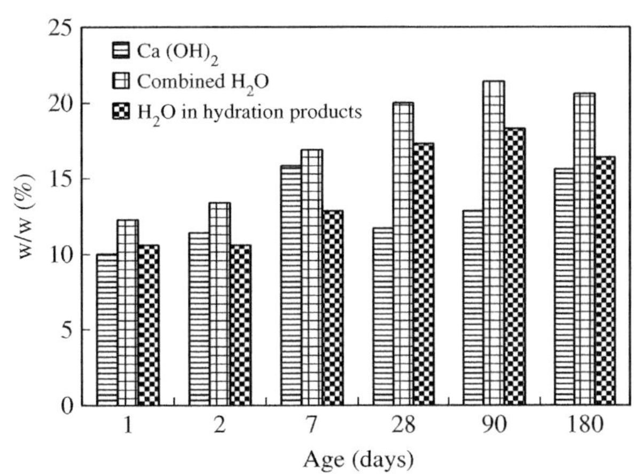

specific mixture. A laboratory investigation on metakaolin (MK) concrete found that the computed efficiency factors varied with replacement level and age. At 28 days, the *k* values ranged from 1.6 to 2.3 for MK, while at 180 days, the k values varied from 1.8 to 4.0 for MK. Generally, the *k* factors increased with age but declined with higher pozzolanic content. It was also observed that change in w/cm ratio from 0.33 to 0.27 did not significantly affect the resultant efficiency factors. Based on the obtained results, it was concluded that a single *k* value for a pozzolan does not exist, even for specific cement–pozzolan content.

Kim et al. [38] studied the XRD patterns measured at 28 days for the ordinary Portland cement and the pastes mixed with 0, 5, 10, 15 and 20% of metakaolin. The XRD patterns of the hydrate products in pastes are presented in Fig. 4.8. It can be seen that the crystalline phase of CH (portlandite) decreases as the replacement ratio of MK and SF increases, while the weak peaks of C–A–H, in pastes replaced by MK, slightly increase. The decrease of the CH peak is related with the consumption by pozzolanic reaction of MK. Further EDS analysis revealed the

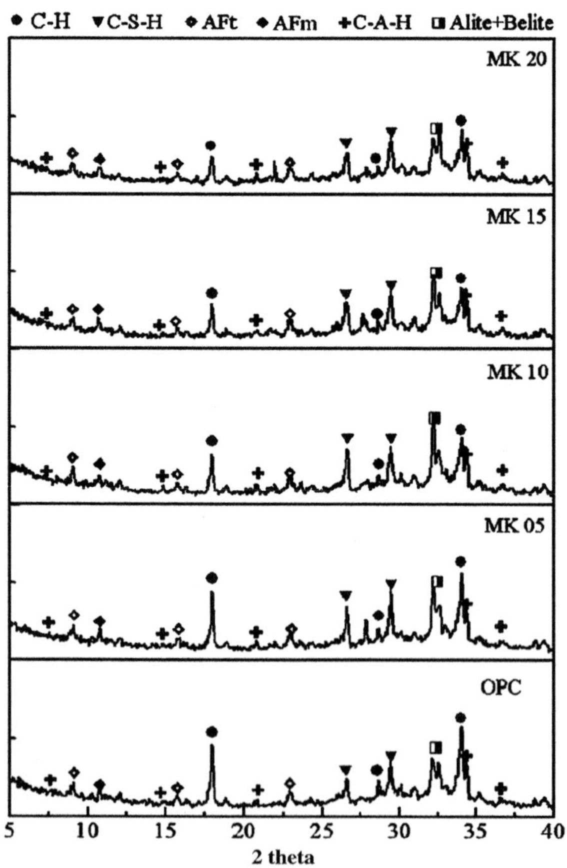

Fig. 4.8 XRD patterns of paste sample with replacement of metakaolin [38]

presence of C–S–H and CH as the main hydrated products in the case of ordinary Portland cement, and C–A–S–H and C–A–H were observed for MK paste.

Lagier and Kurtis [42] examined the influence of Portland cements of varying composition on early age reactions with metakaolin of varying surface area (11,100 and 254,000 m^2/kg). They observed that (1) metakaolins appeared to have a catalyzing effect on cement hydration, leading to an acceleration in the reaction rates, an increase in cumulative heat evolved during early hydration, and—for some cements—apparently an increased intensity in heat evolved during certain periods of early hydration. The surface area of the metakaolin also seemed to influence early hydration behaviour, with the higher surface area material producing a greater rate of heat evolution, greater cumulative heat, and greater intensities during early hydration; (2) strongly exothermic reactions appeared to occur between the cements and metakaolins, particularly in the first 24 h, and the reactions seemed to be most closely associated with the "third peak" observed in calorimetry—that which is most often related to the reaction of calcium aluminate phases; and (3) reaction of metakaolin appeared to be quite sensitive to variations in total alkali content in the cement.

Janotka et al. [31] investigated the hydration of cement paste containing Slovak poor metakaolin sands with different metakaolin content [36.0% (MK-1), 31.5 (MK-2) and 40.0% (MK-3)]. The percentage of metakaolin sands was 10, 20 and 40%. From the calorimetric results (Table 4.9), it was concluded that the addition of MK-1 and MK-2 sands to Portland cement induced a delay up to 2 h of the precipitation of the main hydration products in the blended-cement pastes and decreased the maximum heat evolution rate. On the contrary, the incorporation of 40% of MK-3 sand shortened 6 h its apparition and increased significantly the maximum heat evolution rate. The presence of the metakaolin sands reduced the heat released during the hydration process with respect to non-blended-cement pastes as a consequence of the precipitation of hydration products is less

Table 4.9 Calorimetric data for the Portland cement and metakaolin-blended-cement pastes [31]

Cement system	Peak start time (h)	Peak max. time (h)	Max. rate (J/g h)	Peak end time (h)	Peak duration (h)	Peak heat released (J/g)	Total heat released (J/g)
PC reference	1.87	12.98	9.15	96.86	94.99	272.03	310.12
PCMK-1/10	2.11	13.27	8.61	90.49	88.38	239.98	266.15
PCMK-1/20	2.15	12.98	8.17	114.54	112.39	249.25	296.68
PCMK-1/40	2.29	13.64	5.71	142.44	104.15	211.66	235.09
PCMK-2/10	2.09	13.91	7.84	91.01	88.92	241.77	245.07
PCMK-2/20	2.03	14.20	6.69	108.62	106.59	229.36	226.24
PCMK-2/40	2.22	15.32	4.56	121.17	118.95	168.80	171.28
PCMK-3/10	2.01	15.05	7.26	137.39	135.38	273.16	276.47
PCMK-3/20	2.07	13.07	13.07	123.25	121.18	269.51	273.44
PCMK-3/40	2.85	6.90	38.28	97.96	95.11	195.44	202.56

exothermic in the former and also the lower amount of hydration products formed in metakaolin sands-blended-cement systems.

4.3.1 Temperature Effect

Oriol and Pera [48] investigated the effect of a microwave curing on lime consumption in metakaolin blended cements. The microwave treatment conditions were assessed for plain Portland cement pastes, and then the lime consumption for various metakaolin contents was evaluated by Fourier transform infra-red (FTIR) spectrometry and differential thermal analysis (DTA). The results were compared with those obtained at room temperature. It was concluded that pozzolanic reaction of metakaolin could be consequently promoted by microwave heating. As heat is generated quickly inside the cementitious material, thermal acceleration of the reaction was more efficient, allowing both reduction of the necessary amount of metakaolin (15% instead of 30–40% in normal conditions) and of the water: cement ratio (0.40 instead of 0.50).

Frías et al. [25] discussed the effect of metakaolin on the heat evolution in metakaolin–cement mortars in comparison to the behavior of other traditional pozzolanic materials such as fly ash and silica fume. The results revealed that MK mortars produced a slight increase in heating when compared to a 100% Portland cement mortar, due to the high pozzolanic activity of MK. With respect to the hydration heat, MK-blended mortar showed closer behavior to silica fume than to fly ash. The results obtained for pozzolanic activity are shown in Fig. 4.9. After 2 h, both the SF and MK showed pozzolanic activity, since the samples had fixed significant amounts of calcium ions (lime), while the FA, due to its lesser activity at early stages, hardly showed any reaction with lime before day 28.

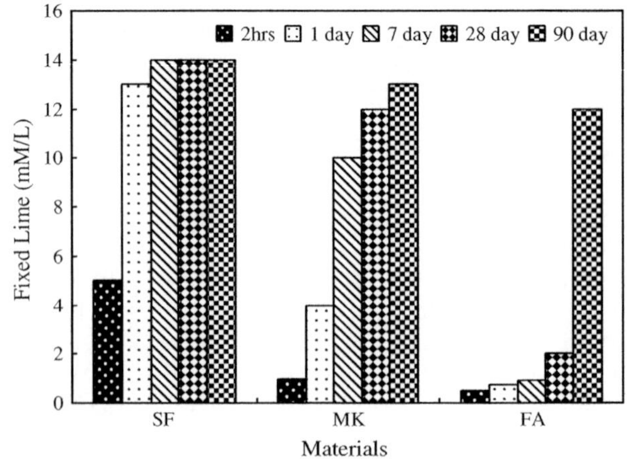

Fig. 4.9 Variation of Pozzolanic activity: fixed lime over time [25]

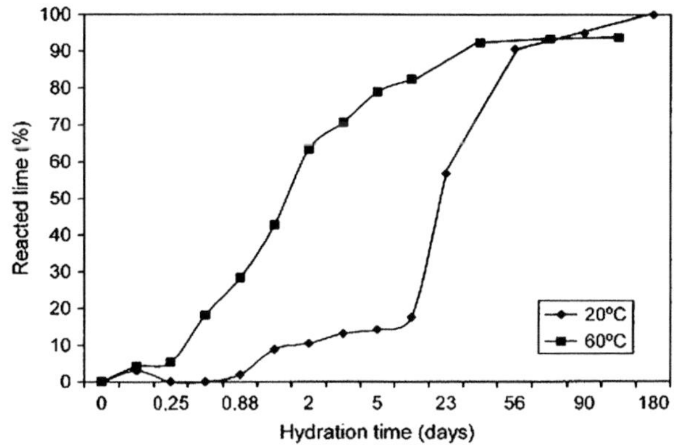

Fig. 4.10 Evolution of reacted lime with curing time at different temperatures [56]

Rojas [58] determined the effect of curing temperature on the reaction kinetics in a metakaolin/lime mixture cured at 60°C and after 60 months of hydration. The stability of hydrated phases formed during the pozzolanic reaction was evaluated. The results exhibited that metastable hexagonal phases (C_2ASH_8 and probably C_4AH_{13}) coexist with stable cubic phase (hydrogarnet) in the absence of lime. Also, there was evidence of the possible presence of a calcium aluminum silicate hydroxide hydrate (vertumnite). DTA curve showed the presence of CSH and the coexistence of metastable and stable phases in a MK/lime system cured at 60°C for 60 months of hydration. The metastable phase was assigned to stratlingite (C_2ASH_8) and probably to C_4AH_{13}. The stable phase was attributed to katoite, a hydrogarnet structured phase. XRD pattern confirmed the presence of stratlingite and katoite as main crystalline compounds present under tested conditions.

Rojas and Sánchez de Rojas [57] carried out an experimental study on kinetics of pozzolanic reaction in MK/lime binder as well as MK-blended cement at 60°C. The sequence, development and crystallinity of hydrated phases up to 123 days of curing time were studied by means of DTA and XRD techniques. It is well known that the pozzolanic reaction between metakaolin and calcium hydroxide produce CSH, stratlingite (C_2ASH_8), C_4AH_{13} and hydrogarnet (C_3ASH_6). However, the presence or absence of these hydrated phases depends on different parameters, such as curing temperature, matrix used, etc. They concluded that (1) the sequence and formation of the hydrated phases was different in both matrices cured at 60°C; (2) in MK/lime system (1:1 in weight), hydrated phases formed at high temperature were essentially CSH, C_2ASH_8, C_4AH_{13} and C_3ASH_6. CSH was detected at 6 h of hydration time, followed by C_2ASH_8 and C_4AH_{13} at 12 h of curing and finally C_3ASH_6 at 30 h of age; (3) in a MK-blended cement system (10, 20 and 25% of MK), the development of hydrated phases presented a different behavior from the MK/lime system. Stratlingite was detected in all cases as the sole reaction product from pozzolanic reaction. There was no evidence of formation of C_4AH_{13} and hydrogarnet (only traces); (4) XRD patterns for the MK/lime systems, stratlingite and hydrogarnet were identified as crystalline hydrated phases but not C_4AH_{13}; (5) in MK-blended cement at 60°C, XRD showed important changes with respect to the crystallinity of hydrated phases. Stralingite did not appear as a crystalline phase, indicating a very low crystallinity of this compound in blended cements. At 20 and 25% of MK, the presence of hydrogarnet traces was observed.

Rojas and Cabrera [56] studied the influence of curing temperature on the kinetics of reaction of a metakaolin ()/lime mixture. Chemical composition of MK was 51.5% of SiO_2, 41.3% of Al_2O_3, 4.64% of Fe_2O_3, 0.1% of CaO, 0.16% of MgO, 0.63% of alkalis and 0.83% of TiO_2. MK and analytical grade $Ca(OH)_2$ were mixed in a ratio of 1:1 by weight with a water/binder ratio of 2.37. Specimens were cured at 20 and 60°C. In the first case, the curing time varied from 2 h up to 180 days and, in the second case, from 2 h up to 123 days. A mathematical model was applied to calculate the rate constant for the hydration reaction. The amount of $Ca(OH)_2$ in cementitious materials can be measured quantitatively by thermal analysis. The results of lime consumption versus time are presented in Fig. 4.10. A different trend of decreasing lime content as the temperature increased was observed. In the samples cured at 60°C, lime was rapidly consumed during the first 120 h. Between 5 and 9 days, the amount of transformed lime at 60°C was approximately 65% higher than that at 20°C. Total lime consumption was 63, 82 and 92% at 2, 9 and 34 days, respectively. Beyond 34 days (between 34 and 123 days) of curing time, there is only an additional 1% of lime reacted. The curve corresponding to the samples cured at 20°C shows that lime is consumed at a slower rate than that at 60°C. The total consumption of lime was 10, 17 and 90% at 2, 9 and 56 days, respectively. However, at 180 days of curing time, lime had totally reacted by pozzolanic reaction. These results revealed an important increase in the activity of MK between 7 and 56 days at 20°C, while in the same samples cured at 60°C, the interval of maximum pozzolanic Activity was detected between 6 h and 2 days.

4.3.2 Effect of Dehydroxylation

Salvador [61] used flash-calcination processes to reduce the calcination time to a few seconds on fine products. Flash-calcination enables the dehydroxylation of powdered kaolinite clay within several tenths of a second, when traditional soak-calcinations require minutes at least. The pozzolanic properties of the metakaolin produced from two different kaolinites, using two different flash calciners increased with the dehydroxylation rate, and rapidly decreased occurrence of recrystallization for temperatures above 900°C. Flash-calcined products revealed structural properties different from soak-calcined products. Two different lime reactivity tests were done to asses the pozzolanic properties of products: the compressive strength and Chapelle test. Processing flash-calcination in a temperature range with a sufficient residence time, led to metakaolin with lime reactivity similar or better than reactivity of standard metakaolin obtained by soak-calcination. By measuring compressive strength, they verified that, after a flash-calcination, the quality of the pozzolana increased with α (degree of dehydroxylation) values. During the few seconds or tenths of seconds of flash-calcination, temperatures above 950°C induced recrystallization from the amorphous phase. Recrystallization was accompanied, as with a soak-calcination, by a rapid drop of lime reactivity. This performance drop was correlated with the decrease of the exothermic band surface area on DTA traces.

Shvarzman et al. [63] investigated the effect of heat treatment parameters on the dehydroxylation/amorphization process of the kaolinite-based materials such as natural and artificial kaolin clays with different amounts of amorphous phase (metakaolin). The process of dehydroxylation/amorphization of kaolinite was characterized by DTA/TGA with mass-spectrometry and X-ray powder diffraction. Based on the study, they concluded that (1) at calcinations temperatures below 450°C, kaolin clays showed relatively low level of the dehydroxylation degree, less then 0.18. In the range from 450 to 570°C, the degree of dehydroxylation sharply increased to 0.95, and finally at the temperatures between 570 and 700°C the kaolinite was fully dehydroxylated since the only moderate change of degree of dehydroxylation was observed in this range (from 0.95 to 1.0); (2) the dehydroxylation was accompanied by kaolinite amorphization, which affected the activity of additives; (3) pastes containing less than 20% of amorphous phase can be considered as inert materials from the standpoint of pozzolanic activity. It was shown that chemical activity is a linear function of amorphous phase content (APC) in its range of 50–100%; (4) activity strength index (ASI) of mature mixes (7, 28 and 90 days) depends significantly on APC, as well as the degree of dehydroxylation. In contrast to the chemical activity, the increase of APC by over 55% did not lead to additional growth in ASI. Therefore, even with the partial dehydroxylation of kaolinite accompanied with ~55% amorphization the material may be considered as very active pozzolanic admixture (according to ASTM 618).

4.4 Fresh Properties of Mortar/Concrete Containing Metakaolin

Caldarone et al. [18] observed that although the slump of concrete containing 10% MK was reduced from that of concrete with Portland cement only, the concrete containing MK required 25–35% less high-range water-reducers (HRWR) than equivalent SF mixtures. This reduction in HRWR demand resulted in the concrete containing MK having less sticky consistency and better finish.

Wild et al. [67] reported the results of workability tests of metakaolin concrete. Portland cement was partially replaced with 0, 5, 10, 15, 20, 25, and 30% metakaolin Control concrete mixture proportion was 1:2.3:3.4 with water–binder ratio (w/b) of 0.45. Workability results are given in Table 4.10.

Brooks and Johari [14] reported decrease in slump values and increase in both initial and final setting times with the increase in MK content in concrete with mixture proportion of 1:1.5:2.5 with w/b ratio of 0.28 (Table 4.11).

Table 4.10 Workability of metakaolin concretes [67]

Metakaolin (MK) (%)	Superplasticizer (%)	Slump (mm)	Compacting factor	Vebe time (s)
0	0	5	0.81	26
5	0.6	10	0.84	15
10	1.2	15	0.88	10
15	1.8	25	0.89	9
20	2.4	75	0.89	7
25	3.0	75	0.89	4
30	3.6	90	0.90	5

Table 4.11 Workability, setting times of MK concretes [14]

Concrete mixes	Slump (mm)	Initial setting time (h)	Final setting time (h)
OPC	100	5	7.7
MK5	30	6.42	8.82
MK10	20	6.98	9.42
MK15	5	6.45	9.31

Qian and Li [53] studied the dosage of superplasticizers on the slump of high-performance concrete made with 0, 5, 10, and 15% metakaolin as partial replacement of cement (Fig. 4.11). It was observed that (1) increase in SP content increased the slump; and (2) slump of the HPC decreased with the increase in MK content.

Li and Ding [45] investigated the consistency and setting times of Portland cement containing metakaolin and ultra-fine slag (S). Four series of pastes were made. Test results of water consistency and setting times are given in Table 4.12. It was observed that (1) standard consistency of cement paste was in the range of 26.3–29.1%. Mixture PO had the lowest water requirement, but MK blended cement M1 had the highest water requirement. After slag was mixed with MK blended cement, the water requirement of cement mixture S2 and S3 was reduced,

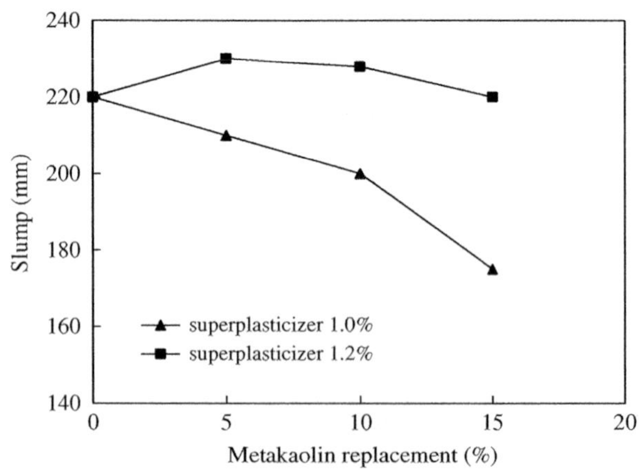

Fig. 4.11 Relationships of slump with metakaolin replacement [53]

Table 4.12 Physical properties of cement paste [45]

Mixture type	Composition (%)			Water required for standard consistency (%)	Setting times (h:min)	
	Cement	MK	Slag		Initial	Final
PO	100	0	0	26.3	1:36	3:27
M1	90	10	0	29.1	1:11	2:12
S2	70	10	20	28.4	1:55	4:34
S3	60	10	30	28.0	2:19	6:15

Table 4.13 Physical properties of blended cements with metakaolin [11]

sample	Water demand	Setting time (min)	
		Initial	Final
PC	27.5	105	140
MK-10	32.5	155	180
MK-20	41.0	205	230
MKC-20	37.5	140	170

MKC commercial metakaolin with high purity, *MK* metakaolin produced from poor Greek Kaolin

but it was still higher than that of PC; (2) initial and final setting times of the cement incorporating 10% MK were shorter than those of PC, while the initial and final times of S2 and S3 cements were longer than those of PC.

Batis et al. [11] concluded that increase in metakaolin content increased the water demand and setting times of blended cements (Table 4.13). The blended cements demand significantly more water than the relatively pure cement and this phenomenon is attributed to the high fineness of metakaolin. The increase of the metakaolin content causes a significant increase of the water demand. PC with 10% MK (MK-10) showed the lowest water demand, compared with the other blended cements. The initial and final setting time of metakaolin cements is higher than the setting time of pure cement.

Badogiannis et al. [9] investigated the effects of five metakaolins (derived from poor Greek kaolins) and a commercial metakaolin (MKC) on the water demand and setting times of cements (Table 4.14). The metakaolinite contents in metakaolins (MK1, MK2, MK3, and MK4) were 36, 37, 71, and 49%, respectively. Commercial metakaolin (MKC) was of high purity (95% metakaolinite). Cement was replaced with 0, 10, and 20% metakaolin. It was concluded that (1) blended cements demanded significantly more water than the relatively pure cement. The increase in water demand could be attributed to the high fineness of metakaolin as well as to their narrow particle size distribution; and (2) the initial and final setting time of metakaolin cements was affected by the metakaolin content. Cements with 10% metakaolin, generally, exhibited similar setting times to that of PC, while for 20% metakaolin content there was a delay in the setting. MK4 showed the greatest effect on the setting delay of the cements.

Kim et al. [38] investigated the effect of metakaolin (5, 10, and 15%) as partial replacement of cement on the flow rate of mortar. It was observed that flow values of mortar mixtures decreased with the increase in MK content. There was

Table 4.14 Properties of metakaolin cements [9]

Sample	Metakaolin (% w/w)	Water demand (% w/w)	Setting time (min)	
			Initial	Final
PC	–	27.5	105	140
MK1-10	10	29.0	75	130
MK2-10	10	29.0	85	130
MK3-10	10	32.0	105	160
MK4-10	10	32.5	155	180
MKC-10	10	31.0	95	130
MK1-20	20	32.0	105	160
MK2-20	20	31.5	110	165
MK3-20	20	38.5	120	160
MK4-20	20	41.0	205	230
MKC-20	20	37.5	140	170

approximately 8, 12, and 24% reduction in the flow value of mortars in comparison with that of the control mortar (0% MK) The results demonstrated the higher water demand with the increase in MK content.

Khatib [37] reported the workability of concrete containing metakaolin (MK) at a low water-to-binder ratio of 0.3. Portland cement (PC) was partially replaced with 0–20% MK. Results of slump, vebe time and compacting factor are given in Table 4.15. All mixes including the control (i.e. 0% MK) exhibited very low workability. Slump values were below 20 mm. Also Vebe times and compaction factors were more than 24 s and less than 0.8, respectively. The presence of MK reduces the workability.

Cachim et al. [17] examined the effect of metakaolin addition on the slump and density of hydraulic lime. Weight of hydraulic lime concrete ingredients were 1,074 kg coarse aggregates, 321 kg and, 550 kg of hydraulic lime, and 247.5 kg water. Two mixtures had a partial replacement of hydraulic lime by metakaolin. This replacement, by weight, was 20 and 30%. It was observed that slump values were 15, 8, and 16 mm for 0, 20 and 30% MK content whereas density of fresh concrete was 2,240, 2,220 and 2,160 kg/m^3.

Janotka et al. [31] investigated the effect of Slovak poor metakaolin sands with different metakaolin content [36.0% (MK-1), 31.5 (MK-2) and 40.0% (MK-3)] on the consistency and setting times of cement paste. The percentage of metakaolin sands in the blended cements was 10, 20 and 40%. Table 4.16 shows the consistency and initial and final setting times of Portland cement and metakaolin

Table 4.15 Workability of concrete mixes containing MK [37]

MK (%)	Slump (mm)	Vebe (s)	Compacting factor
0	17	24	0.87
5	12	25	0.86
7.5	8	27	0.83
12.5	3	30	0.81
15	2	36	0.80
20	0	37	0.78

Table 4.16 Properties of mortar containing metakaolin [31]

Cement system	Specific surface (m^2/kg)	Normal consistency	Setting times	
			Initial set	Final set
PC reference	428.27	29.6	3 h 25 min	4 h 40 min
PCMK-1/10	409.68	32.3	3 h 35 min	4 h 40 min
PCMK-1/20	415.97	38.0	3 h 50 min	4 h 55 min
PCMK-1/40	505.21	41.7	3 h 40 min	5 h 40 min
PCMK-2/10	401.58	30.2	3 h 25 min	4 h 30 min
PCMK-2/20	401.97	30.5	3 h 35 min	4 h 35 min
PCMK-2/40	375.91	32.5	4 h 10 min	5 h 40 min
PCMK-3/10	521.89	32.2	3 h 10 min	4 h 15 min
PCMK-3/20	605.21	36.2	3 h 05 min	4 h 05 min
PCMK-3/40	855.37	42.0	3 h 00 min	4 h 10 min

sand-blended-cement pastes. It was concluded that (1) normal consistency increased with the metakaolin content for all the three types; and (2) the incorporation of metakaolin sands MK-1 and MK-2 induced a delay of the setting times whereas MK-3 sand slightly accelerated setting times. This may be because of higher reactivity of PCMK-3 which is due to the higher specific surface and higher pozzolanic activity of MK-3 sand with respect to the other metakaolin sands.

4.5 Properties of Hardened Mortar/Concrete Containing Metakaolin

4.5.1 Pore Size Distribution

Bredy et al. [12] conducted study on porosity and pore size distribution of pastes containing MK. Pastes were made with MK content of 0–50% at different water-to-binder (w/b) ratios to maintain the same consistency. They concluded that total porosity of paste decreased when the MK content was less than 20%. Beyond 30%, an increase in porosity was found. Larbi and Bijen [43] reported that at 100 days of curing, inclusion of MK in mortar decreased pore volume and threshold diameter.

Khatib and Wild [34] reported the porosity and pore size distribution of OPC-metakaolin (0, 5, 10 and 15%) paste prepared at a constant water/binder (w/b) ratio of 0.55. The intruded pore volume and the pore structure were measured by mercury intrusion proximity. Pore volume of pastes containing metakaolin (MK) and proportion of pores with radii smaller than 20 μm are given in Table 4.17. For the control paste, there was a continuous reduction in total pore volume with age and also the rate of reduction declined with age. It can be seen from the table that total pore volume decreased with an increasing rate up to 14 days. Between 14 and 28 days, pore-volume increased. Beyond 28 days, pore volume again continued to decrease with age, but at a decreasing rate. This trend indicated that there was a

Table 4.17 Pore volume and % of pores of pastes containing MK [34]

Age (days)	Pore volume (mm^3/g)				Percentage of small pores (radii <20 μm)			
	Metakaolin (MK) (%)				Metakaolin (MK) (%)			
	0	5	10	15	0	5	10	15
3	262.0	257.6	284.1	277.6	22.2	28.3	31.0	39.9
7	229.6	261.7	268.8	251.6	26.5	32.1	41.0	50.4
14	209.9	203.4	221.0	212.1	30.3	43.0	53.9	55.7
28	189.1	205.3	237.1	222.7	33.7	43.5	48.7	54.9
90	181.4	180.8	219.6	198.9	37.3	44.7	49.9	57.6

clearly defined minimum point at around 14 days for pastes containing metakaolin. Also these pastes possessed a greater total pore volume than the control paste. Thus, the increase in pore volume between 14 and 28 days appeared to be associated with a renewed increase in CH content and a decline in relative strength. The proportion of pores with radii smaller than 20 μm was increased with the increase in metakaolin content.

Threshold value for paste decreased with the increase in metakaolin content. Figure 4.12 shows the threshold radius for pastes with and without metakaolin at different curing periods. The threshold radius is taken as the radius before which the cumulative pore volume rises sharply. Incorporation of metakaolin in cement paste led to refinement of the pore structure. The threshold value for paste decreased as the metakaolin content in the paste increased.

Ambroise et al. [2] determined the total porosity of blended cement with hydration time. Mortars mixtures (binder: sand = 1:3) were prepared: one with plain OPC, and the others with MK blended cements. The amounts of MK were 10, 20, and 30%, respectively. Total porosity of blended cements is given in Table 4.18. It was observed that porosity increased with the increase in MK content; (2) porosity is not proportional to the water demand: in spite of higher water demand (0.34 instead of 0.25), the blended cement containing 20% MK had nearly the same porosity as plain OPC, due to the efficiency of the pozzolanic reaction.

Frías and Cabrera [24] investigated the effect of hydration time (up to 360 days) on the average pore diameter of cement pastes containing 0, 10, 15, 20 and 25% of MK as partial replacement of cement. The water/binder was 0.55 by weight. To determine the degree of hydration, it is necessary to know the total amount of calcium present in the hydrated MK/OPC systems. Average pore size with hydration time is shown in Fig. 4.13. It was observed that (1) incorporation of MK reduced the pore size to less 100 Å. Gradual increase in the very fine pores was observed with age; (2) a strong reduction on the average pore diameter was observed in the first days. Depending on MK contents, the evolution of this parameter was different. In the case of OPC paste, the average pore diameter decreased from 40 to 14.5 μm for a hydration time between 1 and 56 days. Subsequently, the average pore diameter was almost a constant value. In paste with 10% of MK, the reduction of average pore diameter finished at 28 days while for

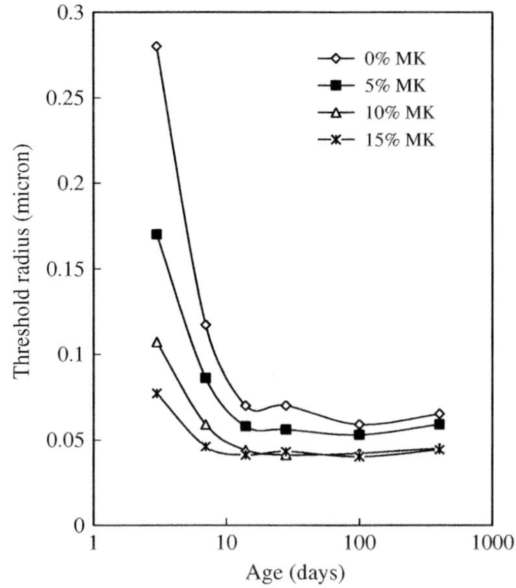

Fig. 4.12 Variation of threshold radius with age and varying MK content [34]

Table 4.18 Total porosity of MK blended cements (%) [2]

Mix No.	OPC: MK (%)	Water:solid ratio (W:S)	Curing time (days)			
			28	90	180	360
1	100:0	0.25	14.4	12.3	12.8	11.0
2	90:10	0.28	15.3	14.3	14.2	14.1
3	80:20	0.34	13.8	13.0	12.9	12.6
4	70:30	0.39	16.8	15.7	15.2	15.0
5	60:40	0.44	25.0	21.6	16.8	16.4
6	50:50	0.54	29.5	29.0	28.0	28.0

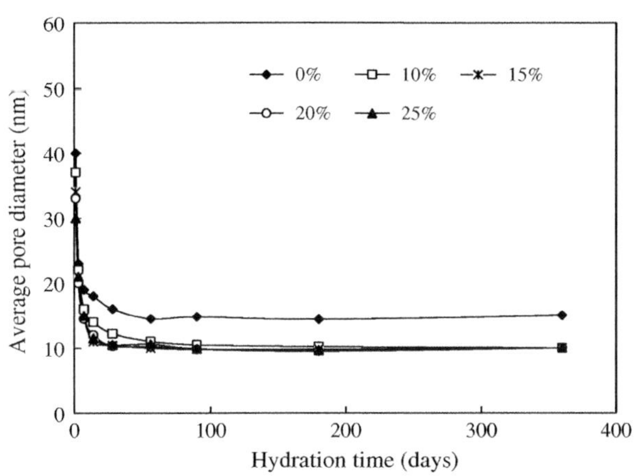

Fig. 4.13 Average pore diameter [24]

the 15, 20, and 25% MK pastes, this phenomenon occurred at 14 days. Above 10% of MK, the average pore diameter curves did not show any difference.

Poon et al. [49] investigated the porosity and pore size distribution of high-performance cement paste blended with metakaolin. The cement pastes were prepared with 5, 10, and 20% MK contents at w/b ratio of 0.3. Pore size and porosity were measured by Mercury Intrusion Porosemetry (MIP), and results are given in Table 4.19. Porosity and average pore size of pastes decreased with age. The pastes containing MK had lower porosity and smaller average pore diameters than the control paste.

Poon et al. [51] measured the total porosity of concrete mixes containing 5, 10, and 20% metakaolin as partial replacement of cement. Concrete mixes were made with w/b ratio of 0.3. The maximum nominal size of the coarse aggregate was 10 mm for the concrete with a w/b of 0.3, and 20 mm for the concrete with a w/b of 0.5. Porosity was measured with MIP, and results are given in Table 4.20. It was observed that the incorporation of MK in the cement pastes resulted in a very dense microstructure of the paste, with a lower total porosity in comparison with the plain Portland cement pastes

Badogiannis and Tsivilis [10] investigated the porosity of concretes made with Greek kaolin of low kaolinite content (49% metakaolinite, 72.1% SiO_2, 30% active SiO_2) and commercial metakaolin (95% metakaolinite, 54.6% SiO_2, 53% active SiO_2) of high purity. Concrete mixtures were made in which metakaolin replaced either cement or sand in percentages 10 or 20% by weight of cement.

Table 4.19 Average pore diameter and total porosity of blended cement pastes [49]

Mix	Average pore diameter (μm)			
	3 days	7 days	28 days	90 days
Control	0.0380	0.0371	0.0362	0.0348
5% MK	0.0357	0.0279	0.0257	0.0243
10% MK	0.0287	0.0251	0.0197	0.0186
20% MK	0.0204	0.0143	0.0122	0.0114
Total porosity of blended cement pastes				
Control	20.11	17.99	15.58	14.04
5% MK	18.17	15.36	13.82	12.51
10% MK	16.84	15.18	12.37	11.68
20% MK	16.30	12.85	10.73	9.21

Table 4.20 Total porosity of concrete mixes with w/b ratio of 0.3 [51]

Mix	MIP measured porosity (% v/v)			
	3 days	7 days	28 days	90 days
Control	8.69 ± 0.11	8.44 ± 0.13	7.92 ± 0.12	6.97 ± 0.28
5% MK	7.22 ± 0.13	7.01 ± 0.15	6.40 ± 0.10	NA
10% MK	6.87 ± 0.14	5.38 ± 0.12	4.75 ± 0.09	4.48[a]
20% MK	6.59 ± 0.08	5.32 ± 0.10	4.66 ± 0.12	NA

NA not available

[a] Only one measurement

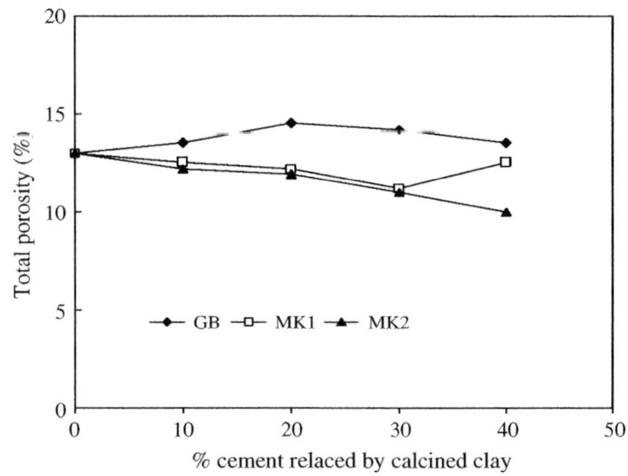

Fig. 4.14 Influence of proportion of cement replaced by each calcined-clay on total porosity [28]

The addition of metakaolin caused reduction in concrete total porosity. The total porosity of metakaolin concrete varied from 7.2 to 11.2%, while control concrete presented a total porosity of 11.1%.

Gonçalves et al. [28] determined the total porosity of mortars prepared with two metakaolin samples and ground calcined-clay brick. Three types of calcined-clays (1) ground calcined-clay brick (GB); (2) commercial metakaolin (MK1) and (3) a calcined-clay produced in the laboratory Metakaolin (MK2). Major constituents of ground calcined-clay brick were SiO_2 (63.89%), Al_2O_3 (25.49%), Fe_2O_3 (7.73%). Major constituents of commercial metakaolin (MK1) were SiO_2 (51.2%), Al_2O_3 (35.3%), Fe_2O_3 (4.0%) whereas laboratory produced metakaolin (MK2) had SiO_2 (52.46%), Al_2O_3 (44.24%), Fe_2O_3 (2.06%). Mortar mixtures 1:1.5, with water/cementitious material ratio of 0.5, were prepared with 10, 20, 30 and 40% of cement was replaced by calcined-clay samples (GB, MK1 or MK2). Total porosity results are shown in Fig. 4.14. It was observed that (1) for all percentages of replacement, the mortar porosity was greater than that of the control sample; and (2) replacement of up to 40% of cement by either one of the metakaolin samples (MK1 or MK2) resulted in a similar continuous reduction in total porosity. This may be attributed, at least in part, to improved packing of the mixtures. Distinct behaviors were found for replacements of 40% cement by the metakaolin samples: whereas replacement by MK1 resulted in a further reduction in porosity, the same was not observed when MK2 was added.

4.5.2 Water Absorption and Sorptivity

Khatib and Mangat [33] demonstrated that sorptivity of concrete; taken from the top surface of a concrete cube can be several times greater than those for concrete taken from the bottom surface. Dias [23] observed that air-cured concrete, over a 4-year period, undergoes an increase in weight and a reduction in sorptivity as a result of carbonation of the surface zone. Also, the greater the initial (6 months) sorptivity, the greater was the reduction in sorptivity (at 4 years) due to carbonation. Equivalent water-cured concrete did not show any drop in sorptivity from carbonation, although sorptivity did decrease with time as a result of continued hydration.

Courard et al. [21] studied the water absorption of mortars containing 0, 10, 15, and 20% metakaolin as partial replacement of cement. One mixture with natural kaolin was also made. The specimens were made with 1:3 with water-cementitious ratio 0.5. Water absorption was measured by total immersion at 28 days and 14 months, and results are given in Table 4.21. It can be observed that (1) water absorption increased with the increase in metakaolin content both at 28 days and 14 months; and (2) at 14 months water absorption was less than that observed at the age of 28 days. Percentage decreased varied from 4.1 to 9.7% for mortar made with MK, and 16.9% with natural kaolin.

Khatib and Clay [36] investigated the effect of metakaolin on the water absorption of concrete. The mixture proportion of control mixture was 1:2.3:3.4 with water–binder ratio of 0.45. Cement was partially replaced with 5, 10, 15 and 20% MK. Tests were conducted up to the age of 90 days from the time of casting. Water absorption (WA) by total immersion is shown in Fig. 4.15. They reported

Table 4.21 Water absorption of mortars made with metakaolin and kaolin [21]

Material	Water absorption (% in mass)		Percentage decrease
	After 28 days	After 14 months	
CEM I 42.5	8.16	7.82	4.1
5% Metakaolin	8.39	8.04	4.1
10% Metakaolin	8.78	8.44	3.9
15% Metakaolin	9.71	8.77	9.7
20% Metakaolin	9.70	8.97	7.5
10% Kaolin	9.51	7.90	16.9

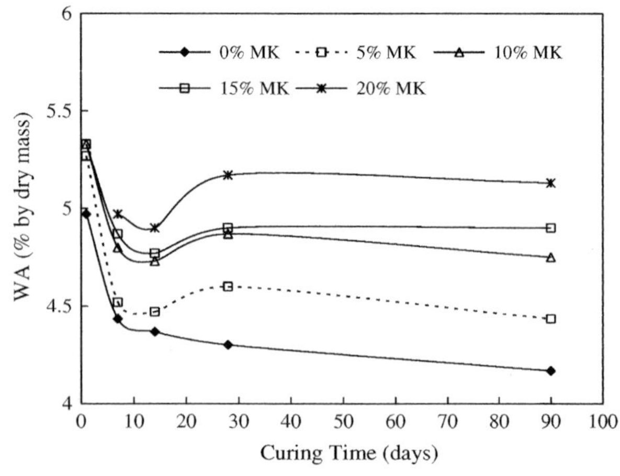

Fig. 4.15 Variation of WA with curing time for concrete containing MK [36]

that (1) water absorption for the control mix decreased with the increase in curing period. The decrease in WA was noticeable during the first 7 days. After 14 days, there was little decrease in WA; (2) a decrease in water absorption capacity was observed for all MK mixes up to about 14 days. Between 14 and 28 days curing period, there was a slight increase in WA caused by the presence of MK. This might be due to the formation of a denser hydration phase. After 28 days, 5 and 10% MK mixes showed a slight decrease in WA, whereas at higher MK replacement levels (15 and 20%), there was little change in WA; and (3) there tends to be a small systematic increase in WA with the increase in MK contents at all curing times. WA values for all mixes ranged between 4.2 and 5.4% by dry mass.

Razak et al. [55] studied the effect of metakaolin on the water absorption and sorptivity of concrete. Cement was replaced with 10% metakaolin. The inclusion of metakaolin greatly reduced the water absorption and sorptivity of concrete when compared with concrete made with OPC.

Badogiannis and Tsivilis [10] determined the sorptivity of concrete mixtures made with Greek kaolin of low kaolinite content (49% metakaolinite, 72.1% SiO_2, 30% active SiO_2) and commercial metakaolin (95% metakaolinite, 54.6% SiO_2, 53% active SiO_2) of high purity. High performance concrete mixtures were made where metakaolin replaced either cement or sand in percentages 10 or 20% by weight of the control cement content. The addition of metakaolin caused a relative decrease in concrete sorptivity. The sorptivity varied from 0.062 to 0.097 mm/min$^{0.5}$, while the PC concrete had sorptivity of 0.114 mm/min$^{0.5}$. Concrete specimens with commercial metakaolin (MKC) demonstrated the best behavior compared to MK concrete specimens and the concrete with MKC as 20% replacement of sand has the lower sorptivity.

4.5.3 Compressive Strength

Wild et al. [67] determined the compressive strength of concrete made with varying percentages of metakaolin (0, 5, 10, 15, 20, 25, and 30%) up to the age of 90 days. Control concrete mixture proportion was 1:2.3:3.4 with water–binder ratio (w/b) of 0.45. Compressive strength results are presented in Table 4.22.

Table 4.22 Compressive strengths and densities of metakaolin concretes [67]

(MK) (%)	Density (kg/m^3)	Compressive strength (MPa)				
		1-day	7 days	14 days	28 days	90 days
0	2,490	19.07	50.23	57.10	62.60	72.43
5	2,440	21.50	53.80	58.97	63.50	71.63
10	2,460	22.43	62.30	69.23	71.00	80.07
15	2,470	20.23	64.80	74.67	76.00	83.70
20	2,480	19.33	66.47	75.73	82.47	85.13
25	2,470	15.73	62.50	69.77	73.93	82.23
30	2,480	14.53	60.53	72.33	76.73	81.80

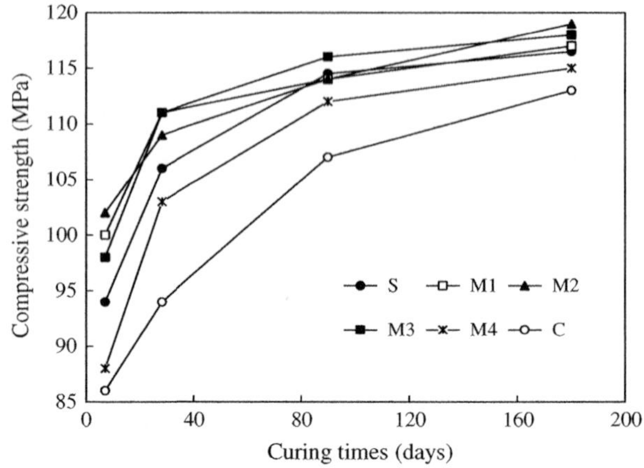

Fig. 4.16 Compressive strength versus curing time [22]

They concluded that inclusion of MK enhanced the compressive strength of concrete at all ages, but the optimum replacement level of OPC by MK to give maximum long term strength enhancement was about 20%.

Curcio et al. [22] determined the compressive strength of high-performance mortars containing 15% metakaolin as replacement of cement with a water/binder ratio of 0.33. Four commercially available MK (M1, M2, M3, and M4) samples were used. Major chemical composition (SiO_2 and Al_2O_3) of M1, M2, M3, and M4 metakaolin were; 51.89 and 43.53%; 51.42 and 44.81%; 55.0 and 40.0%; 53.0 and 44.0%, respectively. Silica fume had 96% SiO_2. Figure 4.16 shows the compressive strength development up to the age of 180 days. It was observed that (1) specimens containing three of the four metakaolin samples had a higher rate of compressive strength development as compared to that of the control at ages below 28 days, a consequence of the higher hydration rate; (2) difference in the compressive strength between the specimens with microfillers and control mortar decreased after 28 days, because of a smaller slowdown of the hydration rate in the control; (3) extent of strength enhancement at short curing ages was influenced by the type of cement; it was lower for early strength Portland as compared to ordinary Portland.

Brooks and Johari [14] concluded that inclusion of metakaolin (5, 10, and 15%) as partial replacement of cement increased its 28-day compressive strength (87 MPa) to 91.5, 104, and 103.5 MPa, with 5, 10, and 15% MK content. Poon et al. [49] observed that the cement pastes containing 5–20% MK had higher compressive strengths than the control at all ages from 3 to 90 days, with the paste containing 10% MK performing the best. Roy et al. [59] also reported increase in compressive strength on addition of MK.

Jin and Li [32] investigated the effect of metakaolin on the mechanical behavior of young concrete under either uniaxial compression or tension. Tests were conducted at ages of 1/2, 1, 2, 3, 7, and 28 days. They reported that (1) stress–strain (deformation) curves for young concretes were different from mature concretes in their capability in resisting the deformation. More ductile behavior was observed for a young concrete. However, the complete stress-deformation curves for young concrete under uniaxial tension showed less ductility as compared to those under uniaxial compression; and (2) metakaolin showed the best enhancement on the mechanical properties of young concrete.

Li and Ding [45] studied the effect of metakaolin (MK) or combination of MK and slag on the 28-day compressive strength of cement mortars. Four series of cement mortars pastes were made. First series (PO) contained 100% cement, 0% MK, and 0% slag; second (M1) 90% cement and 10% MK, 0% slag; third (S2) 70% cement, 10% MK, and 20% slag, and the fourth (S3) 60% cement, 10% MK, and 30% slag. Cement-to-sand ratio was 1:2.5, and w/b ratio was 0.44. Compressive strength results are shown in Fig. 4.17. It was observed that (1) MK enhanced the compressive strength of cement mortar. Concrete containing 10% MK gave the best result; (2) when PC was mixed with MK and ultra-fine slag at the same time, the 3-day compressive strength of S2 and S3 cement mortars was lower than that of PC and cement mortar M1. However, after 7 days, the strengths of S2 and S3 were higher than that of the M1 and PC mortars. After 28 days, the strength of S2 and S3 had greatly surpassed that of the two former cement mortars; and (3) both MK and slag can react with CH release by cement clinker hydration to produce secondary C–S–H gel inside the cement paste. The secondary formed C–S–H gel improved the microstructure of cement paste matrix; therefore, the macroscopic property of cement was also improved. XRD analysis indicated that more $Ca(OH)_2$ was consumed after adding both mineral admixtures.

Poon et al. [50] investigated the performance of metakaolin concrete at elevated temperatures up to 800°C. Normal and high strength concrete (HSC) mixes incorporating 0, 5, 10 and 20% MK were made. They found that (1) after an increase in compressive strength at 200°C; the MK concrete suffered a more severe loss of compressive strength. MK concrete showed a distinct pattern of strength gain and loss at elevated temperatures. A sharp reduction in compressive strength was observed after 400°C followed by severe cracking and explosive spalling.

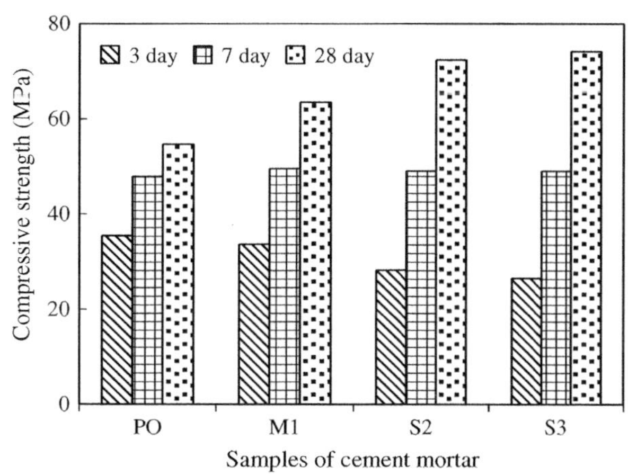

Fig. 4.17 The compressive strength of cement mortars [45]

Within the range 400–800°C, MK concretes suffered more loss and possessed lower residual strengths than the other concretes.

Badogiannis et al. [8] studied the influence of two types of metakaolins; (1) one produced from Poor Greek kaolin (C); and commercial metakaolin (MKC) on the high-performance concrete. Metakaolin was replaced either with cement or sand in percentages of 10 or 20% by weight of the control cement content. The strength of HPC was evaluated using the efficiency factor (*k* value). Both metakaolins exhibited very high *k*-values (close to 3.0 at 28 days) and were characterized as highly reactive pozzolanic materials that can lead to concrete production with an excellent performance.

Lee et al. [44] reported the compressive strength of mortars incorporating with 0, 5, 10 and 15% of MK by mass. The water-cementitious material ratio (w/cm) was fixed at 0.45 by mass. The cementitious material/fine aggregate was invariant at 2.0 in all mortar mixtures. Table 4.23 presents the data on strength development of mortar specimens cured in tap water. These data indicated that MK15 mortar specimens showed a better compressive strength development compared to MK0 mortar specimens during the entire period of tap water curing.

Badogiannis et al. [9] determined the compressive strength of cement containing five metakaolin up to the age of 180 days. The metakaolinite contents in metakaolins MK1, MK2, MK3, and MK4 (derived from poor Greek kaolins) were 36, 37, 71, and 49%, respectively, whereas it was 95% in a commercial high purity metakaolin (MKC). Cement was replaced with 0, 10, and 20% metakaolins. It was observed that metakaolins had a very positive effect on the cement strength after 2 days and specifically at 28 and 180 days (Table 4.24). This may be because produced metakaolins as well as the commercial-one gave similar hydration

Table 4.23 Strength development of mortar specimens cured in tap water [44]

Mixture	Replacement levels of MK	Compressive strength (MPa)			
		7 days	28 days	91 days	360 days
MK0	0	37.0	48.6	56.8	59.2
MK5	5	33.8	46.3	55.5	60.4
MK10	10	43.2	50.4	56.9	62.2
MK15	15	41.8	51.2	63.0	66.8

Table 4.24 Compressive strength of concrete containing metakaolins [9]

Age (days)	Compressive strength (MPa)											
	PC		MK1		MK2		MK3		MK4		MKC	
	10%	20%	10%	20%	10%	20%	10%	20%	10%	20%	10%	20%
1	17	18	17	16	18	16	21	19	20	18	20	19
2	30	30	28	23	30	24	33	29	32	28	33	29
7	44	44	45	44	48	45	50	50	51	50	54	50
28	57	57	64	68	65	67	66	69	68	65	72	70
180	70	70	79	77	80	75	74	72	80	78	76	76

products after 28 days and the pozzolanic reaction was accelerated between 7 and 28 days, accompanied by a steep decrease of $Ca(OH)_2$ content.

Poon et al. [51] determined the compressive strength of concrete mixes containing 5, 10, and 20% metakaolin as partial replacement of cement. Concrete mixes were prepared with two w/b ratios of 0.3 and 0.5. The maximum nominal size of the coarse aggregate was 10 mm for the concrete with a w/b of 0.3 and 20 mm for the concrete with a w/b of 0.5. The results (Table 4.25) showed that the use of 10% gave the best performance, which resulted in the highest strength increase over the control concretes at all ages, particularly at the age of 3 days. This could be attributed to higher rate of hydration in the metakaolin concrete.

Potgieter-Vermaak and Potgieter [52] studied the compressive strength of mortars containing metakaolin between 10 and 30% up to the age of 28 days. Activation temperature of MK ranged from 550 to 850°C for durations of 30 and 60 min. Mortars were made with sand-cementitious proportion of 3:1 by mass and water: cementitious material ratio of 0.375. It was observed that (1) best activation temperature to produce metakaolin from kaolin is >700°C, and should preferably be at least 750°C; (2) compressive strengths increased with the increase in curing times and depended strongly on the activation temperature. Strength enhancements did not depend significantly on the concentration of metakaolin addition; and (3) longer activation time resulted in marginally higher compressive strengths in the mortars containing MK heated for 60 min compared to those having material heated for 30 min.

Khatib [37] investigated the compressive strength of concrete containing metakaolin (MK) at a low water-to-binder ratio of 0.3. Portland cement (PC) was partially replaced with 0–20% MK. Results of compressive strengths and average density are presented in Table 4.26. Generally, the density tends to increase slightly with increasing amounts of MK in concrete, mainly resulting from the filling effect of MK particles. Compressive strength increased with the increase in time and the incorporation of MK increased the strength beyond 1-day of curing. At 1-day of curing, replacing PC with MK beyond 5% slightly reduced the compressive strength. However, beyond 1-day of curing, MK in the mix increased the compressive strength, and attained a maximum value at 15% MK. Moreover

Table 4.25 Compressive strength of control and blended concretes [51]

Series	w/b ratio	Mix	Total charge passed (C)			
			3 days	7 days	28 days	90 days
1	0.30	Control	68.5	81.1	96.5	102.5
		5% MK	73.0	88.2	103.6	112.9
		10% MK	85.9	99.8	116.8	120.3
		20% MK	70.8	87.6	99.6	113.8
2	0.50	Control	28.6	41.2	52.1	60.4
		5% MK	32.6	45.9	57.1	66.5
		10% MK	40.4	55.2	66.2	71.6
		20% MK	30.0	43.2	58.4	69.1

Table 4.26 Compressive strength of concrete mixes containing MK [37]

MK (%)	Density (kg/m^3)	Age (days)				
		1	7	14	28	56
0	2,293	28.9	55.3	60.7	66.1	71.3
5	2,261	30.6	59.2	67.7	72.5	79.0
7.5	2,316	25.1	62.2	74.4	81.4	85.7
12.5	2,329	28.5	68.6	81.6	85.3	90.4
15	2,337	28.6	70.2	83.6	89.7	91.0
20	2,329	22.8	62.8	78.9	83.2	86.1

the maximum contribution of MK occurred at 14 days of curing where more than 35% increase in strength was obtained.

Gonçalves et al. [28] determined the influence of metakaolin and ground calcined-clay brick on the uniaxial compressive strength of mortar (Fig. 4.18). Three types of calcined-clays (1) ground calcined-clay brick (GB); (2) commercial metakaolin (MK1) and (3) a calcined-clay produced in the laboratory Metakaolin (MK2). Major constituents of ground calcined-clay brick were SiO_2 (63.89%), Al_2O_3 (25.49%), Fe_2O_3 (7.73%). Major constituent of commercial metakaolin (MK1) were SiO_2 (51.2%), Al_2O_3 (35.3%), Fe_2O_3 (4.0%) whereas laboratory produced metakaolin (MK2) had SiO_2 (52.46%), Al_2O_3 (44.24%), Fe_2O_3 (2.06%). Mortar mixtures 1:1.5, with water/cementitious material ratio of 0.5, were prepared with 0, 10, 20, 30, and 40% of cement was replaced by calcined-clay samples (GB, MK1 or MK2). They concluded that (1) compressive strength of mortars made with calcined-clay brick decreased with the increase in calcined-clay percentage in comparison to the control mixture; up to 20%, there was marginal loss in strength, but, at higher levels of replacement (30 and 40%), significant reduction in strength was observed. This may be due to the fact that packing density of the dry mix did not change with the replacement of PC by GB, then it is

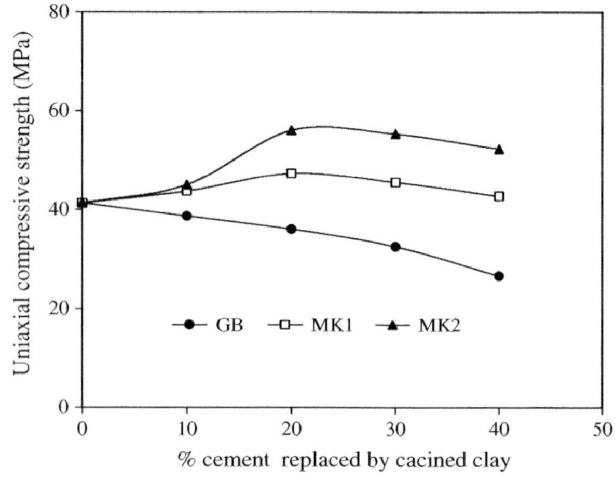

Fig. 4.18 Influence of proportion of cement replaced by each calcined-clay on uniaxial compressive strength of cement mortars [28]

possible to infer that the amount of hydrates formed by the pozzolanic reaction from replacement of up to 20% cement by GB was high enough so that it was able to maintain a constant compressive strength. However, with higher replacement levels, the amount of hydrates resulting from the pozzolanic reaction was probably insufficient to allow maintaining a constant compressive strength, so that the excess of un-reacted GB was only able to contribute in strength due to its physical filler effect; and (2) inclusion of metakaolins (MK1 and MK2) increased the compressive strength of mortar. Optimal results were obtained for 20% replacement of cement by metakaolin, representing increases in 39 and 18% for MK2 and MK1, respectively. This increase may be due to the greater formation of hydrates. At replacement of 30 and 40%, higher compressive strengths than that of the control mixture were observed, but there was reduction in strengths in comparison with 20% replacement. This may probably due to the fact that the more limited availability of calcium hydroxide from the smaller proportion of cement added becomes the limiting factor in the formation of new hydrates. As a result, the unreacted metakaolin can act only as filler.

Janotka et al. [31] determined the compressive strength of cement paste containing Slovak poor metakaolin sands with different metakaolin content [36.0% (MK-1), 31.5 (MK-2) and 40.0% (MK-3)]. The percentage of metakaolin sands was 10, 20 and 40%. It was concluded that (1) control PC paste achieved 43 and 59.6 MPa strength at the ages of 28 and 90 days, respectively. The incorporation of metakaolin sand decreased the strength at 28 days of curing, and decrease in strength was higher as the metakaolin sand content increased. At 28 days, for paste containing MK-1, strength was 41.2, 35.9, and 29.3 MPa whereas strength was 36.8, 28.0, and 14.1 MPa for MK-2 and for MK-3; strength achieved was 32.3, 36.4, and 38.9 MPa; and (2) compressive strength of all paste mixtures increased with age.

4.5.4 Tensile Strength and Elastic Modulus

Qian and Li [53] determined the 28-day direct tensile strength of concrete containing 0, 5, 10, and 15% metakaolin as partial replacement of cement. Specific surface area and average particle diameter of metakaolin were 12,000 m^2/kg and 2.23 μm. Tensile strength results are given in Table 4.27. It was observed that tensile strength increased with the increase in metakaolin content. The average tensile strength increases were 7, 16, and 28% for concrete mixtures made with 5, 10, and 15% MK whereas average ultimate strain increases were 3, 19, and 27%.

Table 4.27 Tensile strength of concrete with different metakaolin replacement [53]

Age (days)	Tensile strength (MPa)			
	MK (0%)	MK (5%)	MK (10%)	MK (15%)
28	3.35	3.58	3.88	4.29

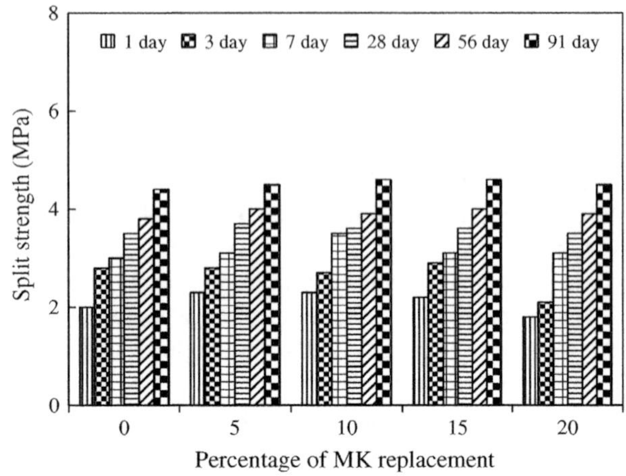

Fig. 4.19 Splitting tensile strength versus replacement of age [38]

The descending area of over-peak stress was less steep when metakaolin replacement was 5 and 10% whereas with 15% metakaolin it was similar to that for concrete without metakaolin. The tensile elasticity modulus for these specimens ranged between 26 and 27 GPa.

Kim et al. [38] presented the results of splitting tensile strength of high-strength concrete made with metakaolin. Cement was replaced with 0, 5, 10, 15 and 20% of metakaolin. Control mixture had 563 kg of cement, 141 kg of fly ash, 532 kg of fine aggregate, 915 kg of coarse aggregate, and 176 kg of water. 28-day compressive strength of control concrete was 60 MPa. Figure 4.19 presents the tensile strength results at the age of 1, 3, 7, 28, 56, and 91 days. This demonstrated the level of tensile strength development according to the replacement ratio of the binder by MK, from 5 to 20%. For binder replacement ratios ranging between 10 and 15%, it was observed that the strength improved with the increase in replacement ratio, while the strength reduced at 20%.

Gonçalves et al. [28] determined the effect of metakaolin samples and ground calcined-clay brick on the modulus elasticity of mortar. Three types of calcined-clays (1) ground calcined-clay brick (GB); (2) commercial metakaolin (MK1) and (3) a calcined-clay produced in the laboratory Mortar mixtures 1:1.5, with water/cementitious material ratio of 0.5, were prepared with 10, and 20% of cement was replaced by calcined-clay samples (GB, MK1 or MK2). Modulus of elasticity of control mortar was approximately 25 GPa. It was observed that increase in replacement of cement by the calcined-clays resulted in a continuous decrease in the elastic modulus of the mortars, reaching reductions up to 10% in comparison to the control mixture. The reduction in the elastic modulus of Portland cement mortars with calcined-clay additions may be probably attributed to the lamellar structure of the clays, which are more deformable under loading. This is associated with the fact that both kaolinite and ilite, even after being calcined at temperatures as high as 550°C and 850°C, respectively, retain their lamellar residual structure.

4.5.5 Bending Strength

Qian and Li [53] investigated the influence of 0, 5, 10, and 15% metakaolin as partial replacement of cement on the bending strength of concrete up to the age of 80 days. Specific surface area metakaolin was 12,000 m^2/kg, with average particle diameter being 2.23 μm. Beam specimens (400 × 100 × 100-mm) were tested under four-point bend on a span of 300 mm. Bending test results are given in Table 4.28. It was observed that (1) 5% metakaolin had little effect on the bending strength of the concrete; (2) at 10 and 15% MK, the 28-day bending strength was increased by 32 and 38%, respectively. At 80 days, increase in bending strength was 13 and 24% for 10 and 15% replacement, respectively. Therefore, significant improvement in bend strength of concrete can be achieved for replacement levels of 10–15% metakaolin.

Courard et al. [21] investigated the role of metakaolin on the bending strength of mortar containing up to 20% metakaolin. The specimens were made with 3 parts of sand, 1 part of cement and 0.5 part of water) in which cement was OPC (CEM I 42.5). Mortar mixtures were made in which cement was replaced by metakaolin (5–20%). One more mixture mix was made with 10% by mass of natural kaolin. Specimens (40 × 400 × 160 mm) were cast and tested up to the age of 28 days. Bending strength results are shown in Fig. 4.20. They concluded that replacement of cement by metakaolin decreased the bending strength during the first days but achieved almost equal strength at the age of 28 days.

Table 4.28 Bending strength of concrete with different metakaolin replacement [53]

Age (days)	Bending strength (MPa)			
	MK (0%)	MK (5%)	MK (10%)	MK (15%)
28	4.65	4.74	6.16	6.40
80	5.70	5.81	6.46	7.06

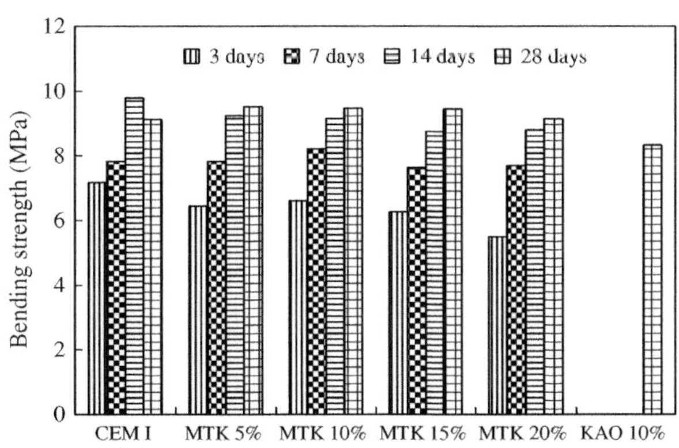

Fig. 4.20 Bending strength of mortars with CEM I 42.5, metakaolin and kaolin [21]

4.5.6 Micro-Hardness

Micro-hardness measurements can contribute to characterization of the properties of the ITZ relative to the bulk cement paste matrix and also helps in estimating the width of the ITZ. Lyubimova and Pinus [46] observed that over a range of water/binder ratios, micro-hardness decreased within 100 μm of the aggregate surface. Whilst reductions in micro-hardness for the ITZ relative to the bulk regions are consistent with the hypothesis that the ITZ is a region of higher than average porosity, such observations are not universal.

Saito and Kawamura [60] reported increase in micro-hardness within 50 μm of the aggregate surface. Igurashi et al. [30] suggested four ITZ/bulk region micro-hardness profiles. For example, in specimens where the microstructure of the ITZ is similar to the bulk region, the close proximity of the aggregate surface can increase micro-hardness values above those recorded for the bulk region by impeding the displacement of material from the indentation site. In contrast, where the ITZ microstructure is weaker than the bulk region, a depression in the micro-hardness profile is expected within the zone of influence of the aggregate, the profile at the aggregate surface also being influenced by the strength of the paste–aggregate bond.

Asbridge et al. [6] determined the variations in the micro-hardness of the hydrated cement matrix as functions of the distance from the aggregate surfaces for specimens in which Portland cement was partially replaced with 0 and 10% of metakaolin. Water/binder ratios were 0.40, 0.50 and 0.60. Cement matrix of mortars were prepared with an aggregate content of 35% by volume. Micro-hardness measurements were made up to distances of 120 μm from the aggregate. Variations in micro-hardness of bulk paste/ITZ are shown in Figs. 4.21, 4.22, 4.23 and 4.24. Figure 4.21 shows that in Portland cement specimens, the ITZ was more porous than the bulk matrix at lower water/binder ratios. The micro-hardness of the ITZ was 22% lower than the bulk matrix at a water/binder ratio of 0.40,

Fig. 4.21 Variation in micro-hardness of bulk paste and ITZ with water/binder ratio (Portland cement binder) [6]

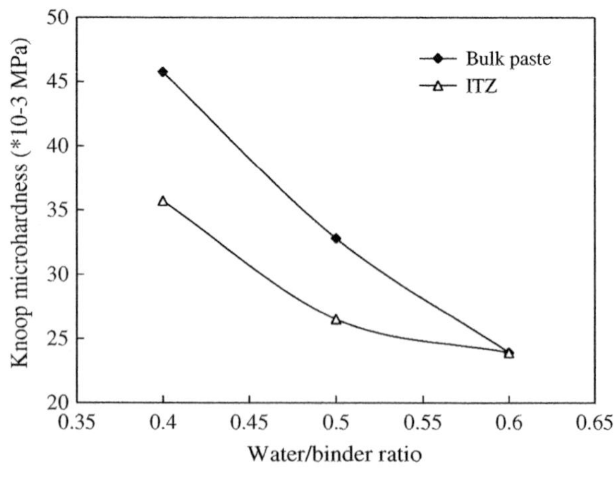

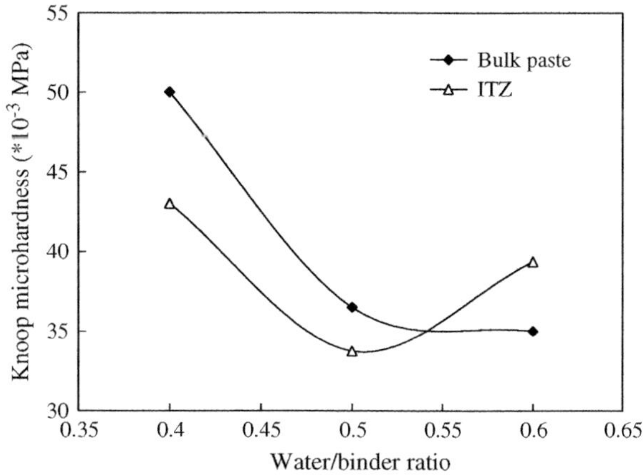

Fig. 4.22 Variation in micro-hardness of bulk paste and ITZ with water/binder ratio (Portland cement/metakaolin binder) [6]

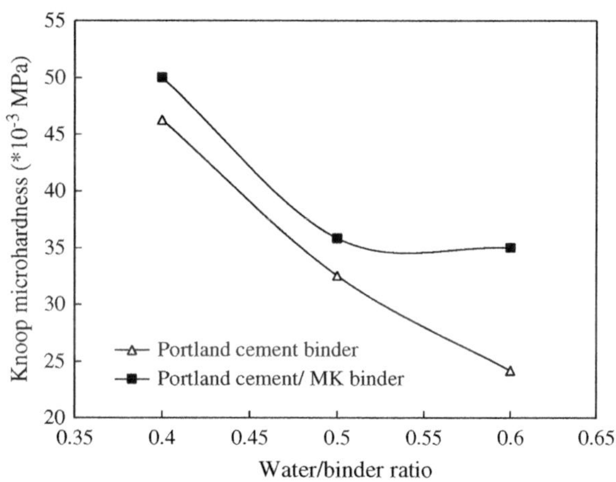

Fig. 4.23 Variation in micro-hardness of bulk paste with water/binder ratio and binder composition [6]

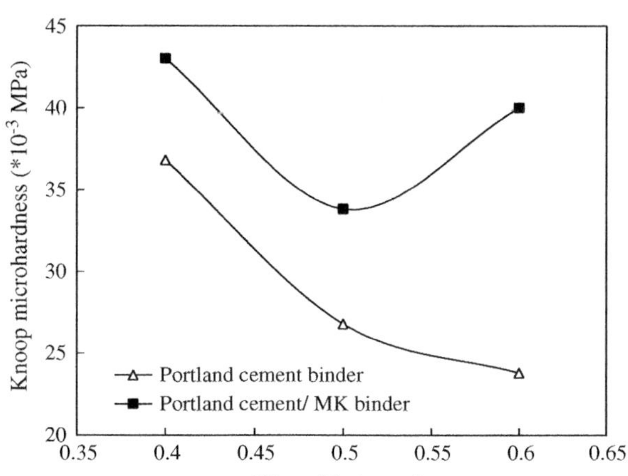

Fig. 4.24 Variation in micro-hardness of ITZ with water/binder ratio and binder composition [6]

and was 18% lower at a water/binder ratio of 0.50. Figure 4.22 indicates that in the specimens prepared with 10% metakaolin, the micro-hardness of the ITZ was similar to that of the bulk matrix. From Figs. 4.23 and 4.24, it is clear that micro-hardness of the bulk matrix and, to a greater degree, the ITZ increased when metakaolin was used.

4.6 Durability Properties of Concrete Containing Metakaolin

4.6.1 Alkali–Silica Reaction

Ramlochan et al. [54] examined the influence of high-reactivity metakaolin (HRM) in controlling expansion due to alkali–silica reaction (ASR) of concrete containing 0–20% HRM as a partial replacement for OPC, up to 2 years. Two types of alkali–silica reactive aggregates: a siliceous limestone (Spratt) and greywacke–argillite gravel (Sudbury) were used. The major chemical compounds in cement were; SiO_2 (19.97%), CaO (62.69%), Al_2O_3 (5.08%), Fe_2O_3 (2.26%), MgO (2.66%), Na_2O (0.30%). Concrete prisms (75 × 75 × 300 mm) were cast with each coarse reactive aggregate, a non-deleteriously reactive sand and high-alkali Portland cement with a 1.02% equivalent soda content (or Na_2O). Prisms containing the Spratt aggregate were cast with 0, 5, 10, 15, and 20% HRM as a partial cement replacement, and prisms containing the Sudbury aggregate were cast with 0, 10, and 15% HRM replacement levels. The accelerated mortar bar method was used. Figures 4.25 and 4.26 show the expansion evolution of concrete prisms stored at 38°C and 100% RH containing different levels of HRM for the two reactive aggregates. It was observed that expansion decreased with the increase in HRM content. Both control specimens (0% HRM) exceeded the expansion limit criterion of 0.04%. The siliceous limestone aggregate (Spratt) reacts very rapidly and control specimens exceeded 0.04% expansion at 56 days.

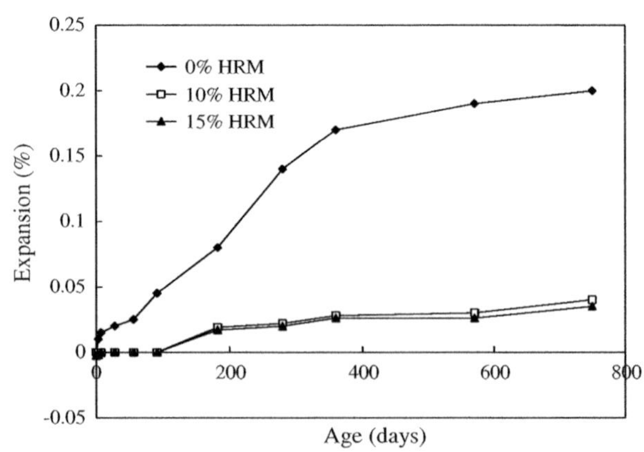

Fig. 4.25 Expansion evolution of CAN/CSA A23.2-14A concrete prisms containing HRM and Sudbury aggregate [54]

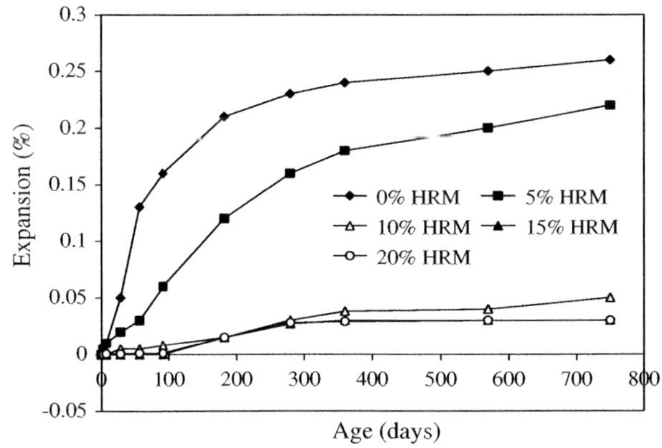

Fig. 4.26 Expansion evolution of CAN/CSA A23.2-14A concrete prisms containing HRM and Spratt aggregate [54]

The greywacke aggregate (Sudbury) is slower reacting, but exceeded 0.04% expansion at about 150 days. Both control specimens also exhibited visual signs of distress early on; surface cracking with exuded gel and an associated white streaking was observed in Spratt control prisms at 90 days and in Sudbury control prisms at 180 days.

Aquino et al. [3] studied the influence of high-reactivity metakaolin (HRM) and silica fume on the alkali–silica reaction of mortar. A high-alkali cement (0.46% Na_2O, 1.06% K_2O, 1.17% Na_2O eq). The oxide composition of HRM ($Al_2O_3 + SiO_2 + Fe_2O_3$) was 95.3%, and silica fume had SiO_2 content of 93%. Four different mortar mixes were prepared with water-to-cementitious materials ratio of 0.56 and aggregate-cementitious ratio of 2.25. Control mixture A contained no mineral admixture. In two other mixtures, cement was replaced with 10% of SF and HRM. Control mixture B contained high-alkali cement and reactive aggregates, but no mineral admixture. Mortar bars of size $25 \times 25 \times 280$ mm were made. The expansion measurements were made at the age of 1, 3, 5, 7, 14, and 21 days of curing in 1 N NaOH solution. Results of expansion measurements are shown in Fig. 4.27. It was observed that (1) control A mixture, which did not incorporate any reactive aggregate, did not show any significant expansion; (2) Control B mixture which contained reactive aggregate but did not incorporate either SF or HRM, expanded at a faster rate and reached expansions of 0.48 and 0.51% at 14 and 28 days; and (3) mortar bars containing HRM or SF showed significantly lower expansion at all ages.

4.6.2 Chloride-Ion Diffusion/Permeability

Zhang and Malhotra [71] measured the chloride penetration resistance (ASTM C1202) of PC concrete and PC-10% MK concrete. The concrete with w/b ratio 0.40 was moist-cured for 28 and 91 days prior to being subjected to electrical conductivity measurements. The PC–MK concrete showed significantly lower

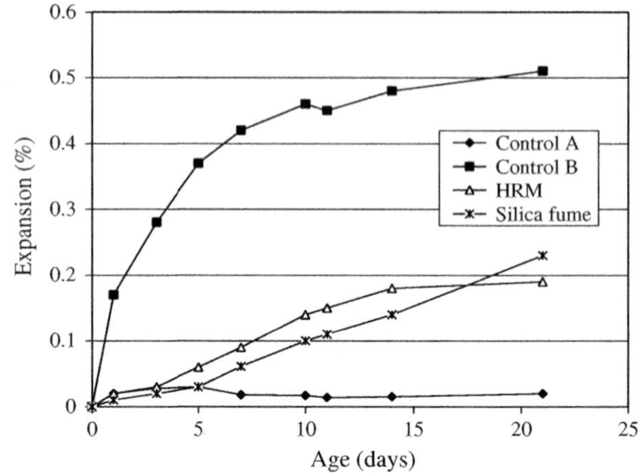

Fig. 4.27 Influence of HRM and SF on mortar bar expansion over a period of 21 days after immersion in 1 N NaOH at 80°C [3]

conductivity values than the PC concrete. Cabrera and Nwaubani [16] measured the effective chloride diffusion coefficients of PC and PC-15% MK (for two different MKs) pastes using a chloride diffusion cell. The pastes of w/b ratio 0.40 were moist-cured (100% RH) for 60 days. PC–MK pastes gave lower chloride diffusion coefficients than the PC paste.

Thomas et al. [66] studied the chloride binding capacity of cement paste with and without 8% HRM by exposing dried paste samples to various concentrations of NaCl solution and determining the amount of chloride removed from solution on establishment of equilibrium. For up to a 1-M solution concentration, the binding capacities of the two pastes were similar but above this concentration, the binding capacity of the HRM paste increased at a much greater rate with increased solution concentration than did the pure PC paste.

Boddy et al. [13] and Gruber et al. [29] measured the chloride diffusion at the age of 365 and 1,095 days. The apparent diffusion coefficients were reduced with age and on decreasing the w/b ratio. It also showed marked decreases with increasing HRM content.

Asbridge et al. [5] studied the effect of metakaolin and aggregate content on the diffusion kinetics of chloride ions in hydrated Portland cement mortars (Table 4.29) under steady- and non-steady-state conditions. In cement mortars, cement was replaced with 10% metakaolin. It was concluded that (1) the steady-state diffusivity of chloride ions in mortar pastes incorporating metakaolin did not vary significantly with aggregate volume fractions. Mortar-paste capillary porosity data showed only some increases at aggregate volume fractions greater than 40%;

Table 4.29 Composition of pastes and mortars [5]

	Steady state	Non-steady state
Binder	PC/MK	PC, PC/MK
Water/binder ratio	0.35	0.40
Aggregate contents (% vol)	0, 15, 30, 40, 50, 55	0, 15, 35, 55

Table 4.30 Chloride diffusion rates for mortars with CEM I 42.5, metakaolin and kaolin [21]

Material	Breakthrough time (days)	Apparent diffusion coefficient (m^2/s)
CEM I 42.5	13	1.29×10^{-12}
5% Metakaolin	45	4.71×10^{-12}
10% Metakaolin	82	3.31×10^{-12}
15% Metakaolin	203	1.23×10^{-12}
20% Metakaolin	Not after one year	–
10% Kaolin	4	1.81×10^{-12}

(2) non-steady-state chloride diffusivities of mortar pastes increased with aggregate volume in samples with only Portland cement as binder; (3) the capillary porosity of the mortar paste increased with aggregate volume, indicating that the ITZ had a higher overall porosity than the bulk paste; and (4) the use of 10% metakaolin as a partial replacement for Portland cement reduced non-steady-state chloride diffusivities by approximately one-order of magnitude relative to those samples with only Portland cement as binder. Mortar-paste diffusivities did not vary significantly with variations in aggregate content. Relative to control samples, metakaolin reduced mortar-paste capillary porosities although the latter still increased with increasing aggregate volumes. The lack of variation in mortar-paste diffusivity and rising capillary porosity with increasing aggregate content supports the hypothesis that the use of metakaolin increases permeation path tortuosity and inhibits percolation.

Courard et al. [21] investigated the effect of metakaolin on the chloride diffusion rates of mortars. Metakaolin percentages as partial replacement of cement were 0, 10, 15, and 20%. One mixture with natural kaolin was also made. Mixture proportion of mortar was 3 parts of sand, 1 part of cement, with w/c ratio of 0.5. Diffusion rates of Cl^- and Na^+ ions into cement mortars were monitored using two-compartment diffusion cells. At periodic intervals, chloride concentration was measured by titration from a 20 cm^3 sample of the solution. The evolution of chloride diffusion was measured for 314 days for cement, metakaolin and kaolin (Table 4.30). The breakthrough time was calculated from the intercept of the concentration versus time date. Breakthrough time represents the time necessary for the initiation of Cl^- ion transfer through the sample. It gives a view of the porous skeleton. Mortar with 20% MK gave the best results as even after 1 year, no diffusion was observed. An increase from 10 to 15% metakaolin content seemed to induce an increase of 150% of occurrence time and a decrease of 170% for diffusion coefficient. Kaolin had no effect and seemed on the contrary to accelerate the phenomenon of diffusion in comparison with the reference mix.

Poon et al. [51] studied the effect of metakaolin on the chloride penetrability of the concrete mixes (Table 4.31). Concrete mixes were prepared with two w/b ratios of 0.30 and 0.50. Cement was replaced with 5, 10, and 20% metakaolin. The maximum nominal size of the coarse aggregate was 10 mm for the concrete with a w/b of 0.30 and 20 mm for the concrete with a w/b of 0.50. It was observed that MK concretes showed lower total charges passed than the control. At the w/b of

Table 4.31 Chloride permeability of control and blended concretes [51]

Series	w/b ratio	Mix	Total charge passed (C)			
			3 days	7 days	28 days	90 days
1	0.30	Control	2,461	2,151	1,035	931
		5% MK	1,327	1,244	862	646
		10% MK	417	347	199	135
		20% MK	406	395	240	124
2	0.50	Control	5,312	4,054	2,971	2,789
		5% MK	4,215	3,765	2,079	1,065
		10% MK	1,580	1,247	918	752
		20% MK	751	740	640	580

0.3, concrete with a 10% MK replacement showed the best performance, while at the w/b of 0.50, concrete with a 20% replacement was the best.

Badogiannis and Tsivilis [10] determined the chloride and gas permeability of concretes made with Greek kaolin of low kaolinite content (49% metakaolinite, 72.1% SiO_2, 30% active SiO_2) and commercial metakaolin (95% metakaolinite, 54.6% SiO_2, 53% active SiO_2) of high purity. Concrete mixtures were made in which metakaolin replaced either cement or sand in percentages 10 or 20% by weight of cement. Chloride and gas permeability results are given in Table 4.32. The addition of metakaolin caused a significant increase of concrete resistance to chloride penetration. The charge passed (in the rapid chloride test) in metakaolin concrete varied from 180 to 820 C, while charge passed in PC concrete was 2,460 C. Concrete with MKC and 10% replacement of sand exhibited higher resistance to chloride penetration. Concrete with metakaolin exhibited lower gas permeability values compared with PC concrete. The gas permeability of the metakaolin concrete varied from 1.35 to 1.85×10^{-16} m^2, while the control concrete presented a gas permeability of 2.94×10^{-16} m^2. The concrete with MK and 10% replacement of cement showed lower gas permeability.

Kim et al. [38] studied chloride permeability of high-strength concrete made with Korean metakaolin. Cement was replaced with 0, 5, 10, 15 and 20% of metakaolin. Control mixture had 563 kg of cement, 141 kg of fly ash, 532 kg of fine aggregate, 915 kg of coarse aggregate, and 176 kg of water. 28-day

Table 4.32 Chloride and gas permeability of concrete mixtures [10]

Mixture	Chloride permeability (C)	Gas permeability ($m^2 \times 10^{-16}$)
Control concrete	2,460	2.94
MKC-CR10	730	1.68
MKC-CR20	240	1.45
MK-CR10	690	1.35
MK-CR20	760	1.60
MKC-SR10	180	1.75
MKC-SR20	530	1.56
MK-RR10	820	1.71
MK-SR20	390	1.85

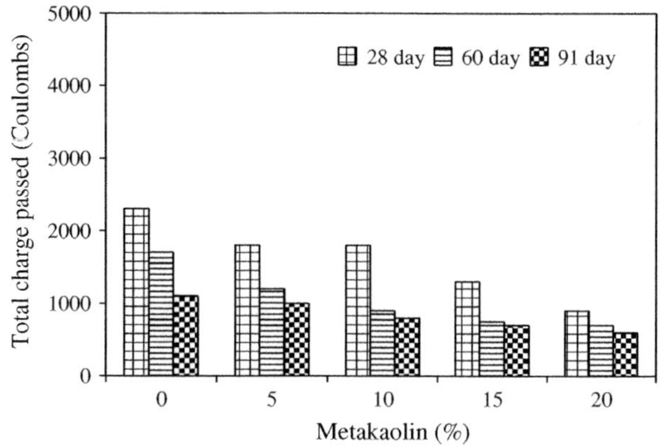

Fig. 4.28 Total charge passed versus replacement level with metakaolin [38]

compressive strength of control concrete was 60 MPa. The resistance of concrete to chloride attack was assessed by rapid chloride permeability test (RCPT) at 28, 56 and 91, and the results are shown in Fig. 4.28. The permeability appeared to decrease as the replacement ratio and curing time increased. Also, all of mixtures with MK revealed very low level of permeability. The increase in resistance to chloride permeability is due to the continued hydration and pozzolanic reaction, accompanied by a decrease in porosity and pore sizes.

Gonçalves et al. [28] investigated the chloride ion penetration of mortars mixtures prepared with metakaolin samples and ground calcined-clay brick. Three types of calcined-clays (1) ground calcined-clay brick (GB); (2) commercial metakaolin (MK1) and (3) a calcined-clay produced in the laboratory Metakaolin (MK2). Major constituents of ground calcined-clay brick were SiO_2 (63.89%), Al_2O_3 (25.49%), Fe_2O_3 (7.73%). Major constituents of commercial metakaolin (MK1) were SiO_2 (51.2%), Al_2O_3 (35.3%), Fe_2O_3 (4.0%) whereas laboratory produced metakaolin (MK2) had SiO_2 (52.46%), Al_2O_3 (44.24%), Fe_2O_3 (2.06%). Mortar mixtures 1:1.5, with water/cementitious material ratio of 0.5, were prepared with 10, and 20% of cement was replaced by calcined-clay samples (GB, MK1 or MK2). They concluded that chloride-ion penetration in control mixture was 13,487 C. The use of calcined-clay promoted a general reduction in the chloride charge. Chloride-ion penetration in mortar mixtures with ground calcined clay brick was 8,460 and 2,111 coulombs; reduction from 1.5 to 6 times for mixtures containing 10 and 20% GB, respectively, whereas this reduction was even more significant, by 10 and 31 times for mixtures containing 10 and 20% MK1, respectively, in comparison to the control mixture. This may be because of the resistance to chloride-ion penetration is more directly associated to the refinement of the pore structure.

4.6.3 Hydroxide Ion Diffusion

Coleman and Page [20] studied the hydroxide ion concentrations of pore solutions of ordinary Portland cement pastes containing 0, 10 and 20% of metakaolin by

Fig. 4.29 Effect of additions of MK501 on pore solution chloride ion concentration for samples containing 1.0% internal chloride ions (by weight of total solids) [20]

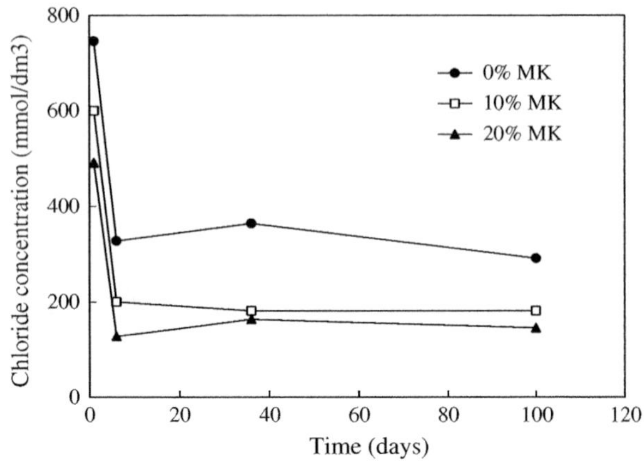

weight of cement at various stages of hydration. Blended cements were prepared from OPC and two grades of metakaolin, labeled as MK501 (major elements; SiO_2 54.2%; Al_2O_3 40.8%) and MK505 (major elements; SiO_2 52.1%; Al_2O_3 41%), of differing composition with respect to minor contaminants. Blended cement pastes were made by partial replacement of OPC with metakaolin having water/solids ratio of 0.5. The effects of metakaolin on the pore solution chloride ion concentrations for samples containing 1.0% of internal chloride ions are shown in Figs. 4.29 and 4.30.

It was observed that both grades of metakaolin caused a substantial reduction in the concentrations of free chloride ions that were retained in the pore solution phase of paste specimens in which 1% chloride (by weight of total solids) had been included as a mix contaminant. This effect was also observed for specimens made from the 501 grade of metakaolin with 0.4% chloride. In specimens with only 0.1% chloride as a mix contaminant, the capacity of the cement hydration products to bind chloride ions was such that the concentrations of chloride remaining in the

Fig. 4.30 Effect of additions of MK505 on pore solution chloride ion concentration for samples containing 1% internal chloride ions (by weight of total solids) [20]

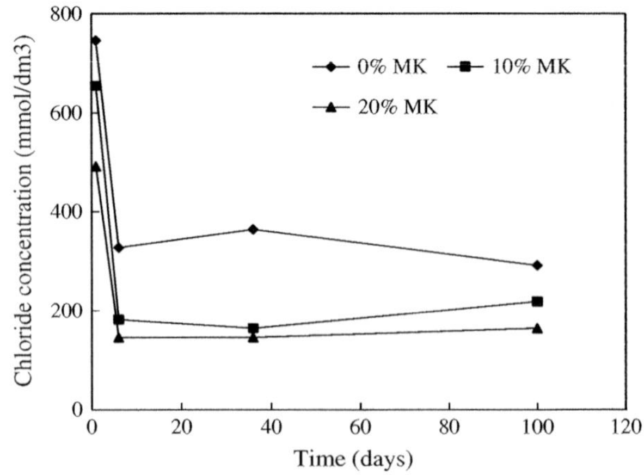

pore solution at ages of 36 days or longer had fallen to around 10 mmol dm^{-3}, irrespective of whether the material contained 0, 10 or 20% metakaolin.

4.6.4 Sulfate Resistance

Sulfate attack is one of the most aggressive environmental deteriorations that affect the durability of concrete structures. The sulfate attack on concrete leads to expansion, cracking, and deterioration of many civil engineering structures exposed to sulfate environment such as piers, bridges, foundations, concrete pipes, etc. The sulfate attack is generally attributed to the reaction of sulfate ions with calcium hydroxide and calcium aluminate hydrate to form gypsum and ettringite. The gypsum and ettringite formed as a result of sulfate attack is significantly more voluminous (1.2 to 2.2 times) than the initial reactants. The formation of gypsum and ettringite leads to expansion, cracking, deterioration, and disruption of concrete structures. In addition to the formation of ettringite and gypsum and its subsequent expansion, the deterioration due to sulfate attack is partially caused by the degradation of calcium silicate hydrate (C–S–H) gel through leaching of the calcium compounds. This process leads to loss of C–S–H gel stiffness and overall deterioration of the cement paste matrix.

Khatib and Wild [35] investigated the effect of metakaolin on the sulfate resistance of mortar. Two types of cements were used; high C_3A and intermediate C_3A content. Cement was replaced with 0, 5, 10, 15, 20, and 25% of metakaolin (MK). Specimens of size 25 × 25 × 285 mm were moist-cured in air for 14 days, and their length was measured before immersing in 5% Na_2SO_4 solution. Test results demonstrated that (1) expansion decreased systematically with the increase in MK content for both types of cement; (2) for mortars containing high C_3A cement, the control mortar and mortars containing 5 and 10% MK showed rapid expansion and deterioration between 40 and 70 days exposure to sulfate solution. Mortars containing 15 and 20% MK showed small but quite sharp expansions also between 40 and 70 days, but subsequently stabilized. By 520 days, mortar containing 20 and 25% MK showed no significant changes in length; (3) with intermediate C_3A content in cement, trend was similar to the mortars made with high C_3A content. The control mortar (0% MK) showed significant expansion followed by 5% MK mortar. Mortar containing 10% MK exhibited little expansion and other mortar specimens containing 15, 20 and 25% were found to be extremely stable, showing little overall change and no sign of deterioration after 520 days of exposure in sodium sulfate solution.

Courard et al. [21] studied the sulfate resistance of mortars containing 0, 10, 15, and 20% metakaolin as partial replacement of cement. One mixture with natural kaolin was also made. The specimens were made with 1:3 with water-cementitious ratio 0.5. Resistance to sulfate was determined by measuring the changes in length of prismatic specimens when stored in a standard sulfate solution. At the age of 28 days, the specimens were measured for length and placed in the sulfate solution having a concentration of 16.0 ± 0.5 g/l SO_4^{2-} and prepared by adding sodium sulfate (Na_2SO_4) to water. Corresponding control specimens were placed in limewater. Changes in length were measured after storage periods of 4, 8, 12, 16, 20, 28, 40 and 52 weeks at 23 ± 2°C. It was observed that OPC mortar exhibited expansion only after few days. Variation in length after only 84 days was 3.7%. Behaviors of metakaolin-modified mortar in comparison with reference mortar, an inhibition of sulfate attack was observed, especially for more than 10% replacement of cement part metakaolin, by consuming $Ca(OH)_2$, has a large positive effect on mortar durability in the sulfate environment.

Lee et al. [44] examined the resistance of mortars made with varying percentages of metakaolin (MK), exposed to magnesium sulfate solutions. The resistance of mortar specimens was evaluated by expansion and compressive measurements. Cement was partially replaced with 0, 5, 10 and 15% of MK by mass. The cementitious material/fine aggregate ratio was maintained at 2 with water-cementitious material ratio as 0.45. All mortar and paste samples were immersed in magnesium sulfate solution ($MgSO_4$) for 360 days. Sulfate solution concentrations were 0.42% (3,380 ppm of $SO4^{2-}$), 1.27% (10,140 ppm of $SO4^{2-}$) and 4.24% (33,800 ppm of $SO4^{2-}$), respectively. Results of expansion of mortars at 360 days and reduction in compressive strength are shown in Figs. 4.31 and 4.32, respectively. It was observed that with the increase in concentration from 0.42 to 4.24% resulted in a large increasing expansion, with increasing the solution concentrations. However, 0.42% had little effect on expansion. This implied the possible presence of the critical concentration influencing the expansion of mortar specimens with or without MK. However, with respect to the reduction in the compressive strength of mortar specimens attacked by magnesium sulfate solution, a general trend indicating an almost linear increase of values in the reduction in compressive strength at 360 days with the increase in concentration of the solutions was observed.

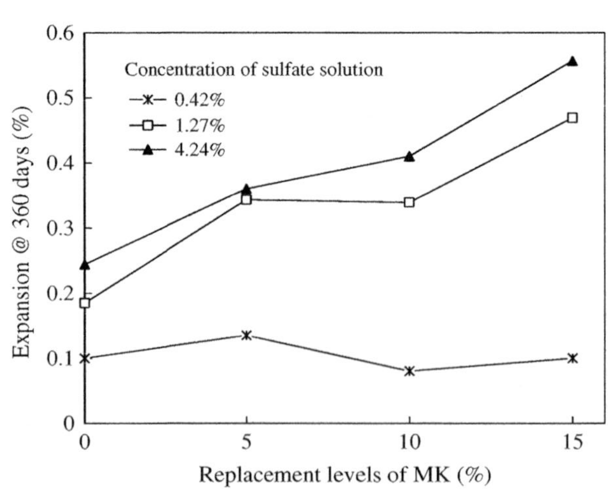

Fig. 4.31 Effect of solution concentrations and MK replacement levels on the expansion of mortar specimens exposed to magnesium sulfate solution [44]

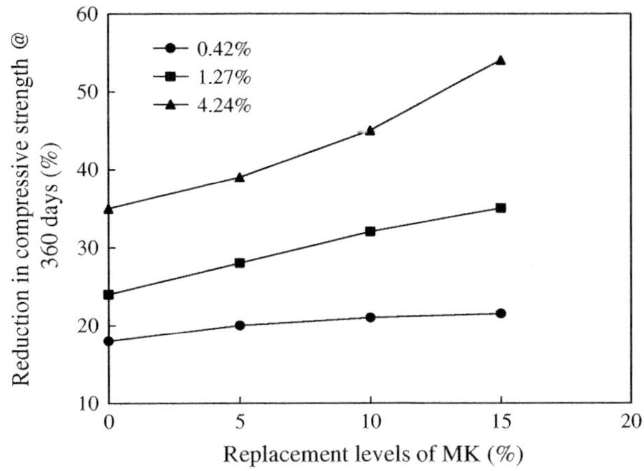

Fig. 4.32 Effect of solution concentrations and MK replacement levels on the reduction in the compressive strength of mortar specimens exposed to magnesium sulfate solutions [44]

Al-Akhras [1] investigated the durability of concrete made with metakaolin (MK) subjected sulfate attack. MK contents were 5, 10, and 15% as partial replacement of cement. Concretes were made with two w/b ratios of 0.50 and 0.60. The other experimental parameters were: initial moist curing period (3, 7, and 28 days), curing type (moist and autoclaving), and air content (1.5 and 5%). The degree of sulfate attack was determined by measuring the expansion of concrete prisms (75 × 75 × 300 mm).

The effect of MK replacement level on the variation of sulfate expansion is shown in Figs. 4.33 and 4.34. It was concluded that (1) the sulfate resistance of MK concrete was higher than that of plain concrete. Sulfate resistance of MK concrete increased with the increase in MK content. Plain concrete was not durable to sulfate attack, reaching maximum sulfate expansion values of 0.4 and 0.45% after 18 months at w/b ratios of 0.50 and 0.60, respectively; (2) the 10 and 15% MK concrete at both w/b ratios showed excellent durability to sulfate attack. The sulfate expansion was 0.10 and 0.07% for concrete at w/b ratio of 0.50, and 0.13

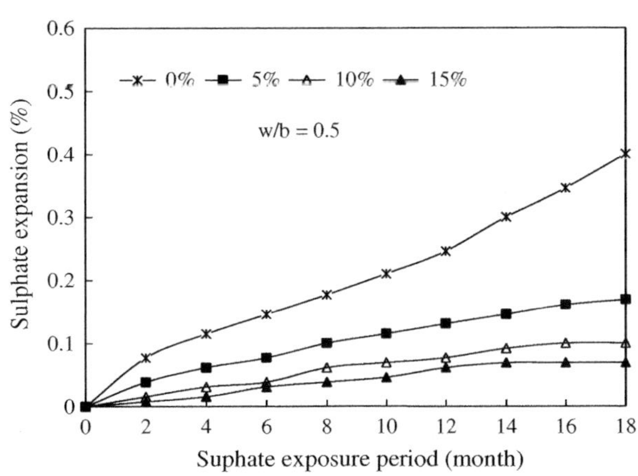

Fig. 4.33 Effect of metakaolin replacement level on the variation of sulfate expansion with sulfate exposure period at w/b ratio of 0.50 [1]

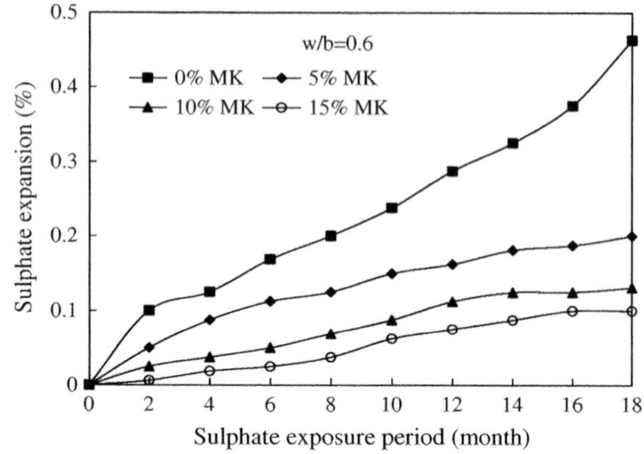

Fig. 4.34 Effect of metakaolin replacement level on the variation of sulfate expansion with sulfate exposure period at w/b ratio of 0.60 [1]

and 0.10% for concrete at w/b ratio of 0.60, respectively; (3) MK (5%) concrete showed intermediate durability to sulfate attack reaching maximum sulfate expansion values of 0.17 and 0.2% for concrete at w/b ratios of 0.50 and 0.60, respectively. The enhancement in the sulfate resistance of concrete with the increase in MK content can be explained by the following mechanisms. First, the replacement of a portion of Portland cement with MK reduces the total amount of tricalcium aluminate hydrate in the cement paste matrix of concrete. The second mechanism is through the pozzolanic reaction between MK and calcium hydroxide released during the hydration of cement, which consumes part of the calcium hydroxides. Thus, the quantity of expansive gypsum formed by the reaction of calcium hydroxide will be less in MK concrete than in plain concrete. Furthermore, the formation of secondary C–S–H by the pozzolanic reaction, although less dense than the primary C–S–H gel, is effective in filling and segmenting large capillary pores into small, discontinuous capillary pores through pore size refinement, decreasing the total permeability of concrete. In addition to the pozzolanic reaction, the filler action of MK due to the fine particle size of MK (1 μm) compared to the particle size of cement (12 μm) further densifies the pore structure of MK concrete to enhance the resistance of MK concrete to sulfate attack.

Kim et al. [38] examined the resistance of mortars containing metakaolin. The resistance to acid attack was assessed by immersion test in 2% sulfuric acid solution until 8 weeks after 28 days of water curing. Seven mix proportions of mortar were made with 0, 5 and 15% MK content having w/b ratio of 0.50. Significant reduction in compressive strength of concrete mixtures containing metakaolin was observed compared control mixture. The compressive strength at 56 days revealed a reduction ratio of about 20% compared to the specimens replaced with 5 and 15% of MK. Test results related to the resistance of mortar to sulfuric acid verified that the reduction rate of the compressive strength at 56 days was similar for replacement rates of 5 and 15% by metakaolin.

4.6.5 Corrosion Résistance

Batis et al. [11] investigated the influence of metakaolin on the corrosion resistance of cement mortar. A poor Greek kaolin with low kaolinite content was thermally treated and the produced metakaolin (MK) was ground to the appropriate fineness (20% residue at 13.6 μm). In addition, a commercial metakaolin (MKC) of high purity was also used. For the corrosion measurements, mortar mixes were made by replacing cement or sand with 10 and 20% of metakaolin. For the corrosion measurements, mortar specimens were then exposed to the corrosive environment of either partial or total immersion in 3.5% w/w NaCl solution. The corrosion tendency of the reinforcing steel bars was estimated in all types of the mortar specimens, by half-cell potential development versus the exposure time in the corrosive environment. Plots of this evolution are shown in Fig. 4.35. Initially the potential values of almost all specimens immersed in chloride environment equal −200 mV versus saturated calomel electrode (SCE). Thereafter decay was observed to more negative values and finally after 8 months of exposure it differentiates between the various specimens. It is well established (C876, 1991) regarding the significance of the numerical value of the potentials that, if potentials over an area are numerically less than 0.200 V versus SCE, there is a greater than 90% probability that there is no corrosion of the reinforcing steel bars at the time of measurement, whereas if potentials over the examined area fluctuate between −0.200 and −0.350 V versus SCE, the corrosion activity of the reinforcing steel is uncertain. Finally, if potentials are numerically greater than −0.350 V versus SCE, there is a greater than 90% probability that corrosion of reinforcing steel bars occurs.

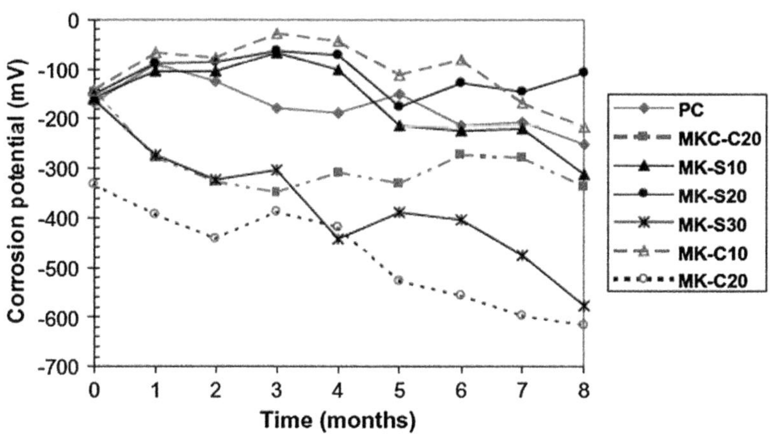

Fig. 4.35 Corrosion resistance of cement mortar [11]

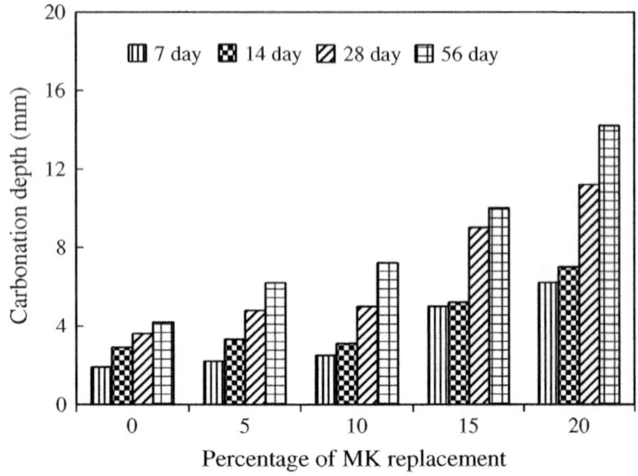

Fig. 4.36 Carbonation depth versus replacement level of metakaolin [38]

4.6.6 Carbonation

Kim et al. [38] studied the effect of metakaolin on the carbonation of high-strength concrete. Cement was replaced with 0, 5, 10, 15 and 20% of metakaolin. Control mixture had 563 kg of cement, 141 kg of fly ash, 532 kg of fine aggregate, 915 kg of coarse aggregate, and 176 kg of water. The resistance of concrete to carbonation was assessed by the phenolphthalein indicator method. The accelerated conditions were 5% CO_2, 60% RH and 30°C. The carbonation depth of concrete was measured at 7, 14, 28 and 56 days, and the results are shown in Fig. 4.36. It was observed that the carbonation depth of concrete mixtures increased with increase in metakaolin content, and also with age. At 28 days, carbonation depth of concrete increased by 20–30% with MK content of 5 and 10%, whereas carbonation depth increased by 40 and 70% at 56 days. On the other hand, the carbonation depth increases to about 100–370% for replacement ratios of 15 and 20% regardless of the age of concrete. It is probably due to the fact that the replacement of cement by MK decreases the content of portlandite in hydrate products due to pozzolanic reaction.

4.6.7 Creep and Shrinkage

Wild et al. [69] showed that, for cement–MK pastes (w/b = 0.55), autogenous shrinkage increased for MK content up to a maximum of 10%; then, it appeared to be comparable to that of the control cement paste for MK content above 15%. An expansion up to 14 days for all the compositions (0–25% MK) was also observed, except for 10% MK. Kinuthia et al. [39] also found, in cement–MK pastes (w/b = 0.50), that 5 and 10% MK increased the autogenous shrinkage of cement pastes, while at 15 and 20% MK, they observed a significant decrease. The effect of MK on autogenous shrinkage of cement pastes can be the consequence of four phenomena: (1) cement dilution by MK, less cement generating less shrinkage, (2) heterogeneous nucleation of hydrates on the surface of MK particles, accelerating cement hydration and, consequently, increasing shrinkage, (3) pozzolanic reaction of MK with CH produced by cement and, (4) increase of capillary tension, due to the refinement of pore size distribution, leading to an increase in autogenous shrinkage.

Brooks and Johari [14] studied the effect of metakaolin (MK) on creep and shrinkage of concrete mixtures containing 0, 5, 10, and 15% of MK. Control concrete mixture proportion was 1:1.5:2.5 with a water-to-cement ratio of 0.28. Autogenous shrinkage was measured from the time of initial set to 24 h and from 24 h till 200 days. Total shrinkage consists of autogenous shrinkage from 24 h and drying shrinkage due to water loss to environment. The results of autogenous, total shrinkage and creep are given in Tables 4.33 and 4.34. It was concluded that (1) at very young ages, autogenous shrinkage was reduced with the increase in MK content. After 24 h, autogenous shrinkage of MK15 concrete was reduced by 65% in comparison with OPC concrete. The reduction of autogenous shrinkage could be due to the dilution effect which is caused by reduction in cement content; (2) at long-term, autogenous shrinkage increased with the increase in MK content. At 200 days, there were 91, 80, and 56% increase in autogenous shrinkage of concrete with 5, 10, and 15% MK content, respectively; and (3) inclusion of MK reduced both total and basic creep of concrete, with a greater reduction in creep at higher replacement levels. The reduction in creep could be attributed to a denser pore structure, stronger paste matrix and improved paste aggregate interface of MK concretes as a result of the formation of additional hydrate phases from secondary

Table 4.33 Results for the 200-day total and autogenous shrinkage of concrete [14]

Item/concrete mixtures	OPC	MK5	MK10	MK15
Total shrinkage from initial set (10^{-6})	861	713	618	516
Total shrinkage from 24 h (10^{-6})	558	499	455	410
Total autogenous shrinkage from initial set (10^{-6})	445	485	419	327
Autogenous shrinkage 24 h (10^{-6})	142	271	256	221
Percentage of A.S.–T.S. from initial set	52	68	68	63
Percentage of A.S.–T.S. from 24 h	25	54	56	54
Weight loss of the exposed specimens (%)	0.85	0.67	0.73	0.76

Table 4.34 Results of creep after 200 days [14]

Concrete mixture	Creep (10^{-6})		Stress (MPa)	Specific Creep (10^{-6}/MPa)		Drying creep (10^{-6}/MPa)
	Dry	Sealed		Dry	Sealed	
OPC	358	285	13.9	25.8	20.5	5.3
MK5	312	235	16.8	18.5	14.0	4.5
MK10	201	126	17.7	11.3	7.2	4.1
MK15	171	115	17.9	9.5	6.4	3.1

pozzolanic reaction of MK and its filler effect. The 200-day specific drying creep which is taken as the difference between the specific total creep and basic creep was also reduced for MK concretes.

Gleize et al. [27] investigated the effects of high-purity metakaolin (MK) on the autogenous shrinkage of pastes. Pastes were made with two water/binder ratios (0.3 and 0.5), and cement was partially replaced with 5, 10, 15 and 20% metakaolin. The MK was high-purity MK with a surface area of 23,000 m^2/kg and an average particle size of 4 μm. Figure 4.37 shows the autogenous shrinkage with time for pastes containing up to 20% MK, at water–solid ratios of 0.3 and 0.5. They concluded that (1) no overall expansion of pastes was observed at early ages; and (2) long-term autogenous shrinkage of cement–MK paste, with w/b ratios of 0.3 and 0.5, decreased with the increase in MK content.

Khatib [37] studied the length change of concrete containing metakaolin (MK) at a low water to binder ratio of 0.30. Portland cement (PC) was partially replaced

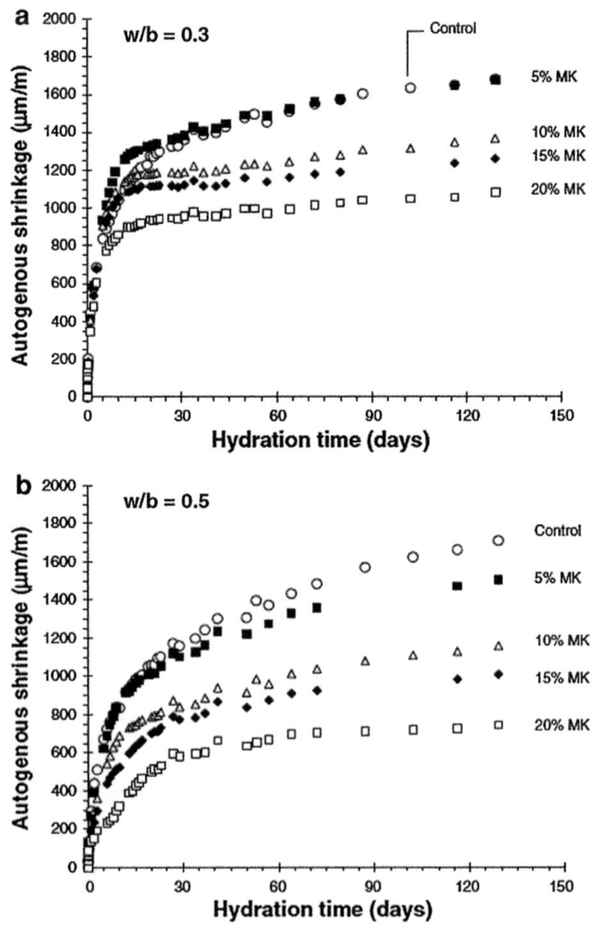

Fig. 4.37 Autogenous shrinkage of cement paste [27]

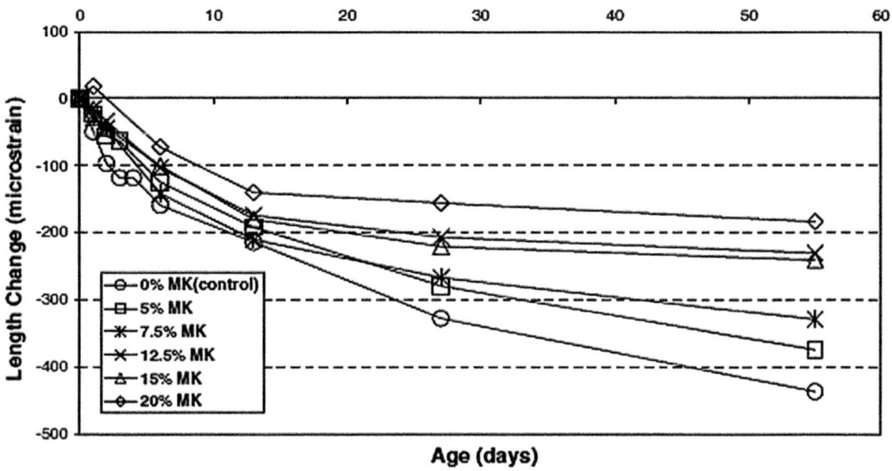

Fig. 4.38 Length change for MK concrete subjected to air curing at 20°C, 55% RH [37]

with 0–20% MK. Results of length change for MK concrete subjected to air curing at 20°C, 55% RH is shown in Fig. 4.38. There appeared to be systematic decrease in shrinkage as the MK content increases. Partial replacement of PC with 20% MK can reduce the long-term shrinkage by more than half compared with that of the control.

References

1. Al-Akhras, N.M.: Durability of metakaolin to sulfate attack. Cem. Concr. Res. **36**(9), 1727–1734 (2006)
2. Ambroise, J., Maximilien, S., Pera, J.: Properties of metakaolin blended cements. Adv. Cem. Based Mater. **1**(4), 161–168 (1994)
3. Aquino, W., Lange, D.A., Olek, D.A.: The influence of metakaolin and silica fume on the chemistry of alkali–silica reaction products. Cem. Concr. Compos. **23**(6), 485–493 (2001)
4. Asbridge, A.H., Walters, G.V., Jones, T.R.: Ternary blended concretes-OPC/GGBFS/metakaolin, pp. 547–557. Concrete across Borders, Denmark (1994)
5. Asbridge, A.H., Chadbourn, G.A., Page, C.L.: Effects of metakaolin and the interfacial transition zone on the diffusion of chloride ions through cement mortars. Cem. Concr. Res. **31**(11), 1567–1572 (2001)
6. Asbridge, A.H., Page, C.L., Page, M.M.: Effects of metakaolin, water/binder ratio and interfacial transition zones on the micro hardness of cement mortars. Cem. Concr. Res. **32**(9), 1365–1369 (2002)
7. ASTM C 618.: Standard Specification for Coal Fly Ash and Raw or Calcined Natural Pozzolan for Use as a Mineral Admixture in Concrete. Annual Book of ASTM Standards, Philadelphia (1993)
8. Badogiannis, E., Papadakis, V.G., Chaniotakis, E., Tsivilis, S.: Exploitation of poor Greek kaolins: strength development of metakaolin concrete and evaluation by means of k-value. Cem. Concr. Res. **34**(6), 1035–1041 (2004)

9. Badogiannis, E., Kakali, G., Dimopoulou, G., Chaniotakis, E., Tsivilis, S.: Metakaolin as a main cement constituent: exploitation of poor Greek kaolins. Cem. Concr. Compos. **27**(2), 197–203 (2005)
10. Badogiannis, E., Tsivilis, S.: Exploitation of poor Greek Kaolins: durability of metakaolin concrete. Cem. Concr. Compos. **31**(2), 128–133 (2009)
11. Batis, G., Pantazopoulou, P., Tsivilis, S., Badogiannis, E.: The effect of metakaolin on the corrosion behavior of cement mortars. Cem. Concr. Compos. **27**(1), 125–130 (2005)
12. Bredy, P., Chabannet, M., Pera, J.: Microstructural and porosity of metakaolin blended cements. Mater. Res. Soc. Symp. **137**, 431–436 (1989)
13. Boddy, A., Hooton, R.D., Gruber, K.A.: Long-term testing of the chloride-penetration resistance of concrete containing high-reactivity metakaolin. Cem. Concr. Res. **31**(5), 759–765 (2001)
14. Brooks, J.J., Johari, M.M.A.: Effect of metakaolin on creep and shrinkage of concrete. Cem. Concr. Compos. **23**(6), 495–502 (2001)
15. Cabrera, J., Rojas, M.F.: Mechanism of hydration of the metakaolin–lime–water system. Cem. Concr. Res. **31**(2), 177–182 (2001)
16. Cabrera, J.G., Nwaubani, S.O.: The microstructure and chloride ion diffusion characteristics of cements containing metakaolin and fly ash. In: V.M. Malhotra (ed.), Sixth CANMET/ACI/JCI International Conference on Fly Ash, Silica Fume, Slag and Natural Pozzolans in Concrete, vol. 1, pp. 385–400. Bangkok, Thailand (1998)
17. Cachim, P., Velosa, A.L., Rocha, F.: Effect of Portuguese metakaolin on hydraulic lime concrete using different curing conditions. Construct. Build. Mater. **24**(1), 71–78 (2010)
18. Caldarone, M.A., Gruber, K.A., Burg, R.G.: High reactivity metakaolin: a new generation mineral admixture. Concr. Int. **16**(11), 37–40 (1994)
19. Changling, H., Osbaeck, B., Makovicky, E.: Pozzolanic reaction of six principal clay minerals: activation reactivity assessments and technological effects. Cem. Concr. Res. **25**(8), 1691–1702 (1995)
20. Coleman, N.J., Page, C.L.: Aspects of the pore solution chemistry of hydrated cement pastes containing metakaolin. Cem. Concr. Res. **27**(1), 147–154 (1997)
21. Courard, L., Darimont, A., Schouterden, M., Ferauche, F., Willem, X., Degeimbre, R.: Durability of mortars modified with metakaolin. Cem. Concr. Res. **33**(9), 473–1479 (2003)
22. Curcio, F., Deangelis, B.A., Pagliolico, S.: Metakaolin as pozzolanic micro filler for high-performance mortars. Cem. Concr. Res. **28**(6), 803–809 (1998)
23. Dias, W.P.S.: Reduction of concrete sorptivity with age through carbonation. Cem. Concr. Res. **30**(8), 1255–1261 (2000)
24. Frías, M., Cabrera, J.: Pore size distribution and degree of hydration of MK–cement pastes. Cem. Concr. Res. **30**(4), 561–569 (2000)
25. Frías, M., Sánchez de Rojas, M.I., Cabrera, J.: The effect that the pozzolanic reaction of metakaolin has on the heat evolution in metakaolin–cement mortars. Cem. Concr. Res. **30**(2), 209–216 (2000)
26. Frías, M., Cabrera, J.: Influence of MK on the reaction kinetics in MK/lime and MK-blended cement systems at 20°C. Cem. Concr. Res. **31**(4), 519–527 (2001)
27. Gleize, F.J.P., Cyr, M., Escadeillas, G.: Effects of metakaolin on autogenous shrinkage of cement pastes. Cem. Concr. Compos. **29**(2), 80–87 (2007)
28. Gonçalves, J.P., Tavares, L.M., Filho, R.D.T., Fairbairn, E.M.R.: Performance evaluation of cement mortars modified with metakaolin or ground brick. Construct. Build. Mater. **23**(5), 1971–1979 (2009)
29. Gruber, K.A., Ramlochan, T., Boddy, A., Hooton, R.D., Thomas, M.D.A.: Increasing concrete durability with high-reactivity metakaolin. Cem. Concr. Res. **23**(6), 479–484 (2001)
30. Igurashi, S., Bentur, A., Mindess, S.: Micro-hardness testing of cementitious materials. Adv. Cem. Based Mater. **4**, 48–57 (1996)
31. Janotka, I., Puertas, F., Palacios, M., Kuliffayova, M., Varga, C.: Metakaolin sand-blended-cement pastes: rheology, hydration process and mechanical properties. Construct. Build. Mater. **24**(5), 791–802 (2010)

32. Jin, X., Li, Z.: Effects of mineral admixture on properties of young concrete. J. Mater. Civil Eng. **15**(5), 435–442 (2003)
33. Khatib, J.M., Mangat, P.S.: Absorption characteristics of concrete as a function of location relative to casting position. Cem. Concr. Res. **25**(5), 999–1010 (1995)
34. Khatib, J.M., Wild, S.: Pore size distribution of metakaolin paste. Cem. Concr. Res. **26**(10), 1545–1553 (1996)
35. Khatib, J.M., Wild, S.: Sulfate resistance of metakaolin mortar. Cem. Concr. Res. **28**(1), 83–92 (1998)
36. Khatib, J.M., Clay, R.M.: Absorption characteristics of metakaolin concrete. Cem. Concr. Res. **34**(1), 19–29 (2004)
37. Khatib, J.M.: Metakaolin concrete at a low water to binder ratio. Construct. Build. Mater. **22**(8), 1691–1700 (2008)
38. Kim, H.S., Lee, S.H., Moon, H.Y.: Strength properties and durability aspects of high strength concrete using Korean metakaolin. Construct. Build. Mater. **21**(6), 1229–1237 (2007)
39. Kinuthia, J.M., Wild, S., Sabir, B.B., Bai, J.: Self-compensating autogenous shrinkage in Portland cement–metakaolin-fly ash pastes. Adv. Cem. Res. **12**(1), 35–43 (2000)
40. Klimesch, D.S., Ray, A.: Autoclaved cement-quartz pastes with metakaolin additions. Adv. Cem. based Mater. **7**(3–4), 109–117 (1998)
41. Kostuch, J.A., Walters, G,V., Jones, T.R.: High performance concrete incorporating metakaolin—a review, pp. 1799–1811. Concrete 2000, University of Dundee (1993)
42. Lagier, F., Kurtis, K.E.: Influence of Portland cement composition on early age reactions with metakaolin. Cem. Concr. Res. **37**(10), 1411–1417 (2007)
43. Larbi, J.A., Bijen, J.M.: Influence of pozzolans on the Portland cement paste–aggregate interface in relation to diffusion of ions and water absorption in concrete. Cem. Concr. Res. **22**, 551–562 (1992)
44. Lee, S.T., Moon, H.Y., Hooton, R.D., Kim, J.P.: Effect of solution concentrations and replacement levels of metakaolin on the resistance of mortars exposed to magnesium sulfate solutions. Cem. Concr. Res. **35**(7), 1314–1323 (2005)
45. Li, Z., Ding, Z.: Property improvement of Portland cement by incorporating with metakaolin and slag. Cem. Concr. Res. **33**(4), 579–584 (2003)
46. Lyubimova, T.Y., Pinus, E.R.: Crystallization structure in the contact zone between aggregate and Cem.. Concr. Kolloidn. Z. (USSR) **24**(5), 578–587 (1962)
47. Murat, M.: Hydration reaction and hardening of calcined clays and related mineral-I Preliminary investigations on metakaolin. Cem. Concr. Res. **13**(2), 259–266 (1983)
48. Oriol, M., Pera, J.: Pozzolanic activity of metakaolin under microwave treatment. Cem. Concr. Res. **25**(2), 265–270 (1995)
49. Poon, C.S., Lam, L., Kou, S.C., Wong, Y.L., Wong, R.: Rate of pozzolanic reaction of metakaolin in high-performance Cem. pastes. Cem. Concr. Res. **31**(9), 1301–1306 (2001)
50. Poon, C.S., Azhar, S., Anson, M., Wong, Y.L.: Performance of metakaolin concrete at elevated temperatures. Cem. Concr. Compos. **25**(1), 83–89 (2003)
51. Poon, C.S., Kou, S.C., Lam, L.: Compressive strength, chloride diffusivity and pore structure of high performance metakaolin and silica fume concrete. Construct. Build. Mater. **20**(10), 858–865 (2006)
52. Potgieter-Vermaak, S.S., Potgieter, J.H.: Metakaolin as an extender in South African cement. J. Mater. Civil Eng. **18**(4), 619–623 (2006)
53. Qian, X., Li, Z.: The relationships between stress and strain for high-performance concrete with metakaolin. Cem. Concr. Res. **31**(11), 1607–1611 (2001)
54. Ramlochan, T., Thomas, M., Gruber, K.A.: The effect of metakaolin on alkali–silica reaction in concrete. Cem. Concr. Res. **30**(3), 339–344 (2000)
55. Razak, H.A., Chai, H.K., Wong, H.S.: Near surface characteristics of concrete containing supplementary cementing materials. Cem. Concr. Compos. **26**(7), 883–889 (2004)
56. Rojas, M.F., Cabrera, J.: The effect of temperature on the hydration rate and stability of the hydration phases of metakaolin–lime–water systems. Cem. Concr. Res. **32**(1), 133–138 (2002)

57. Rojas, M.F., Sánchez de Rojas, M.I.: The effect of high curing temperature on the reaction kinetics in MK/lime and MK-blended cement matrices at 60°C. Cem. Concr. Res. **33**(5), 643–649 (2003)
58. Rojas, M.F.: Study of hydrated phases present in a MK–lime system cured at 60°C and 60 months of reaction. Cem. Concr. Res. **36**(5), 827–831 (2006)
59. Roy, D.M., Arjunan, P., Silsbee, M.R.: Effect of silica fume, metakaolin, and low-calcium fly ash on chemical resistance of concrete. Cem. Concr. Res. **31**(12), 1809–1813 (2001)
60. Saito, M., Kawamura, M.: Resistance of the cement–aggregate interfacial zone to the propagation of cracks. Cem. Concr. Res. **16**(5), 653–661 (1986)
61. Salvador, S.: Pozzolanic properties of flash-calcined kaolinite: a comparative study with soak-calcined products. Cem. Concr. Res. **25**(1), 102–112 (1995)
62. Sha, W., Pereira, G.B.: Differential scanning calorimetry study of ordinary Portland cement paste containing metakaolin and theoretical approach of metakaolin activity. Cem. Concr. Compos. **23**(6), 455–461 (2001)
63. Shvarzman, A., Kovler, K., Grader, G.S., Shter, G.E.: The effect of dehydroxylation/amorphization degree on pozzolanic activity of kaolinite. Cem. Concr. Res. **33**(3), 405–416 (2003)
64. Sun, W., Zhang, Y.-S., Lin, W., Liu, Z.-Y.: In situ monitoring of the hydration process of K-PS geopolymer cement with ESEM. Cem. Concr. Res. **34**(6), 935–940 (2004)
65. Tafraoui, A., Escadeillas, G., Lebaili, S., Vidal, T.: Metakaolin in the formulation of UHPC. Construct. Build. Mater. **23**(2), 669–674 (2009)
66. Thomas, M.D.A., Gruber, K.A., Hooton, R.D.: The use of high reactivity metakaolin in high performance concrete. 1st Engineering Foundation Conference on High Strength Concrete, Hawaii **1997**, 517–530 (1997)
67. Wild, S., Khatib, J.M., Jones, A.: Relative strength, pozzolanic activity and cement hydration in superplasticised metakaolin concrete. Cem. Concr. Res. **26**(10), 1537–1544 (1996)
68. Wild, S., Khatib, J.M.: Portlandite Consumption in metakaolin cement pastes and mortars. Cem. Concr. Res. **27**(1), 137–146 (1997)
69. Wild, S., Khatib, J., Roose, J.L.: Chemical and autogenous shrinkage of Portland cement–metakaolin pastes. Adv. Cem. Res. **10**(3), 109–119 (1998)
70. Wong, H.S., Razak, H.A.: Efficiency of calcined kaolin and silica fume as cement replacement material for strength performance. Cem. Concr. Res. **35**(4), 696–702 (2005)
71. Zhang, M.H., Malhotra, V.M.: Characteristics of a thermally activated alumino-silicate pozzolanic material and its use in concrete. Cem. Concr. Res. **25**(8), 1713–1725 (1995)

Chapter 5
Rice Husk Ash

5.1 Introduction

Rice husk is an agricultural residue obtained from the outer covering of rice grains during milling process. Current rice production in the world is more than 700 million tons. Rice husk constitutes about 20% of the weight of rice. It contains about 50% cellulose, 25–30% lignin, and 15–20% of silica.

Rice husk ash (RHA) is generated by burning rice husk. On burning, cellulose and lignin are removed leaving behind silica ash. The controlled temperature and environment of burning yields better quality of rice-husk ash as its particle size and specific surface area are dependent on burning condition. The ash produced by controlled burning of the rice husk between 550°C and 700°C incinerating temperature for 1 h transforms the silica content of the ash into amorphous phase. The reactivity of amorphous silica is directly proportional to the specific surface area of ash. The ash so produced is pulverized or ground to required fineness and mixed with cement to produce blended cement.

The production of rice husk ash is primarily in areas where rice crops are abundant. Fully burned rice husk ash could be gray, purple or white, depending on the impurities present and the burning conditions. In open field burning or in uncontrolled combustion environments, the ash will remain mostly un-reactive because of the unfavorable mineralogical composition. Partially burned rice husk ash contains carbon, and is therefore black in colour. The silica in rice husk ash can is amorphous or crystalline, depending on the manner in which it is burned and cooled. If the ash is formed in open field burning or in uncontrolled combustion environments, it will retain a large proportion of non-reactive silica in the form of cristobalite and tridymite, and would require grinding to develop pozzolanic activity.

Rice husk ash reactivity is attributed to its high content of amorphous silica, and to its very large surface area, governed by the porous structure of the particles [19, 36]. Generally, reactivity is favored also by increasing the fineness of the

R. Siddique and M. Iqbal Khan, *Supplementary Cementing Materials*,
Engineering Materials, DOI: 10.1007/978-3-642-17866-5_5,

pozzolanic material. However, Mehta [35] reported that grinding of RHA to a high degree of fineness should be avoided, since it derives its pozzolanic activity mainly from the internal surface area of the particles.

The form of silica obtained after combustion of rice husk depends on the temperature and duration of combustion of rice husk. Mehta [35] reported that amorphous silica can be produced by maintaining the combustion temperature below 500°C under oxidizing conditions for prolonged periods or up to 680°C with a hold time less than 1 min. Yeoh et al. [53] reported that RHA can remain in the amorphous form at combustion temperatures of up to 900°C if the combustion time is less than 1 h, while crystalline silica is produced at 1,000°C with combustion time greater than 5 min. Chopra et al. [18] observed that at burning temperatures up to 700°C, the silica is in amorphous form. Hwang and Wu [29] studied the effect of burning temperatures and the chemical composition of rice husk ash. They observed that at 400°C, polysaccharides begin to depolymerize. Above 400°C, dehydration of sugar units occurs. At 700°C, the sugar units decompose. At temperatures above 700°C, unsaturated products react together and form a highly reactive carbonic residue. The X-ray data and chemical analyses of RHA produced under different burning conditions given by Hwang and Wu [29] showed that the higher the burning temperature, the greater the percentage of silica in the ash. K, S, Ca, Mg as well as several other components were found to be volatile.

Della et al. [20] presented the processing and characterization of high specific surface area silica from RHA. They reported that a 95% silica powder could be produced after heat-treatment at 700°C for 6 h. And specific surface area of particles was increased after wet milling from 54 to 81 m^2/g.

5.1.1 Advantages of Using RHA

Rice-husk ash is a very fine pozzolanic material. The utilization of rice husk ash as a pozzolanic material in cement and concrete provides several advantages such as

- Improved strength
- Enhanced durability properties,
- Reduced materials costs due to cement savings, and
- Environmental benefits related to the disposal of waste materials and to reduced carbon dioxide emissions.

5.1.2 Applications of Rice Husk Ash

Rice husk ash can is used in following applications:

- Blended cements
- Green concrete

- High performance concrete
- Refractory
- Roofing shingles
- Ceramic glaze
- Insulator
- Waterproofing chemicals
- Oil spill absorbent

5.2 Properties of RHA

5.2.1 Physical Properties

Completely burnt rice-husk is grey to white in color, while partially burnt rice-husk ash is blackish.

Della et al. [20] presented the macroscopies of RHA as-received and after-burning out at 700°C for 6 h, and wet-grinding for 80 min in a jar mill. The as-received RHA samples were black with some gray particles, resulting from different stages of the carbon combustion during burning of rice husk (Fig. 5.1a). The active silica obtained after the heating and grinding presents reduced size of particles and grey coloration due to the lower content carbonaceous material (Fig. 5.1b). Level of carbon detected in RHA before thermal treatment was 18.60%. The level of carbon decreased considerably after the thermal treatment; 0.14% for the RHA 700° for 6 h sample (Fig. 5.1b). The amount of black particles decreased with increase in calcination temperature and time. At 700°C for 6 h, the thermal treatment yielded bright white silica.

Rice husk ash is a very fine material. Average particle size of rice-husk ash ranges from 3 to 10 μm. Physical properties values as reported by few authors are given in Table 5.1.

5.2.2 Particle Size Distribution

Della et al. [20] determined the particle size distribution of as-received RHA sample (SiO_2 72.1%), and RHA (SiO_2 94.95%) after burning out at 700°C for 6 h.

Table 5.1 Physical properties of RHA

Property	Value		
	Mehta [36]	Bui et al. [11]	Ganesan et al. [23]
Mean particle size (μm)	–	5	3.80
Specific gravity	2.06	2.10	2.06
Fineness: passing 45 μm (%)	99	–	99
Specific surface (m^2/g)	–	–	36.47

Fig. 5.1 Colour of RHA

The particle mean size of RHA after burning out at 700°C for 6 h was around 33 μm (Fig. 5.2), being all particles lower than 112 μm. After milling for 80 min, the mean size was reduced to 0.68 μm, being 100% of particles lower than 6 μm (Fig. 5.3). As raw material, RHA presented a specific surface area around 177 m^2/g. After burning out at 700°C for 6 h, it was changed to 54 m^2/g. This decrease in specific area is proportional to heating temperature and time, which causes an agglomeration effect, diminishing porosity. After wet grinding for 80 min, the particle's specific area increased to 81 m^2/g.

Agarwal [9] reported the particle size distribution (Fig. 5.3) of rice husk ash collected from paper mill using rice husk as fuel (90.52% SiO_2).

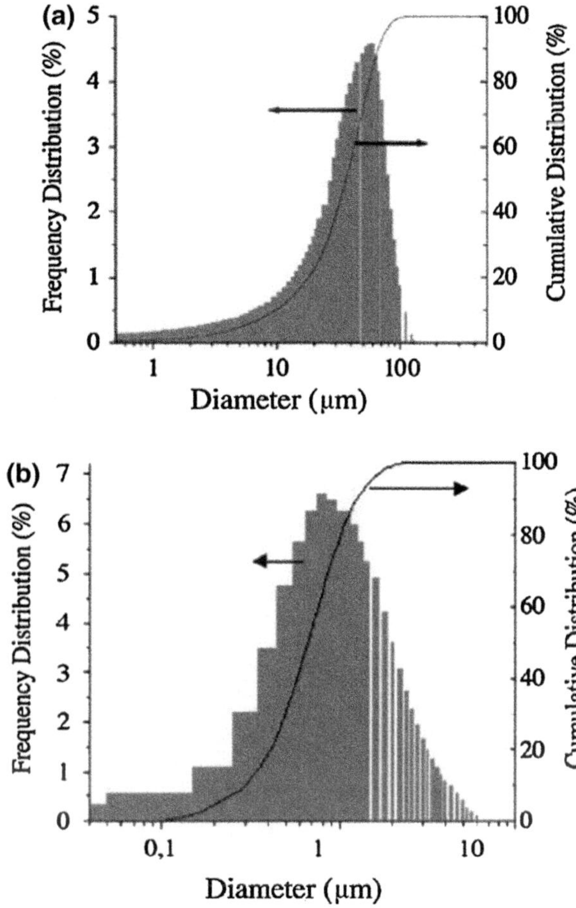

Fig. 5.2 Particle size distribution of RHA: **a** as received; **b** after wet grinding for 80 min in a jar mill [20]

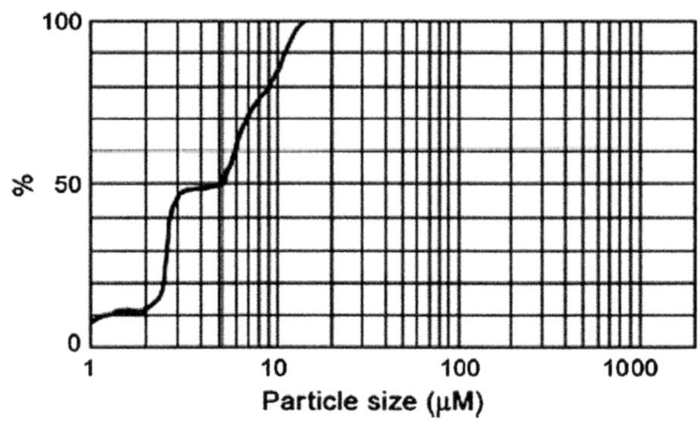

Fig. 5.3 Particle size analysis of RHA [9]

Habeeb and Fayyadh [26] examined the rice husk ash sample obtained after burning rice husk at 700°C. Electron microscope showed that RHA samples were multilayered and microporous surface (Fig. 5.4).

Fig. 5.4 SEM for RHA particle [26]

Fig. 5.5 SEM of ground rice husk ash [15]

Chindaprasirt et al. [15] examined the SEM of ground rice husk ash. SEM picture (Fig. 5.5) revealed that the rice husk ash maintains its cellular structure. After being ground, RHA consists of very irregular-shaped particles with porous cellular surface.

5.2.3 Chemical Composition

Rice husk ash is very rich in silica content. Silica content in rice husk ash is generally more than 80%. Typical chemical composition of RHA is given in Table 5.2. As per ASTM C 618 [2], the combined proportion of silicon dioxide (SiO_2), aluminium oxide ($A1_2O_3$) and iron oxide (Fe_2O_3) in the ash should be not be less than 70%, and LOI should not exceed 12% as stipulated in ASTM requirement.

Table 5.2 Chemical composition of RHA

Constituents	Percentage		
	Mehta [36]	Bui et al. [11]	Ganesan et al. [23]
Silica (SiO_2)	87.2	86.98	87.32
Alumina (Al_2O_3)	0.15	0.84	0.22
Iron oxide (Fe_2O_3)	0.16	0.73	0.28
Calcium oxide (CaO)	0.55	1.40	0.48
Magnesium oxide (MgO)	0.35	0.57	0.28
Sodium oxide (Na_2O)	1.12	0.11	1.02
Potassium oxide (K_2O)	3.68	2.46	3.14
Sulfur oxide (SO_3)	0.24		–
LOI	8.55	5.14	2.10

5.3 Pozzolanic Activity

Rice husk contains high amount of SiO_2. Well burnt and well-ground rice husk ash is very active and considerably improves the strength and durability of cement and concrete. The sensitivity of burning conditions is the primary reason that prevents the widespread use of this material as pozzolan [28, 42].

Zhang et al. [54] investigated the influence of RHA on the hydration, microstructure and interfacial zone between the aggregate and cement paste. It was observed that (1) calcium hydroxide and calcium silicate hydrates were the major hydration products in the RHA paste. Due to the reaction, the paste incorporating RHA had lower $Ca(OH)_2$ content than the Portland cement paste; (2) inclusion of RHA in concrete reduced porosity, $Ca(OH)_2$, amount in the interfacial zone, width of the interfacial zone between aggregate and cement paste.

Jauberthie et al. [33] reported that concentration of silica is high on the external face of the husk, much weaker on the internal face and practically non-existent within the husk as given in Table 5.3. It was confirmed that the presence of amorphous silica is concentrated at the surface of the rice husk and not within the husk itself. Amorphous silica concentrated on the interior and exterior surfaces of the uncalcinated husk promote a pozzolanic action on the surface of the husk and therefore enable its use in cement/concrete.

Table 5.3 Micro analysis of rice husk [33]

Element	External surface of rice husk		Internal surface of rice husk		Interior surface of rice husk	
	% (by weight)	% (by atomic)	% (by weight)	% (by atomic)	% (by weight)	% (by atomic)
C	6.91	11.11	30.20	40.93	62.54	69.54
O	47.93	57.84	42.53	43.27	35.19	29.38
Si	45.16	31.05	27.27	15.80	2.27	1.08
Total	100	100	100	100	100	100

Feng et al. [22] investigated the effect of hydrochloric acid pretreatment and heating on the pozzolanic activity of RHA. Three methods were used to estimate the pozzolanic activity; (1) rapid evaluation method—conductivity measurement, the conductivity method proposed by Luxan et al. [34]. The greater the change in the conductivity of the saturated $Ca(OH)_2$ solution added with rice husk ash, the more active rice husk ash is. This method is valid to evaluate the pozzolanic activity of rice husk ash (1996); (2) measuring the rate of consumption of $Ca(OH)_2$ that reacted with rice husk ash. Rice husk ash/$Ca(OH)_2$ = 1:1, W/C = 1.0, 20°C, the $Ca(OH)_2$ used was analytical grade $Ca(OH)_2$; and (3) comparing the strength of mortar made with and without rice husk ash. Hydrochloric acid-treated rice husk was prepared by immersing rice husk in 1 N HCl aqueous solution. The husks were washed repeatedly with water until hydrochloric acid was undetected in the filtrate and then air-dried at room temperature. Two kinds of rice husk ash were obtained by heating rice husk in a batch furnace under oxidizing atmosphere by which rice husk ash can be produced about 300 g at a time. One kind of rice husk ash (ADR) was obtained by heating hydrochloric acid treated rice husk, and another kind of rice husk ash (RHA) was obtained by heating untreated rice husk. The heating temperature ranged from 350 to 1,100°C and the maintaining time was 4 h, and then the rice husk ash was ground in a ball mill for 60 min by adding grinding agent. Based on the study, they concluded that (1) with hydrochloric acid pretreatment of rice husks, the pozzolanic activity of rice husk ash was stabilized and enhanced as well; the sensitivity of the pozzolanic activity of the rice husk ash to burning conditions was reduced. The pozzolanic activity of ADR (pretreated) was slightly affected by the change of maintaining time, but the maintaining time had a great affect on the pozzolanic activity of RHA (no pretreatment); (2) the two kinds of rice husk ashes had the same rate of lime consumption at a very rapid rate in the initial period of reaction. The mechanism of reaction was consistent with diffusion control and ADR had a faster reaction rate with lime than RHA. The main reaction product was C–S–H gel; (3) during the first 12 h, RHA and ADR showed similar behavior in the increase of hydration heat of the cement. The pozzolanic activity of ADR is higher than that of RHA; and (4) because of the high of amorphous SiO_2 content in ADR with high activity, a significant increase in the strength of ADR specimen was observed compared with the strength of control mortar and that made with RHA. The cement mortar added with ADR had lower $Ca(OH)_2$ content after 7 days. The pore size distribution of the mortar with ADR showed a tendency to shift towards the smaller pore size.

Agarwal [9] studied the accelerated pozzolanic activity of rice husk ash. The RHA was collected from paper mill using rice husk as fuel (90.52% SiO_2). The pozzolanic activity of RHA was determined by compressive strength of mortar cubes cured in water at 7 and 28 days. The pozzolanic activity of the control mortar (1 part cement and 2.75 part sand) was found to be 310 kg/cm^2. Accelerated pozzolanic activity index of RHA is given in Table 5.4. It was observed that (1) RHA as received from the plant had 16% lesser activity, while the remaining fractions 150 μm passing (retaining 75 μm), 45 μm retaining and 45 μm passing had shown 10, 35 and 48% gain in activity over the control. The RHA with 100%

Table 5.4 Accelerated pozzolanic index of RHA [9]

System	Accelerated pozzolanic index (kg/cm^2)	Percentage of control	7 day (RT) kg/cm^2	28 day (RT) kg/cm^2
43 Grade OPC	310	100	250	360
RHA (1.2% LOI) (as received)	260	84	140	260
150 μm passing, retaining 75 μm	340	110	180	340
45 μm retaining	420	135	240	390
Less than 45 μm	460	148	240	415
RHA (13% LOI) <75 μm	390	126	260	400
RHA (20% LOI) <45 μm	380	123	260	410

passing 45 μm demonstrated maximum accelerated activity and (2) RHA with 13 and 20% carbon content has shown 26 and 23% higher activity, respectively.

The X-ray diffraction (Fig. 5.6) of RHA shows a peak of cristobalite and tridymite [9]. Mehta [36] has indicated that the activity of RHA was reduced. However, in the present case, increase in activity from 10 to 48% was observed. It is possible that a fraction of RHA present is showing the peaks of cristobalite and tridymite, which, on grinding, has no significant effect on the activity.

Cizer et al. [17] carried out the XRD analysis of RHA and indicated its amorphous phase with a broad band between 15 and 30 $2\theta°$ (Fig. 5.7). It contains certain amount of crystalline silica in the form of cristobalite and tridymite. This indicated that RHA was obtained by burning at relatively high temperatures

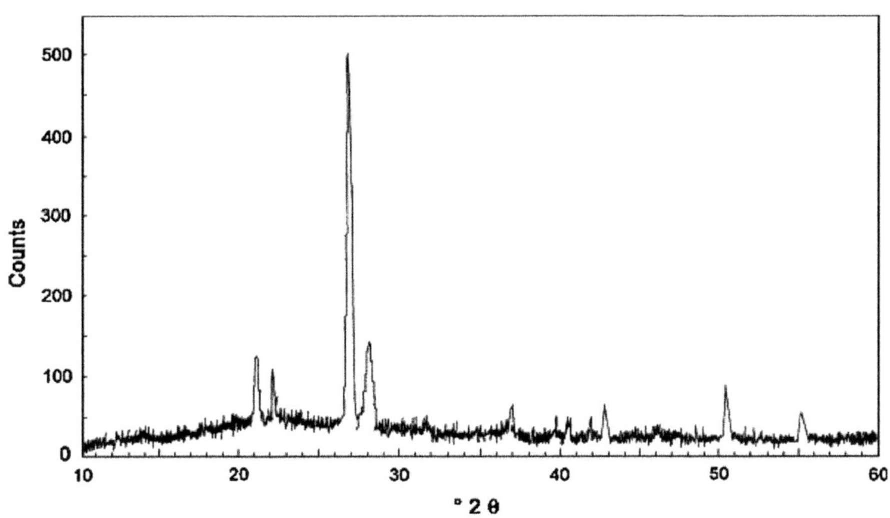

Fig. 5.6 XRD pattern of RHA [9]

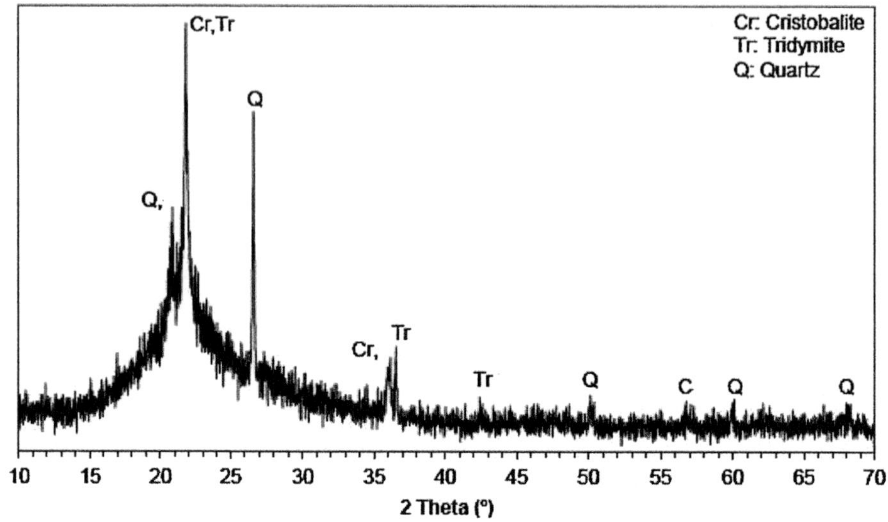

Fig. 5.7 XRD pattern of the rice husk ash [17]

(around 800–1,000°C) so that the crystallization of the amorphous silica took place.

Ramezanianpour et al. [43] determined the pozzolanic activity of rice husk ash. Rice husk was burnt at 550, 600, 650, 700 and 750°C. Time of burning was at 30, 60 and 90 min. Results showed that temperature of 650°C and 60 min burning time are the best combination. The results of pozzolanic activity are shown in Table 5.5. It can be seen that RHA demonstrated high pozzolanic activity over that of the control in accordance with ASTM C-618 test method. On the other hand, produced rice husk ash is a high reactive pozzolanic material, and entirely satisfies other requirements.

Table 5.5 Comparison in chemical and physical specifications of produced RHA with ASTM C618 [43]

Chemical requirement	ASTM	RHA results
$SiO_2 + Al_2O_3 + Fe_2O_3$, min (%)	70	89.9
SO_3, max (%)	4	0.15
Moisture content, max (%)	3	0.23
Loss on ignition, max (%)	6	5.9
Physical requirements		
Fineness: amount retained when wet-sieved on 45 μm sieve, max (%)	34	8
Strength activity index (20% RHA) at 3-day, min (% of control)	–	102
Strength activity index (20% RHA) at 7-day, min (% of control)	75	106
Strength activity index (20% RHA) at 28-day, min (% of control)	75	110

5.4 Fresh Properties of Paste/Concrete Containing RHA

5.4.1 Workability

Ikpong and Okpala [30] reported the workability of concrete made with RHA. Cement was partially replaced with 0, 20, 25 and 30% of RHA. Concretes were designed to have 28-day strength of 20, 25, 30 and 40 N/mm^2. Workability results are given in Table 5.6. It was observed that to attain the same level of workability, the mixes made with RHA required higher water than those made with only ordinary Portland cement. This was reflected in the water–cement ratios of the three mixes (0, 30 and 40% RHA contents).

Ismail and Waliuddin [31] reported slump and density of high-strength concrete (HSC) containing RHA (10–30%) as partial replacement of cement. Control mixture had proportion of 1:1.07:1.90 with cement content of 571 kg/m^3. Slump and density results are given in Table 5.7. Slump and density decreased with the increase in RHA content. However, fineness of RHA did not have significant effect on these properties.

Bui et al. [11] studied the effect of RHA on the slump of concrete mixtures. Two types (PC 30 and PC 40) of ordinary Portland cement were used. Cement PC 30 and PC 40 had Blain specific surface area of 2,700 and 3,759 cm^2/g, respectively. Cement was replaced with 10, 15 and 20% RHA by mass. Three water-to-binder ratios (0.30, 0.32, and 0.34) were used, and superplasticizer was added to all mixtures for obtaining high workability. Figure 5.8 shows the influence of RHA content on the slump of gap-graded mixtures made with w/b ratio of 0.34. It was observed that slump decreased with the increase in RHA content for same level of superplasticizer.

Sensale et al. [48] investigated the influence of partial replacements of Portland cement by rice-husk ash (RHA) on flow diameter and setting times of cement paste. Pastes with water/binder ratio 0.30 and substitutions of 5 and 10% cement

Table 5.6 W/C ratio, slump and compaction factor values of mixes [30]

Strength (MPa)	RHA (%)	W/C ratio	Compaction factor	Slump (mm)
20	0	0.80	0.926	40
	30	0.83	0.93	33
	40	0.85	0.92	35
25	0	0.69	0.89	35
	30	0.72	0.87	35
	40	0.75	0.92	37
30	0	0.59	0.92	50
	30	0.64	0.90	55
	40	0.65	0.89	35
40	0	0.54	0.90	50
	30	0.57	0.87	45
	40	0.61	0.91	45

Table 5.7 Fresh concrete properties [31]

Mix type	RHA (%)		w/(C + RHA)	Density (kg/m^3)	Slump (mm)
	#200	#325			
A	–	–	0.24	2,425	70
Aa-10	10	–	0.31	2,405	30
Aa-20	20	–	0.33	2,400	60
Aa-30	30	–	0.36	2,398	30
Ab-10	–	10	0.30	2,403	30
Ab-20	–	20	0.32	2,396	45
Ab-30	–	30	0.34	2,390	32

by RHA were used. Two sources of ash were considered; a residual RHA (RRHA) from the common rice paddy milling industries in Uruguay and a homogeneous ash produced by controlled incineration from the United States (CRHA). RRHA had SiO_2 content of 87.2% whereas CRHA had SiO_2 content of 88%. Flow diameter and setting times results are given in Table 5.8. It was observed that (1) setting times decreased with increase in RHA content; and (2) flow diameter increased with 5% RHA, but at 10% RHA, flow diameter was less than what was obtained with 5% RHA.

Habeeb and Fayyadh [26] investigated the influence of RHA on the slump and fresh density of concrete mixtures. Cement was replaced with three grades of RHA (F1, F2 and F3, i.e. 180, 270 and 360 min of grinding, respectively). Fineness of F1, F2 and F3 RHA were 27.4, 29.1, and 30.4 m^2/g. The fresh properties of concrete mixtures are given in Table 5.9. The slump was in the range of 210–230 mm. Fresh density of concrete varied between 2,253 and 2,347 kg/m^3, the lowest density values were for 20F1 mixture. This was due to the low specific gravity of

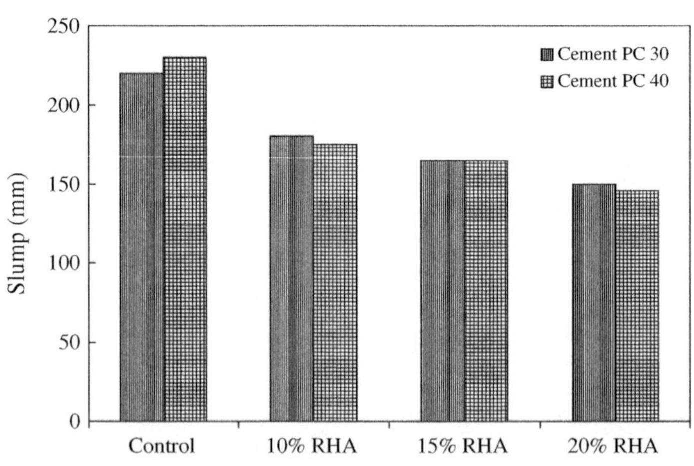

Fig. 5.8 Slump variation of gap-graded concretes made with different fineness at constant superplasticizer content [11]

Table 5.8 Flow diameter and setting times of pastes [48]

Paste	Superplasticizer (%)	Flow diameter (mm)	Setting time (minutes)	
			Initial	Final
Control	0.95	206	410	490
5% RRHA	0.925	220	284	359
10% RRHA	0.83	217	255	305
5% CRHA	1.0	213	316	373
10% CRHA	1.6	205	413	495

RHA which lead to reduction in the mass per unit volume. The concrete incorporating finer RHA resulted in denser concrete matrix. The SP content had to be increased along with the RHA fineness and percentage, this due to the high specific surface area of RHA which would increase the water demand. Therefore, to maintain high workability, SP content rose up to 2.00% for the 20F3 mixture.

5.4.2 Air-Entrainment

Zhang and Malhotra [55] investigated the influence of RHA on the air-entraining admixture (AEA) requirement of concrete mixtures made with RHA (0, 5, 8, 10 and 15%) as partial replacement of cement (Fig. 5.9). It was observed that AEA requirement increased with the increase in RHA content possibly because of high specific surface area of RHA in comparison to cement.

5.4.3 Consistency and Setting Times

Singh et al. [49] examined the effect of lignosulfonate (LS) and $CaCl_2$ on the consistency of blended cement made with RHA (Table 5.10). I is evident that in the presence of 1% LS, the water/binder ratio was reduced considerably as is generally expected in the presence of a superplasticizer. In the presence of 2% $CaCl_2$ (an accelerator), the water/binder ratio was lower than that of control but higher than that in the presence of 1% LS. However, in the presence of a mixture of 1% LS and 2% $CaCl_2$, the water/binder ratio was the lowest. This indicated that

Table 5.9 Fresh concrete properties [26]

Mix	w/b ratio	RHA content (%)	SP content (%)	Slump (mm)	Fresh density (kg/m^3)
CM	0.53	0	0.63	230	2,347
20F1	0.53	20	1.83	220	2,253
20F2	0.53	20	1.90	215	2,270
20F3	0.53	20	2.00	210	2,294

Table 5.10 Variation of water consistency with admixtures [49]

Composition	Water–binder
90% OPC + 10% RHA	0.42
90% OPC + 10% RHA + 2% $CaCl_2$	0.40
90% OPC + 10% RHA + 1% LS	0.38
90% OPC + 10% RHA + 2% $CaCl_2$ + 1% LS	0.37

water reduction was enhanced in the presence of a mixture of a superplasticizer (1% LS) and an accelerator (2% $CaCl_2$).

Jaturapitakkul and Roongreung [32] investigated consistency and setting times of cementing material made from rice husk ash (RHA) and calcium carbide residue (CCR). Calcium carbide residue (CCR) is a by-product of acetylene production process. CCR consists mainly of calcium hydroxide, $Ca(OH)_2$, and is obtained in a slurry form. Chemical composition of CCR was 51.94% CaO, 3.36% SiO_2, 2.56% Al_2O_3, 0.33% Fe_2O_3, 0.46% MgO, 0.03% K_2O, 0.22% SO_3 and 41.72% LOI. The major compound in RHA was SiO_2 (78.22%). Normal consistency and setting times of various combinations of CCR-RHA paste are given in Table 5.11. It was observed that (1) normal consistency of CCR-RHA pastes needed more water than that of cement paste because of the high porosity of the two materials as well as the high LOI of CCR. The normal consistency of cement paste was 23.9% while those of CCR-RHA pastes were between 43.7 and 62.0%, depending on RHA content. The higher the RHA content in the paste, the higher was the water requirement to maintain the same normal consistency; and (2) initial and final setting times of the cement paste were 107 and 195 min, respectively. Setting times of CCR-RHA pastes was more than that of the paste made with OPC. In CCR-RHA paste, the $Ca(OH)_2$ from CCR reacted with SiO_2 from RHA to form CSH while Portland cement reacted with water and formed CSH directly. The long setting times of calcium carbide residue-rice husk ash paste was because the pozzolanic reaction between pozzolan and lime was usually much slower than the hydration of cement.

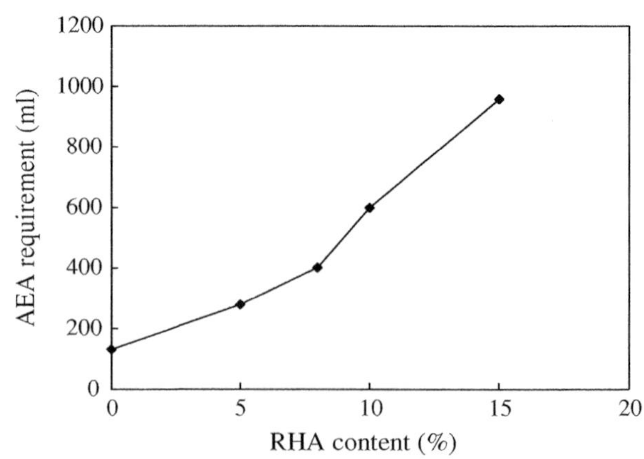

Fig. 5.9 Relationship between requirement of air-entraining admixture and RHA content [55]

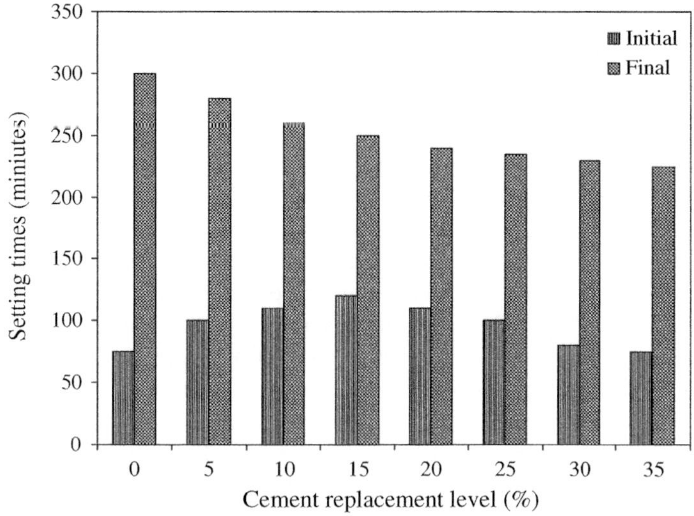

Fig. 5.10 Initial and final setting times of RHA blended cements [23]

Ganesan et al. [23] reported the effect of cement replacement with RHA on the consistency and setting times of cement. Percentages of cement replacement were 0, 5, 10, 15, 20, 20, 25, 30, and 35. They observed that (1) consistency of control mixture was approximate 32%, however, water required for standard consistency linearly increased with an increase in RHA content. The standard consistency with 35% RHA content was 44%. As ashes are hygroscopic in nature and the specific surface area of RHA is much higher than cement, it needs more water. The results of initial and final setting times are shown in Fig. 5.10. It was observed that up to 15%, RHA level increased the initial setting time. At 20, 25, 30 and 35%, there was reduction in initial setting time. The initial setting time measured for RHA blended cements up to 35% was higher than that of control OPC. On the other hand, the final setting time decreased with the increase in RHA up to 35%.

Table 5.11 Normal consistency and setting times of Portland cement and calcium carbide-rice husk ash pastes [32]

Paste	Normal consistency (%)	Setting times (min)	
		Initial	Final
Cement	23.9	107	195
20C80R	62.0	422	775
35C65R	58.8	397	660
50C50R	55.8	345	635
65C35R	50.6	420	670
80C20R	43.7	502	680

5.5 Properties of Hardened Concrete Containing RHA

5.5.1 Porosity and Water Absorption Capacity

Tashima et al. [50] determined the water absorption capacity of concrete mixture prepared with 0, 5, and 10% RHA as partial replacement of cement. RHA had SiO_2 content of 92.99% and blain specific surface of 16,196 cm^2/g. The water absorption results are given in Table 5.12. The results revealed that higher quantity of RHA resulted in lower water absorption values, due to the RHA being finer than cement. Adding 10% of RHA to the concrete, a reduction of 38.7% in water absorption was observed when compared to control mixture.

Cizer et al. [17] determined the porosity of cement mortar blended with RHA and lime. Reference cement mortar (Cref) was prepared in 1:3 cement/sand ratio by weight. In RHA-cement mortars, cement was replaced with 30, 50, and 70% RHA and were designated as (RHA-C.3-7), (RHA-C.5-5) and (RHA-C.7-3), respectively. Two types of ternary blended mortars were prepared with RHA, cement and lime. The ratio of the cement was kept 10% for both mixtures. While one (RHA-C-L.7-1-2) was composed of 70% RHA and 20% lime, the other (RHA-C-L.5-1-4) contained 50% RHA and 40% lime. Open porosity of the mortars at 60, 90 and 120 days is given in Fig. 5.11. Cref had the lowest open porosity. Concerning RHA-cement mortars, the open porosity increased as the content of the cement in the mortar decreased. RHA-C-L.7-1-2 and RHA-C-L.5-1-4 mortars were more porous than the rest as they contained 20 and 40% lime respectively and only 10% cement by weight.

Saraswathy and Song [46] studied the influence of partial replacement of cement with (RHA) on the porosity and water absorption of concrete. Cement was replaced with 0, 5, 10, 15, 20, 25, and 30% RHA. Proportion of control (with out RHA) mix was 1:1.5:3 with w/c ratio of 0.53. Porosity and water absorption test was carried out as per ASTM C 642 [1]. Table 5.13 shows the results of porosity and coefficient of water absorption. It was observed that (1) porosity values decreased with increase in RHA content because small RHA particles improved the particle packing density of the blended cement, leading to a reduced volume of larger pores; and (2) coefficient of water absorption for RHA-Concrete was found to be less when compared to control concrete.

Ganesan et al. [23] investigated the effect of RHA on the water absorption and sorptivity of concrete. Cement was replaced with 0, 5, 10, 15, 20, 25, 30,

Table 5.12 Water absorption of concrete mixtures [50]

Mixture	Water absorption (%)		
	7 days	28 days	91 days
Control (0% RHA)	2.67	1.76	1.54
5% RHA	2.64	1.64	1.35
10% RHA	2.35	1.38	1.11

Table 5.13 Porosity and water absorption of rice husk replaced concrete [46]

RHA (%)	Porosity (%)	Water absorption coefficient (m^2/s)
0	18.06	3.5571×10^{-10}
5	18.18	6.7587×10^{-11}
10	13.82	1.0320×10^{-11}
15	13.80	1.0644×10^{-11}
20	13.54	1.2122×10^{-10}
25	13.04	1.4548×10^{-10}
30	11.89	1.3030×10^{-10}

and 35% RHA. Control concrete mixture was made with 383 kg of cement, 575 kg of sand, 1,150 kg of coarse aggregate per cubic meter with water–binder ratio of 0.53. Water absorption and sorptivity results are given in Table 5.14. It was observed that (1) at 28-day curing, the percentage of water absorption increased with RHA content up to 35%. This is due to the fact that RHA is finer than OPC and also it is hygroscopic in nature. When the curing time was increased to 90 days, percentage of water absorption decreased considerably with increase in RHA content up to 25%. Even at 30% RHA, the value was lower compared to that of control; and (2) sorptivity at 28 days of curing, progressively decreased with increase in RHA content up to 25%. At 30 and 35% RHA there was an increase in sorptivity and these values are also lower than that of control concrete. At 90 days of curing, the sorptivity values up to 35% RHA were quite lower than that of control concrete.

Chindaprasirt and Rukzon [16] studied the porosity of mortars made with blends of ordinary Portland and ground rice husk ash (RHA). Cement was replaced with 0, 10, 20 and 40% RHA. Sand-to-binder ratio of 2.75 by weight, water to binder ratio (w/b) of 0.5 was used. Porosity results of mortars at 7, 28 and 90 days are given in Table 5.15. It was observed that (1) at the age of 7-day, the porosities

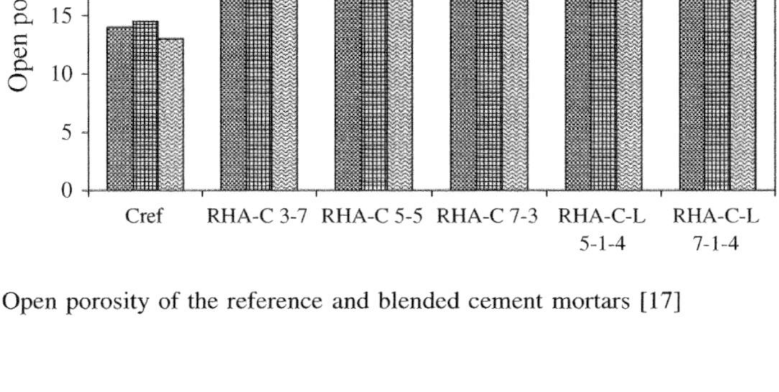

Fig. 5.11 Open porosity of the reference and blended cement mortars [17]

Table 5.14 Water absorption sorptivity of RHA blended cement [23]

Mix	RHA (%)	Saturated water absorption (%)		Sorptivity × 10^{-6} (m/s$^{1/2}$)	
		28 days	90 days	28 days	90 days
R0	0	4.71	3.76	11.05	9.76
R1	5	4.83	3.21	10.6	7.09
R2	10	5.02	3.20	9.16	4.86
R3	15	5.58	3.11	7.37	4.09
R4	20	5.81	2.20	6.00	3.61
R5	25	6.09	2.80	5.53	2.28
R6	30	6.35	3.05	6.08	3.38
R7	35	6.92	3.98	10.30	4.04

of mortar containing 10 and 20% RHA are lower than that of the control mix. The addition of fine particles of RHA causes segmentation of large pores and increases nucleation sites for precipitation of hydration products in cement paste. This results in pore refinement and a reduction of calcium hydroxide in paste; (2) at replacement level of 40%, the porosity of the mortars containing RHA increased in comparison with that of the control. At the age of 7 days, porosity of 40 RHA mix was 21.8% which was significantly larger than 17.8% of the OPC mortar. The increases in porosity with a relative large amount of pozzolans are resulted from the reduced amount of OPC. This results in less hydration products especially at the early age where the pozzolanic reaction is small.

5.5.2 Compressive Properties

Rahman [39] determined the compressive strength of various mix proportions of sand, cement and rice husk ash (RHA) for use in sandcrete block at the age of 7, 28 and 60 days, and the results are given in Table 5.16. It is evident that (1) optimum water/(cement + RHA) ratio increased with rice husk ash contents; (2) up to 40% RHA could be added as a partial replacement for cement without any significant change in compressive strength at 60 days.

Ikpong and Okpala [30] studied the effect of replacement of cement (0, 20, 25 and 30%) with RHA on the compressive strength of concrete (Table 5.17). It was concluded that compressive strength continued to increase with age for each of the mixes. The control mix (0% RHA) attained a higher strength than the OPC/RHA

Table 5.15 Porosity of blended mortars [16]

Mix	Porosity (%)		
	7 days	28 days	90 days
OPC	17.8	13.7	12.8
10RHA	17.0	13.3	12.6
20 RHA	17.8	13.1	12.6
40 RHA	21.8	15.0	12.8

Table 5.16 Compressive strength at optimum water/(cement + RHA) ratio for different mix proportions [39]

(Cement + RHA): sand	Optimum water/ (cement + RHA) ratio	7-day strength (MPa)	28-day strength (MPa)	60-day strength (MPa)
(1.0 + 0.0):4	0.66	9.07	13.15	16.70
(1.0 + 0.1):4	0.70	8.52	12.70	16.50
(1.0 + 0.2):4	0.75	6.58	10.20	16.35
(1.0 + 0.3):4	0.81	6.00	9.72	16.76
(1.0 + 0.4):4	0.88	5.62	9.44	16.37
(1.0 + 0.0):5	0.75	7.68	11.75	14.40
(1.0 + 0.1):5	0.81	6.58	10.40	14.70
(1.0 + 0.2):5	0.88	5.93	9.84	14.63
(1.0 + 0.3):5	0.94	4.99	8.48	14.60
(1.0 + 0.4):5	1.00	4.65	8.18	14.35

mixes at all ages. At the age of 28 days, all the OPC/RHA concretes, except the designed mix of 40 N/mm^2, had attained their 28-day design strengths. For the designed mix of 40 N/mm^2, the mix having 30% RHA content reached 98.5% of its design strength, while the one having 40% RHA content reached 86.5%.

Zhang and Malhotra [55] investigated the effect of RHA (10%) as partial replacement of cement on the compressive strength of concrete and compared it with the compressive strength of concrete containing 10% silica fume (SF). Water-to-cementitious material ratio was maintained at 0.40. Compressive strength results up to the age of 730 days are shown in Fig. 5.12. It was observed that in general RHA concrete achieved higher strength than control concrete mixture but lower than that of silica fume concrete. At 28 days, the RHA concrete had a compressive strength of 38.6 MPa compared with 36.4 MPa for the control concrete and 44.4 MPa for SF concrete. At 180 days, RHA concrete exhibited

Table 5.17 Compressive strength development with age [30]

Design strength (N/mm^2)	RHA (%)	Compressive strength (N/mm^2)			
		7 days	14 days	28 days	90 days
20	0	17.5	23.9	33.1	37.8
	30	15.8	18.2	29.1	36.1
	40	13.27	15.2	27.2	33.9
25	0	22.9	27.2	35.4	39.2
	30	18.7	23.8	31.1	37.5
	40	14.8	21.0	28.6	34.0
30	0	28.4	37.6	43.5	46.3
	30	22.6	32.0	38.3	43.2
	40	16.7	27.5	31.8	37.1
40	0	32.4	38.8	44.2	47.6
	30	24.0	33.5	39.4	44.9
	40	1.5	29.3	34.6	38.9

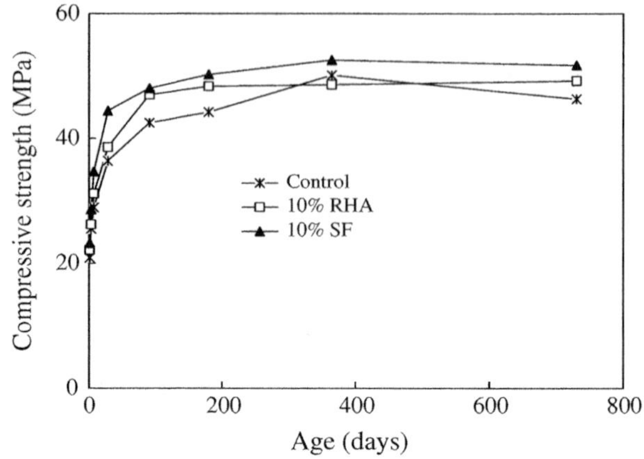

Fig. 5.12 Development of compressive strength of concrete with RHA and silicafume [55]

compressive strength of 48.3 MPa compared with 44.2 MPa for the control concrete and 50.2 MPa for SF concrete.

Ismail and Waliuddin [31] investigated the influence RHA (passing #200 and #325 sieves) as partial replacement of cement on the compressive strength of high-strength concrete. Control mixture had mixture proportion of 1:1.07:1.90 with cement content of 571 kg/m^3. Compressive strength results are given in Table 5.18. It was observed that (1) high-strength concrete could be made using RHA; (2) optimum replacement of cement with RHA was around 10 to 20% with finely ground RHA; (3) rate of hydration in concrete made with RHA was slow as compared to control concrete. This fact was very dominant during the initial three days of age of concrete. This rate of slow hydration also affects the 150-day strength of concrete.

Singh et al. [49] studied the effect of lignosulfonate (LS) and $CaCl_2$ on the compressive strength of blended cement made with RHA. Pastes of 10% RHA-blended Portland cement were made with the addition of 2% $CaCl_2$, 1% LS, and (1% LS + 2% $CaCl_2$). Compressive strength was determined at the age of 7 and 28 days (Fig. 5.13). They concluded that (1) with 2% $CaCl_2$, compressive strength was higher at 7 days and lower at 28 days of hydration than that of the control; (2) with 1% LS, the values were lower than that of control on all days of hydration.

Table 5.18 Compressive strength of RHA concrete [31]

Mix type	RHA (%)		w/(C + RHA)	Compressive strength (MPa)			
	#200	#325		3 days	7 days	28 days	150 days
A	–	–	0.24	54.3	62.3	72.4	85.0
Aa-10	10	–	0.31	46.2	56.0	68.1	71.1
Aa-20	20	–	0.33	35.3	46.8	57.3	57.4
Aa-30	30	–	0.36	31.5	39.3	47.7	48.8
Ab-10	–	10	0.30	47.0	61.0	71.0	72.4
Ab-20	–	20	0.32	46.7	56.0	70.2	70.3
Ab-30	–	30	0.34	43.1	51.7	63.0	63.2

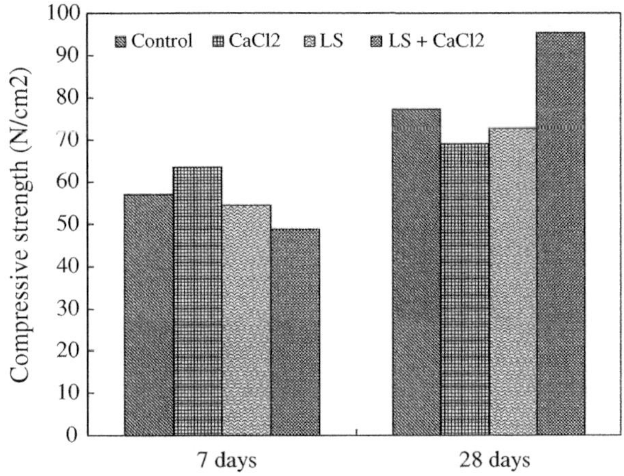

Fig. 5.13 Effect of admixtures on the compressive strength of blended cement [49]

However, in the presence of a mixture of 1% LS and 2% $CaCl_2$, the compressive strength was maximum at 28 days of hydration; and (3) when mixture of a superplasticizer (1% LS) and an accelerator (2% $CaCl_2$) was added to the RHA blended OPC, the superplasticizer reduced the water demand and the accelerator enhanced the compressive strength. Thus the role of the two admixtures becomes supplementary to each other.

Jaturapitakkul and Roongreung [32] determined the compressive strength of mortar made with a cementing material (combination of rice husk ash and calcium carbide residue). Calcium carbide residue consists mainly of calcium hydroxide, $Ca(OH)_2$, and is obtained in a slurry form. Major chemical composition of CCR was 51.94% CaO, and 41.72% LOI. The major compound in RHA was SiO_2 (78.22%). Control mortar was made using Type I Portland cement. The mortar contained 1.0 part of cementitious material to 2.75 parts of river sand by weight, having water-cementitious material ratio of 0.65. Superplasticizer was employed in order to maintain the flow of mortar between 110 ± 5 mm. Compressive strengths results are given in Table 5.19. It was observed that (1) mortar 50C50R gave the highest compressive strength whereas mortar 50C50R had lower compressive strength than that of the control mortar; (2) difference in compressive

Table 5.19 Compressive strength of Portland cement and calcium carbide-rice husk ash pastes [32]

Mortar	Compressive strength (MPa)							
	1 day	3 days	7 days	14 days	28 days	60 days	90 days	180 days
Cement	8.5	17.6	23.0	28.3	30.9	32.8	33.7	34.7
20C80R	0.7	5.4	7.0	7.9	9.0	9.7	10.1	10.4
35C65R	0.4	5.8	9.4	12.6	14.6	15.6	15.8	16.7
50C50R	0.9	5.8	10.0	13.6	15.6	17.5	18.6	19.1
65C35R	0.4	3.1	7.0	9.9	10.9	11.5	11.9	12.6
80C20R	0.1	1.3	4.0	5.4	6.3	6.7	7.0	7.2

Table 5.20 Axial compressive strength of concrete [50]

Mixture	Compressive strength (MPa)		
	7 days	28 days	91 days
Control (0% RHA)	45.9	48.1	58.3
5% RHA	52.9	60.4	62.0
10% RHA	45.8	54.2	60.9

strength of mortar 50C50R and control mortar at an early age was very high, and reduced to about two times after 28 days. It is important to mention the mix proportion of the mortar 50C50R did not contain Portland cement, and the strength of mortar was solely contributed from the pozzolanic reaction of the two materials; and (3) mortar 50C50R contained 50% of CCR and 50% of RHA as cementitious material gave the highest compressive strength. Increasing or decreasing of CCR content from 50% of the total cementitious material resulted in lower compressive strength of mortar.

Tashima et al. [50] investigated the influence of 0, 5, and 10% RHA as partial replacement of cement on the compressive strength of concrete. RHA had SiO_2 content of 92.99% and blain specific surface of 16,196 cm^2/g. Compressive strength results up to the age of 91 days are given in Table 5.20. It was observed that (1) addition of RHA caused an increase in compressive strength due to the pozzolanic action of RHA; (2) compressive strength increased with age; and maximum increase (25%) in strength was found to be with 5% RHA.

Bui et al. [11] investigated the effect of rice husk ash (10–30%) on the compressive strength of concrete mixtures made with two types (PC30 and PC40) of ordinary Portland cement having Blain specific surface area of 2,700 and 3,759 cm^2/g, respectively. Strength measurements were made up to the age of 90 days. They concluded that (1) rice husk ash can be used as a highly reactive pozzolanic material to improve the microstructure of the interfacial transition zone (ITZ) between the cement paste and the aggregate in high-performance concrete; (2) in addition to the pozzolanic reactivity of RHA, the particle grading of cement and RHA mixtures exerted significant influences on the blending efficiency and; (3) relative strength increase was higher for coarser cement.

Cizer et al. [17] investigated the effect of RHA on the compressive strength of cement mortars up the age of 120 days. Several cement mortar compositions blended with RHA were prepared. Reference cement mortar (Cref) was prepared in 1:3 cement/sand ratios by weight. In RHA-cement mortars, cement was replaced with 30, 50, and 70% RHA and were designated as (RHA-C.3-7), (RHA-C.5-5) and (RHA-C.7-3), respectively. Compressive strength results are shown in Fig. 5.14. Compressive strength of the Cref and RHA-C.3-7 mortars increased until 28 days but after that period no significant increase was observed. RHA-C.5-5 and RHA-C.7-3 mortars composed of lower cement content than RHA-C.3-7 yielded a gradual increase in their compressive strength between 7 and 120 days of hardening.

Fig. 5.14 Compressive strength of the reference cement mortar and RHA-cement mortars [17]

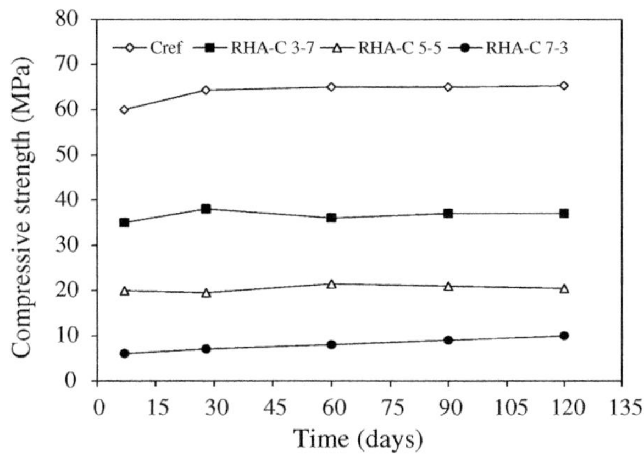

Sensale [47] investigated the influence of RHA (10 and 20%) on the compressive and splitting tensile strength of concretes. Two sources of ash were considered; a residual RHA from rice paddy milling industry in Uruguay and another produced by controlled incineration from the USA. The percentage of reactive silica contained in USA RHA was 98.5% and in the UY RHA was 39.55%. Strengths test results are given in Table 5.21. It was concluded that (1) at lower age (7 days), concretes with UY-RHA exhibited higher compressive strength that concretes with USA-RHA; (2) at higher age (91 days), the RHA concrete had higher compressive strength in comparison with that of concrete without RHA, and the highest compressive strengths were achieved in concretes with 20% USA-RHA; (3) long term compressive strength of the concretes with UY-RHA was not as high as the one obtained with USA-RHA, which also

Table 5.21 Compressive and splitting tensile strength results of RHA concretes [47]

w/(c + RHA)	RHA		Compressive strength (MPa)			28-day splitting tensile strength (MPa)
	Type	%	7 days	28 days	91 days	
0.32	–	0	48.4	55.5	60.6	3.63
	UY	10	51.1	60.4	64.3	3.57
		20	44.3	54.8	62.7	3.54
	USA	10	39.5	51.4	64.5	3.62
		20	30.5	47.4	68.5	3.54
0.40		0	35.8	42.3	45.6	–
	UY	10	41.1	50.4	54.9	–
		20	27.9	40.7	51.4	–
	USA	10	29.7	40.8	51.5	–
		20	23.6	39.4	57.3	–
0.50		0	24.6	32.9	35.9	2.85
	UY	10	24.1	31.5	35.5	2.32
		20	24.9	34.9	37.9	2.63
	USA	10	22.7	34.5	44.4	2.92
		20	20.8	35.9	52.9	3.00

Table 5.22 Physical and mechanical properties of aggregates [45]

Properties	Ilmenite	Baryte	Gravel	Sand
Specific gravity (kg/m^3)	4,200	4,000	2,700	2,650
Unit weight (kg/m^3)	2,800	2,750	1,720	1670
Max nominal size (mm)	20	20	20	–
Void ratio (%)	33	27	37	–
Water absorption (%)	2.2	1.7	0.8	–
Crushing value (%)	38	43	16	–
Fineness modulus	1.85	2.4	–	2.65

increased with the increase in RHA content; (4) the results of splitting tensile revealed the significance of the filler and pozzolanic effect for the concretes with RHA. On the one hand, the results were consistent with the compressive strength development at 28 days for the USA-RHA. On the other hand, in the concretes with UY-RHA, lower splitting tensile strengths were observed, which can be due to the fact that with residual RHA, the filler effect of the smaller particles in the mixture was higher than the pozzolanic effect.

Sakr [45] examined effect of RHA on the mechanical properties of heavyweight concrete. The RHA had a specific surface area of 5.6×10^6 mm^2/g, and unit weight was 2.06×10^3 kg/m^3. The chemical composition of the RHA was 87.0% SiO_2, 1.75% Al_2O_3, 2.5% Fe_2O_3, 2.5% CaO, 2.3% MgO, and 2.5% K_2O. Percentage replacement of RHA was 0, 5, 10, 15, and 20% by weight of cement. Properties of aggregates are given in Table 5.22. Results of compressive strength, indirect tensile strength, flexural strength, bond strength, and modulus of elasticity at the age of 28 days are given in Table 5.23. They concluded that (1) with the

Table 5.23 Mechanical properties of heavyweight concrete with RHA at 28 days [45]

Type of concrete	Replacement (%)	Compressive strength (MPa)	Indirect tensile strength (MPa)	Flexural strength (MPa)	Bond strength (MPa)	Modulus of elasticity (GPa)
Gravel	0	44.5	2.7	4.9	5.5	21.5
	5	45.5	2.9	5.5	6.7	22.0
	10	49.5	3.1	6.0	7.0	23.0
	15	50	3.4	7.5	7.5	24.2
	20	43.0	3.4	7.4	7.4	23.8
Baryte	0	45.0	2.9	4.9	6.2	29.6
	5	46.0	3.1	6.0	7.1	30.1
	10	52.0	3.3	6.8	7.7	32.0
	15	54.0	3.6	7.9	8.0	33.5
	20	52.0	3.6	7.8	7.9	31.5
Ilmenite	0	46.5	3.7	5.9	6.9	35.5
	5	47.0	3.9	7.1	7.9	37.0
	10	50.0	4.0	8.5	8.0	37.0
	15	52.0	4.5	8.9	8.3	38.0
	20	51.0	4.0	8.5	8.1	36.5

increase in RHA content up to 15%, generally, all the mechanical properties increased. Beyond 15%, reduction in strength properties was observed. This may be due to the high specific surface areas of RHA, which consume more water to have the same workability; (2) ilmenite concrete had the highest compressive strength when compared to the corresponding barite concrete or gravel concrete. This increase may be attributed to the shape, specific surface area, and hardness of the ilmenite aggregates. It may also be due to the effect of titanium oxide (TiO_2) in ilmenite, which may be combined with water to form a gelatinous material that fills the voids in the cement matrix, consequently increasing bonding between the cement particles and thus increasing its compressive strength; and (3) 15% of RHA mixed with baryte concrete showed a higher tensile strength when compared to gravel concrete but a lower tensile strength than ilmenite concrete mixed with 15% of RHA. The ilmenite concrete exhibited the highest flexural strength, bond strength, and static modulus of elasticity in compression as compared to the baryte or gravel concrete.

Saraswathy and Song [46] investigated the effect of RHA on the compressive, splitting tensile strength and bond strength of concrete. Proportion of control (with out RHA) mix was 1:1.5:3 with w/c ratio of 0.53. Percentage replacements were 0, 5, 10, 15, 20, 25, and 30%. 28-day results of compressive strength, splitting tensile strength and bond strength are given in Table 5.24. Based on the results, it was concluded that (1) compressive strength increased with increase in RHA content. At 28 days, all RHA concretes exhibited higher compressive strength than the control concrete; (2) inclusion of RHA (up to 25%) as partial replacement of cement did not affect the splitting tensile strength of concrete. After 25% replacement level, a slight decrease in split tensile strength was observed; (3) all the rice husk replaced concretes showed higher bond strength than the conventional concrete, indicating that replacement of rice husk ash does not affect the bond strength properties.

Gastaldini et al. [24] studied the effect of chemical activators on the compressive strength of concrete made with 20% of rice husk ash as partial replacement of cement. Water/binder ratios were 0.35, 0.50 and 0.65, and binder/aggregate ratios were 1:3.75, 1:5.25 and 1:6.9. Potassium sulfate (K_2SO_4), sodium sulfate (Na_2SO_4) and sodium silicate (Na_2SiO_3) were used as chemical activators in concentrations of 1% by weight of cement. Compressive strength results are

Table 5.24 28-day strengths of RHA concretes [46]

RHA (%)	Compressive strength (MPa)	Splitting tensile strength (MPa)	Bond strength (MPa) at 0.25 mm slip
0	36.45	4.49	3.32
5	36.49	4.57	4.11
10	37.43	4.65	4.31
15	37.38	4.92	3.79
20	37.71	4.60	3.43
25	39.55	4.58	4.07
30	37.80	3.67	3.87

Table 5.25 Compressive strength of RHA concretes [24]

Mixture	w/b ratio	Compressive strength (MPa)		
		7 days	28 days	91 days
REF (control)	0.35	58.3	64.2	76.1
	0.50	36.4	47.7	53.5
	0.65	24.6	28.0	31.9
20 RHA	0.35	54.2	69.7	83.4
	0.50	36.4	48.1	53.9
	0.65	17.7	27.0	33.6
20 RHA 1% Na_2SO_4	0.35	59.2	73.3	82.8
	0.50	39.5	50.7	56.5
	0.65	25.3	36.8	42.5
20 RHA 1% K_2SO_4	0.35	65.9	77.4	91.0
	0.50	47.2	54.3	71.8
	0.65	28.9	38.9	48.3
20 RHA 1% Na_2SiO_3	0.35	54.2	74.4	77.5
	0.50	37.4	48.7	53.8
	0.65	29.1	40.3	45.1

given in Table 5.25. It was concluded that (1) mixture with 20% rice husk ash and w/b ratio (0.50) showed compressive strength values equal to that of the reference mixture (REF) at 7 days. The same mixture with different w/b ratios (0.35 and 0.65) showed lower strength when compared with the reference mixture. At 28 and 91 days, compressive strength values were higher than those of the reference samples for all w/b ratios; (2) the three activators yielded a huge increase in strength at 7 days. The mixture activated with K_2SO_4 showed the highest increase when compared with the sample without activator. The percentage increase were 21, 30 and 63% at 7 days, 11, 13 and 44% at 28 days and 9, 33 and 44% at 91 days for the mixtures with w/b ratio = 0.35, 0.50 and 0.65, respectively. When compared with the reference sample, the increases for the same w/b ratios and age ranged from 13 to 52%; and (3) significant growth in the samples with w/b = 0.65, mainly in the sample activated with K_2SO_4, can be attributed to the increase in pH and the changes in the RHA structure caused by the activator, which improved the hydration. In addition, more space is available for the deposition of hydration products when compared with lower w/b ratios. This composition showed higher strength at 7 days (28.9 MPa) when compared with the sample without activator (17.7 MPa) and the reference sample (24.6 MPa), which illustrates the benefits of the activator.

Chindaprasirt et al. [14] determined the compressive strength of blended cements containing RHA. RHA content was 20 and 40% as partial replacement of cement. RHA contained high silica content of 90% and low loss on ignition (LOI) of 3.2%. The Blaine fineness of RHA was 14,000 cm^2/g. Mortars were made with sand-to-binder ratio of 2.75 and adjusted water contents to achieve similar flow of 110 ± 5%. Table 5.26 shows water-to-binder (w/b) ratios and compressive strengths of the mortar mixes. Incorporation of RHA resulted in significant

Table 5.26 Compressive strength of blended cements [14]

Mix	w/b ratio	Compressive strength (MPa)			
		7 days	28 days	90 days	180 days
OPC	0.55	44	51	57	60
RHA20	0.68	31	54	61	62
RHA40	0.80	17	32	43	53

increase due to the high surface area of RHA. Use of RHA resulted in a reduction of 7- and 28-day compressive strength. For RHA mixes, the low initial strength was due to the high water-to-binder (w/b) of the mixes. For 20% RHA replacement level, although the w/b ratio was increased, the strength at 28 days was higher than that of PC mix, suggesting that RHA was quite reactive. Although possessing high fineness, RHA contribution to the strength development was limited by the high water demand associated with its high surface areas.

Sensale et al. [48] determined the compressive strength of cement paste made with RHA. Pastes with water/binder ratio 0.30 and substitutions of 5 and 10% cement by RHA were used. Two sources of ash were considered; a residual RHA (RRHA) from the common rice paddy milling industries in Uruguay and a homogeneous ash produced by controlled incineration from the United States (CRHA). RRHA had SiO_2 content of 87.2% whereas CRHA had SiO_2 content of 88%. Compressive strength results are given in Table 5.27. It was observed that compressive strength of cement paste decreased with the inclusion of RHA.

Chindaprasirt et al. [15] determined the compressive strength of blended Portland cement mortar containing ground rice husk ash (RHA). Ordinary Portland cement (OPC) was partially replaced with 20 and 40% RHA. RHA had silica content of 93.2%. Compressive strength results are given in Table 5.28 It was observed that (1) strength development of OPC mortar was good. At 20% replacement, the strengths of mortars containing RHA were between 102 and 104% of those of OPC mortar at the same age. For 40% replacement level, 7-day strength was 77% of that of OPC mortar at the same age. At the age of 90 days, strength of RHA mortar was 103% of that of OPC mortar. The low early strengths and later age strength development is the common feature of pozzolanic materials.

Table 5.27 Compressive strength of pastes [48]

Paste	Compressive strength (MPa)		
	1 day	2 days	3 days
Control	52.5	88.9	100.5
5% RRHA	41.1	80.8	100.5
10% RRHA	48.3	86.7	105.4
5% CRHA	37.6	70.6	85.9
10% CRHA	39.7	83.3	97.5

Table 5.28 Compressive strength of mortars [15]

Mix	Compressive strength (MPa)		
	7 days	28 days	90 days
OPC	43.5 (100%)	57.0 (100%)	60.0 (100%)
20 RHA	44.5 (102%)	58.5 (103%)	62.5 (104%)
40 RHA	33.5 (77%)	55.0 (97%)	62.0 (103%)

Ramezanianpour et al. [43] investigated the compressive strength of concrete mixtures containing rich husk ash (RHA) up to the age of 90 days. Concrete mixtures were made with 0, 7, 10 and 15% RHA as partial replacement of cement.

Compressive strength results are shown in Fig. 5.15. It was observed that RHA concrete achieved higher compressive strengths up to 90 days when compared with the control concrete. The results showed that it was possible to obtain a compressive strength as high as 46.9 MPa after 28 days. In addition, strengths up to 63.2 MPa were obtained at 90 days.

Gastaldini et al. [25] reported the compressive strength of concrete made with 10, 20, and 30% RHA as partial replacement of cement (Table 5.29). For concrete containing 10% RHA, increase in strength for w/b ratios of 0.35, 0.50 and 0.65 ranged from 15 to 27% and from 10 to 21% at 28 and 91 days, respectively in comparison with the reference concrete. For substitutions of 20% RHA and the same w/b ratios and test ages, the increases in strength ranged from 11 to 34% and from 19 to 26%. For substitutions of 30%, the increases in strength ranged from 6 to 26% and from 7 to 27% for each age.

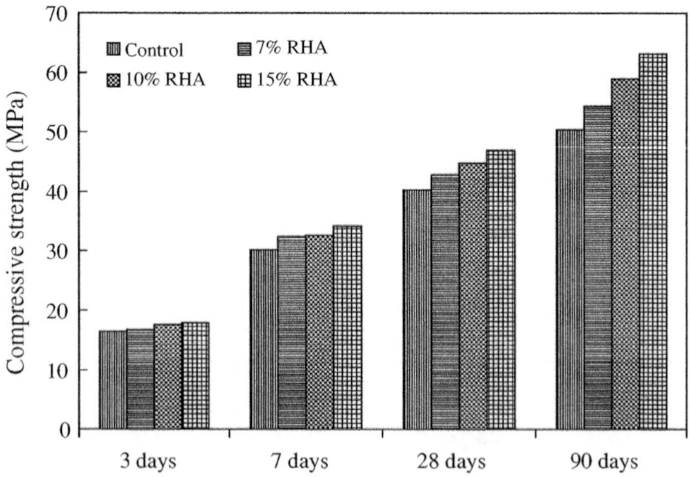

Fig. 5.15 Compressive strength of control and RHA mixtures [43]

Table 5.29 Compressive strength of concrete mixtures containing RHA [25]

Mixture	w/b	Compressive strength (MPa)	
		28 days	91 days
Ref	0.35	54	67.9
	0.50	47	51.4
	0.65	28	35
10 RHA	0.35	68	76.4
	0.50	47	62.1
	0.65	32	38.6
20 RHA	0.35	72	85.6
	0.50	51.5	62.9
	0.65	32	41.7
30 RHA	0.35	67	78.9
	0.50	50	65.1
	0.65	30	37.3

5.5.3 Tensile Strength and Modulus of Elasticity

Zhang and Malhotra [55] reported the mechanical properties of concrete made with 10% RHA and 10% silica fume (SF). Tests were conducted at the age of 28 days, and results are given in Table 5.30. These results indicated that splitting tensile strength, flexural strength and modulus of elasticity of control and concrete incorporating RHA and SF were comparable.

Tashima et al. [50] studied the effect of RHA on the splitting tensile strength and modulus of elasticity of concrete. Cement was replaced with 0, 5, and 10% RHA. RHA had SiO_2 content of 92.99% and blain specific surface of 16,196 cm^2/g. Results of splitting tensile strength and elastic modulus are given in Table 5.31. It was observed that (1) addition of RHA did not had significant effect on the tensile strength, however at 5% RHA, there was marginal increase in strength; and (2) elastic modulus decreased with the increase in RHA content.

Cizer et al. [17] determined the flexural strength of cement mortars up the age of 120 days. Several cement mortar compositions blended with RHA were prepared. Reference cement mortar (Cref) was prepared in 1:3 cement/sand ratio by weight. In RHA-cement mortars, cement was replaced with 30, 50, and 70% RHA and were designated as (RHA-C.3-7), (RHA-C.5-5) and (RHA-C.7-3), respectively. Standard mortar beams (40 × 40 × 160 mm) were made and flexural strength results are shown in Fig. 5.16. Flexural strength of the reference cement

Table 5.30 Mechanical properties of concrete mixtures Zhang and Malhotra [55]

Mix	RHA (%)	SF (%)	W/Cm ratio	Splitting tensile strength (MPa)	Flexural strength (MPa)	Modulus of elasticity (GPa)
1	0	0	0.40	2.7	6.3	29.6
2	10	–	0.40	3.5	6.8	29.6
3	–	10	0.40	2.8	7.0	31.1

Table 5.31 Splitting tensile strength and elastic modulus of concrete [50]

Mixture	Splitting tensile strength (MPa)			Modulus of elasticity (GPA)		
	7 days	28 days	91 days	7 days	28 days	91 days
Control (0% RHA)	4.85	5.37	5.41	30.08	40.85	45.04
5% RHA	4.94	5.79	5.9	40.72	40.76	41.84
10% RHA	4.82	5.78	5.4	40.23	40.21	40.03

mortar and RHA-cement mortars increased gradually upon hardening. Increase in the flexural strength of the RHA-C.3-7 mortar between 7 and 28 days was more than that of RHA-C.5-5 and RHA-C.7-3 mortars having lower cement content. Flexural strength of the RHA-C.3-7 mortar did not change between 28 and 60 days but afterwards it reached a higher value at 120 days.

Ahmadi et al. [10] studied the flexural strength and modulus of elasticity of self-compacting concrete made with 10 and 20% RHA as partial replacement of cement. Results were compared with ordinary concrete. Flexural strength and modulus of elasticity results up to the age of 180 days are shown in Figs. 5.17 and 5.18. It was observed that (1) flexural strength of SCC mixes were 12 to 20% higher than normal concrete. Mixes containing rice husk ash indicated lower compressive strength until 60 days rather than samples with no replacement and by increasing the rate of pozzolanic reactions of rice husk ash in the matrix, strength of composite mixes goes up. Also the mixes containing 20% rice husk ash achieved the highest flexural strength in all cases. Moreover, by increasing the amount of replacement, flexural strength increased. This increase was greater in normal concrete than SCC; (2) with aging and hardening of concrete mixes, the module of elasticity of concrete mixes increased. Normal concrete mixes showed higher module of elasticity around 9 to 17% more than of SCC ones. Also by increasing the amount of rice husk ash in the matrix, module of elasticity of all mixes reduced.

Ganesan et al. [23] measured the 28-day splitting tensile strength of concrete mixtures containing 0, 5, 10, 15, 20, 25, 30, and 35% RHA as partial replacement

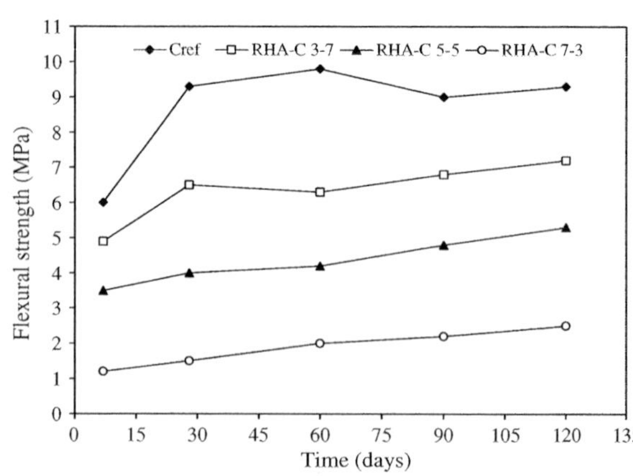

Fig. 5.16 Flexural strength of the reference cement mortar and RHA-cement mortars [17]

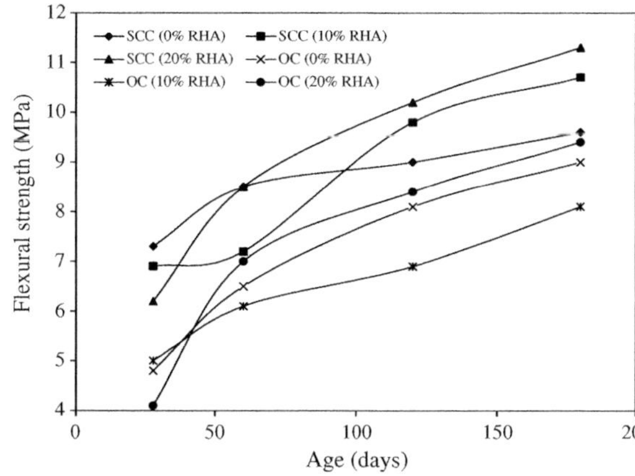

Fig. 5.17 Flexural strength with water–binder ration of 0.40 [10]

of cement. Control concrete mixture was made with 383 kg of cement, 575 kg of sand, 1,150 kg of coarse aggregate per cubic meter with water–binder ratio of 0.53. Splitting tensile strength of control concrete was 4.5 MPa. Splitting tensile strength marginally increased with RHA content up to 20% and then decreased marginally at 25 and 30% RHA, even then, value of splitting tensile strength was equal to of OPC concrete (4.5 MPa). Splitting tensile strength of concrete with 5, 10, 15, and 20 RHA was 4.7, 4.8, 4.9 and 5 MPa, respectively.

Ramezanianpour et al. [43] studied the influence of RHA on the splitting tensile strength and modulus of elasticity of concrete. Concrete mixtures were made with 0, 7, 10 and 15% RHA as partial replacement of cement. Results of splitting tensile strength and modulus of elasticity are shown in Figs. 5.19 and 5.20, respectively. It was concluded that (1) concrete containing RHA achieved greater splitting tensile strength than that the control concrete at all ages. It is clear that with the increase in RHA content, strength increased up to 20%. For instance, at 90 days 15% RHA concrete had splitting tensile strength of 5.62 MPa compared with 4.58 MPa for the control concrete; and (2) with the increase in RHA content modulus of elasticity increased at 28 and 90 days. At 90 days, mixture containing 15% of RHA

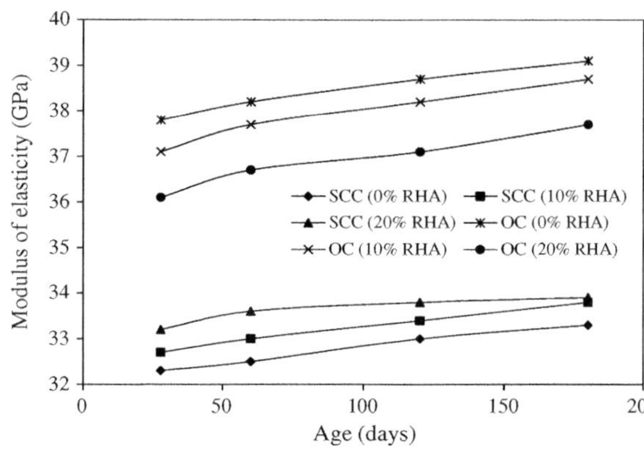

Fig. 5.18 Modulus of elasticity [10]

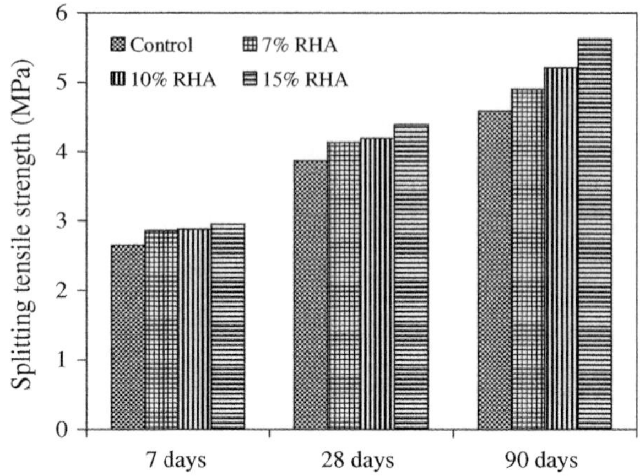

Fig. 5.19 Splitting tensile strength of control and RHA mixtures [43]

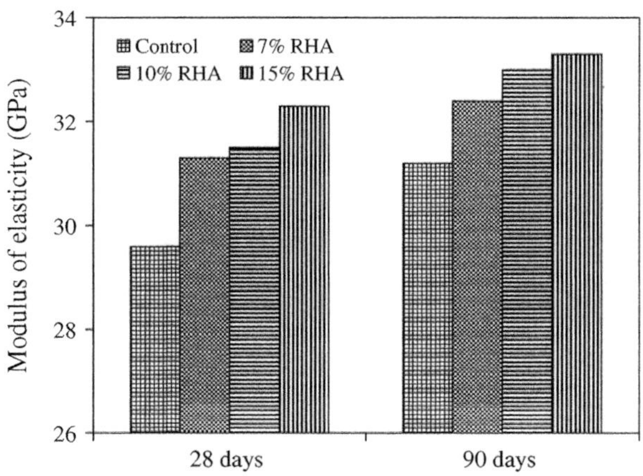

Fig. 5.20 Modulus of elasticity (GPa) of control and RHA mixtures [43]

showed 7% increase in modulus of elasticity in compression to the control concrete.

Habeeb and Fayyadh [26] studied the effect of RHA on the flexural strength, splitting tensile strength, and modulus of elasticity of concretes. Cement was replaced with three grades of RHA having fineness 27.4, 29.1, and 30.4 m^2/g. Results of mechanical properties are given in Table 5.32. It was observed that (1) flexural strength values were in the range of 4.5–6.1 MPa. Addition of RHA to concrete exhibited an increase in the flexural strength and higher strength was obtained for the finer RHA mixture due to the increased pozzolanic reaction and the packing ability of the RHA fine particles; (2) splitting tensile strength values were in the range of 2.6–3.9 MPa. Tensile property was enhanced by adding RHA to the mixture. The coarse RHA mixture (20F1) showed the least improvement; and (3) modulus of elasticity of concrete mixtures was in the range of 29.6–32.9 GPa. Addition of RHA exhibited marginal increase in the elastic properties of concrete; the highest value was recorded for (20F3) mixture due to the increased reactivity of the RHA.

Table 5.32 Mechanical properties of concrete [26]

Mix	Flexural strength (MPa)			Tensile strength (MPa)			Elastic modulus (GPa)		
	28	90	180	28	90	180	28	90	180
CM	4.5	4.9	5.1	2.6	2.8	2.9	29.6	30.5	31.0
20F1	4.9	5.4	5.5	2.9	3.0	3.2	30.1	30.8	31.4
20F2	5.0	5.4	5.7	3.2	3.3	3.5	30.2	31.4	31.7
20F3	5.2	5.7	6.1	3.2	3.5	3.9	30.5	32.3	32.9

5.5.4 Drying Shrinkage

Zhang and Malhotra [55] studied the drying shrinkage strain of concretes made with 10% RHA and 10% silica fume (SF). Figure 5.21 shows the drying shrinkage strain of concretes after 7 days of initial curing in lime-saturated water. Results indicated that RHA concrete had a drying shrinkage of 638×10^{-6} after 448 days, which was similar to the strains for the control and silica fume concretes.

Sensale et al. [48] studied the effect of partial replacements of Portland cement with rice-husk ash (RHA) on the autogenous shrinkage of cement paste. Pastes with water/binder ratio 0.30 and substitutions of 5 and 10% cement by RHA were used. Two sources of ash were considered; a residual RHA (RRHA) from the common rice paddy milling industries in Uruguay and a homogeneous ash produced by controlled incineration from the United States (CRHA). RRHA had SiO_2 content of 87.2% whereas CRHA had SiO_2 content of 88%. Autogenous deformation was measured up to 28 days (Fig. 5.22). It was observed that RHA decreased the autogenous deformation. With no RHA, a considerable autogenous deformation (600 μm/m) was obtained during 4 weeks of sealed hardening. Replacement of 5 and 10% of Portland cement by RHA led to a successive reduction of this autogenous deformation. 10% RHA decreased the autogenous deformation by 250 μm/m after 4 weeks.

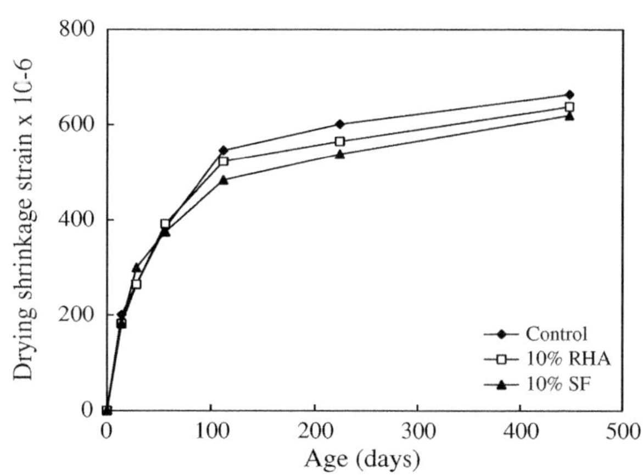

Fig. 5.21 Drying shrinkage of RHA and SF concretes [55]

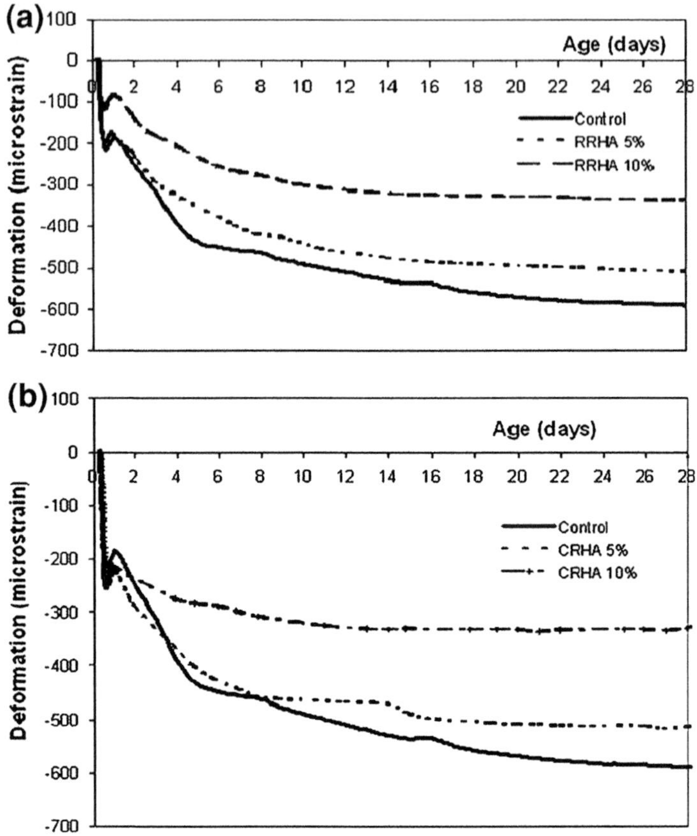

Fig. 5.22 Autogenous deformation versus age of Portland and RHA cement pastes [48]

Habeeb and Fayyadh [26] studied the shrinkage of concrete mixtures made with RHA. Cement was replaced with three grades of RHA (F1, F2 and F3 i.e. 180, 270 and 360 min of grinding, respectively). Fineness of F1, F2 and F3 RHA were 27.4, 29.1, and 30.4 m^2/g. Drying shrinkage was investigated at the age of 7, 14, 28, 42, 56, 90, 180 days under water curing for initial 7 days, and then samples were left in the air (Fig. 5.23). The results showed that average particle size of RHA had a significant effect on the drying shrinkage, the 20F3 concrete mixture exhibited higher shrinkage than the control. 20F2 concrete was comparable, while the shrinkage for 20F1 was lower compared to the control. The reduction in the RHA particle size increased the pozzolanic activity and contributed to the pore refinement of the RHA concrete matrix. Some authors have concluded that that concretes incorporating pore refinement additives will usually show higher shrinkage and creep [37]. On the other hand, others showed that using pozzolanic materials as cement replacement will reduce the shrinkage [13, 55]. These contradictory results about shrinkage are probably due to interpretational differences based on

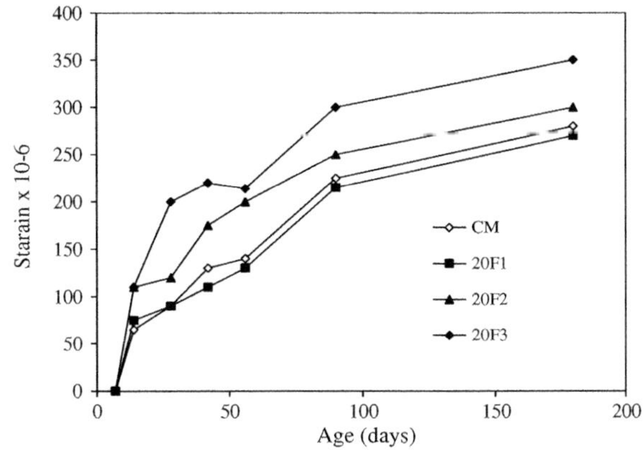

Fig. 5.23 Shrinkage values at various ages [26]

deferent concepts, definitions and measuring techniques [44]. And that may also be because the deferent characteristics and degree of reactivity of the pozzolanic materials used.

5.5.5 *Electrical Resistivity and Conductivity*

Gastaldini et al. [25] studied the electrical resistivity and conductivity in concrete mixes made with 10, 20, and 30% RHA as partial replacement of cement. It was determined using the four-electrode method. Specimens of size 100 × 100 × 170 mm were tested at the age of 91 days (Table 5.33). It was observed that (1) with the increase in RHA content from 10 to 20%, electrical resistivity increased substantially. For a content of 30% RHA, the increase in electrical resistivity when compared with the mix with 20% RHA amounted to 71, 47 and 47% for w/b ratios of 0.35, 0.50 and 0.65, respectively. The mix with 30% RHA was the one that

Table 5.33 Apparent electrical resistivity of concrete mixtures containing RHA [25]

Mixture	w/b	Electrical resistivity (Ω m)	Electrical conductivity (1/Ω cm)
Ref	0.35	316	2.834
	0.50	154	2.504
	0.65	128	2.101
10 RHA	0.35	446	2.525
	0.50	291	2.294
	0.65	229	2.076
20 RHA	0.35	813	2.371
	0.50	569	2.137
	0.65	440	1.929
30 RHA	0.35	1,395	2.204
	0.50	837	1.933
	0.65	646	1.987

showed the highest electrical resistivity for the three w/b ratios: 1,395, 836.8 and 646.2 Ω m, which corresponds to increases of 340, 442 and 404% for the w/b ratios of 0.35, 0.50 and 0.65 when compared with the reference sample; (2) for the reference mixture (REF) and mixes 10 RHA, 20 RHA and 30 RHA, conductivity values ranged from 2.101/Ω cm to 2.834/Ω cm, from 2.076Ω cm to 2.525/Ω cm, from 1.929/Ω cm to 2.371/Ω cm and from 1.837/Ω cm to 2.204/Ω cm, respectively, when the w/b ratio changed from 0.65 to 0.35.

5.6 Durabiliy Properties of Concrete Containing RHA

5.6.1 Permeability

Zhang and Malhotra [55] studied the chloride-ion penetration resistance of concretes made with 10% RHA and 10% silica fume (SF). Details of the concrete mixture along with the chloride-ion penetration test conducted as per ASTM C1202 [7] are given in Table 5.34. It can be seen that use of RHA and SF has drastically reduced the chloride-ion penetration at both the ages. These values were less than 1,000. As per ASTM C1202, when charge passed through concrete is less than 1,000 C, the concrete has very high resistance to chloride-ion penetration.

Nehdi et al. [38] made a comparative study of the rapid chloride permeability of concrete mixtures made with (1) 0% rice husk ash; (2) Egyptian rice husks; EG-RHA (A), EG-RHA (B) and EG-RHA (C); (3) a raw rice husk ash (RAW-US-RHA) and a high quality RHA (US-RHA) from USA, produced using fluidized bed technology; and (4) silica fume (SF). Three different percentages (7.5, 10, and 12.5%) of Egyptian rice husk ashes, SF, and two percentages (7.5 and 10%) of raw US rice husk ashes were used. Chloride permeability results are shown in Fig. 5.24. They concluded that (1) non-ground RHA did not significantly change the rapid chloride penetrability classification of concrete; (2) finely ground RHA reduced the rapid chloride penetrability of concrete from a moderate rating to low or very low ratings depending on the type and addition level of RHA. Such reductions are comparable to those achieved by SF.

Coutinho [21] investigated the rapid chloride permeability of concrete made with RHA (10, 15, and 20%) as partial replacement of cement and when using controlled permeability formwork (CPF). Controlled permeability formwork

Table 5.34 Test results of resistance of concrete to chloride-ion penetration [55]

Mix no.	Type of concrete	W/Cm	Unit weight (kg/m^3)	Compressive strength (MPa)	Chloride-ion resistance (C)	
					28 days	90 days
CO-D	Control	0.40	2,320	36.5	3,175	1,875
R10-D	10% RHA	0.40	2,340	45.5	875	525
SF10-D	10% SF	0.40	2,310	42.8	410	360

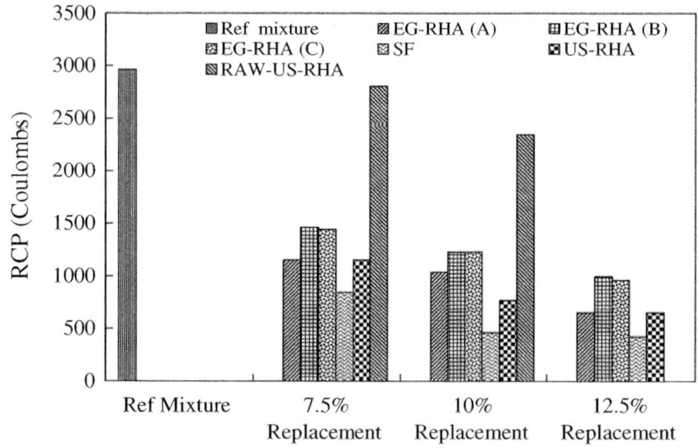

Fig. 5.24 Rapid chloride permeability at 28 days for various concrete mixtures made with rice husk ashes from Egypt and US [38]

(CPF) is the technique developed for directly improving the concrete surface zone. This technique reduces the near-surface water/binder ratio and reduces the sensitivity of concrete to poor site curing. CPF consists of using a textile liner on the usual formwork, allowing air bubbles and surplus water to drain out but retaining the cement particles and so enabling the water–cement ratio of the outer layer to become very low and the concrete to hydrate to a very dense surface skin as the filter makes enough water available at the right time to activate optimum hydration. So CPF enhances durability by providing an outer concrete layer which is richer in cement particles, with a lower water/binder ratio, less porous and so much less permeable than when ordinary formwork is used. Resistance to chloride penetration was assessed with the AASHTO T277-83 test method up to the age of 100 days (Table 5.35). It is evident from these results that inclusion of RHA significantly reduced the charge passed. Further more, when CPF was used, it greatly reduced the permeability of concrete mixtures.

Sensale [47] studied the air-permeability of concretes made with two sources of rice-husk ash; a residual RHA from rice paddy milling industry in Uruguay and

Table 5.35 Rapid chloride permeability results [21]

Type of mixture	Average charge passed (C)
Control	2349.3
Control + CPF	1916.3
10% RHA	435.0
10% RHA + CPF	384.7
15% RHA	322.0
15% RHA + CPF	245.0
20% RHA	260.0
20% RHA + CPF	202.0

Table 5.36 Permeability results of RHA concretes [47]

w/(c + RHA)	RHA		Permeability coefficient (m^2)
	Type	%	
0.32	–	0	1.08×10^{-16}
	UY	10	0.23×10^{-16}
		20	0.05×10^{-16}
	USA	10	0.08×10^{-16}
		20	0.03×10^{-16}
0.50		0	28.20×10^{-16}
	UY	10	71.82×10^{-16}
		20	49.10×10^{-16}
	USA	10	26.36×10^{-16}
		20	14.20×10^{-16}

another from USA. Two (10 and 20%) replacement percentages of cement by RHA, and two water/cementitious material ratios (0.32 and 0.50) were used. The percentage of reactive silica contained in the USA RHA was 98.5% and in the UY-RHA was 39.55%. Air-permeability for concrete was measured with the "Torrent permeability tester" method [51, 52]. Permeability results of RHA concretes are given in Table 5.36. It can be seen that (1) for a particular water-cementitious ratio, permeability of UY-RHA concrete was more than that USA-RHA concrete; (2) with the increase in water-cementitious ratio, permeability increased for both types of RHA. The results of air permeability revealed the significance of the filler and pozzolanic effect for the concretes with RHA. On the one hand, the results are consistent with the compressive strength development at 28 days for the USA RHA. On the other hand, in the concretes with UY RHA, lower air permeability was observed, which can be due to the fact that with residual RHA, the filler effect of the smaller particles in the mixture is higher than the pozzolanic effect.

Saraswathy and Song [46] studied the effect of rice husk ash (RHA) on the chloride permeability of concrete. Proportion of control (with out RHA) mix was 1:1.5:3 with w/c ratio of 0.53. Cement was replaced with 0, 5, 10, 15, 20, 25, and 30% RHA. 28-day rapid chloride permeability [8] results are given in Table 5.37. It was observed that replacement of rice husk ash drastically reduced Coulomb values. As the replacement level increased, the chloride penetration decreased.

Table 5.37 Chloride diffusivity of rice husk replaced concrete [46]

RHA (%)	Charge passed (C)
0	1,161
5	1,108
10	653
15	309
20	265
25	213
30	273

Table 5.38 Chloride-ion permeability of RHA concretes [24]

Mixture	w/b ratio	Total charge passed (C)	
		28 days	91 days
REF (control)	0.35	1,727	1,288
	0.50	3,166	2,136
	0.65	3,681	2,866
20 RHA	0.35	999	452
	0.50	1,557	692
	0.65	2,677	1,176
20 RHA 1% Na_2SO_4	0.35	933	515
	0.50	1,393	630
	0.65	2,004	760
20 RHA 1% K_2SO_4	0.35	820	326
	0.50	1,312	552
	0.65	2,242	818
20 RHA 1% Na_2SiO_3	0.35	704	342
	0.50	914	578
	0.65	1,470	732

As per ASTM C1202, RHA reduced the rapid chloride penetrability of concrete from a low to very low ratings from higher to lower replacement levels.

Gastaldini et al. [24] studied the influence of chemical activators on the chloride-ion permeability of concrete made with 20% of rice husk ash as partial replacement of cement. Water/binder ratios were 0.35, 0.50 and 0.65 whereas binder/aggregate ratios were 1:3.75, 1:5.25 and 1:6.9. Potassium sulfate (K_2SO_4), sodium sulfate (Na_2SO_4) and sodium silicate (Na_2SiO_3) were used as chemical activators in concentrations of 1% by weight of cement. Results of chloride-ion penetration are given in Table 5.38. They concluded that (1) at 28 days, RHA concrete exhibited significant reduction in the total charge passed. This reduction amounted to 42, 51 and 27% for w/b 0.35, 0.50 and 0.65, respectively. At 91 days, the same w/b ratios showed reductions of 65, 68 and 59%; (2) concrete mixtures with activators showed lower total charge passed when compared with the mixture without activator. The mixtures activated with K_2SO_4 showed the best results. At 28 days, the mixture activated with Na_2SiO_3 showed the lowest charge passed. Overall, the best results at 91 days were seen in the sample activated with K_2SO_4; and (3) at 91 days, all mixtures with chemical activators showed very low charge passed (100–1,000 C), even for w/b ratios as high as 0.65 which can be rated as very low as per ASTM C1202.

Ganesan et al. [23] examined the influence of RHA on the chloride permeability of concrete. Cement was replaced with 0, 5, 10, 15, 20, 25, 30, and 35% RHA. Control concrete mixture was made with 383 kg of cement, 575 kg of sand, and 1,150 kg of coarse aggregate per cubic meter with water–binder ratio of 0.53. The rapid chloride permeability test results for RHA blended concrete specimens are shown in Fig. 5.25. It was observed that the chloride permeability reduced considerably by partial replacement of OPC with RHA up to 30%. The total charge

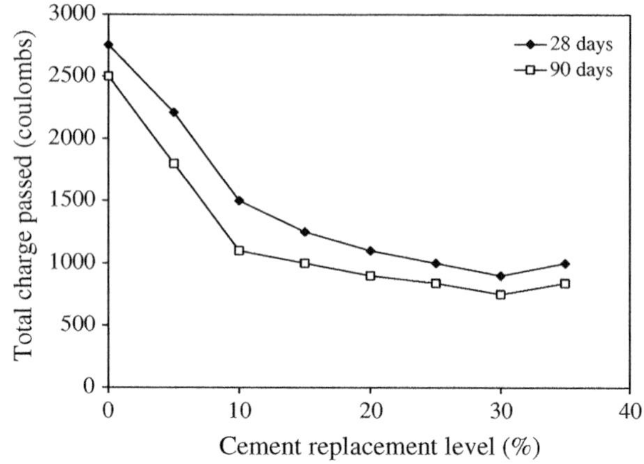

Fig. 5.25 Chloride permeability of RHA Concrete [23]

passed for 30% RHA blended concrete was considerably reduced (more than 70% reduction) both at 28 and 90 days. Since the total charge passed through the concrete depends on the electrical conductance, the lower unburnt carbon content (loss on ignition value 2.1%) present in RHA might have contributed to the significant reduction in the electrical charge passed. It is worth mentioning that the unburnt carbon particles may contribute to the conductivity of the medium and a reduction in the unburnt carbon content may be beneficial from the chloride permeability point of view.

Chindaprasirt et al. [15] determined the chloride penetration resistance of blended Portland cement mortar containing ground rice husk ash (RHA) at the age of 28 days. Ordinary Portland cement (OPC) was partially replaced with 20 and 40% RHA by weight of cementitious materials. RHA had silica content of 93.2%. The 100 × 50 mm cylinders were tested at the age of 28 days for rapid chloride penetration test (RCPT) in accordance with ASTM C1202. The results of the RCPT test are shown in Fig. 5.26. The charge passed was substantially reduced with incorporation of RHA as compared to 7,450 C of normal OPC mortar.

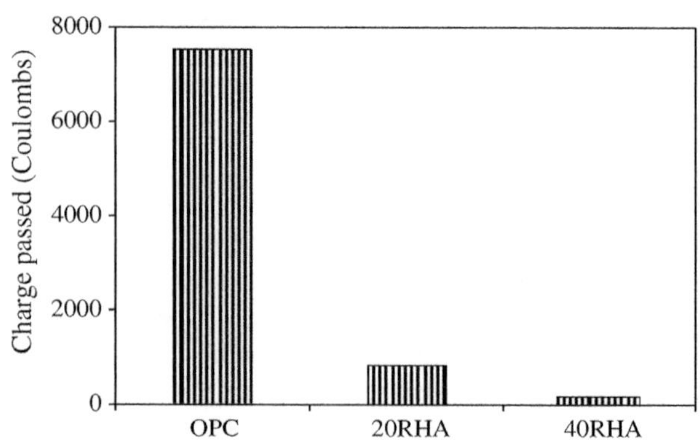

Fig. 5.26 Charge passed of rapid chloride penetration test [15]

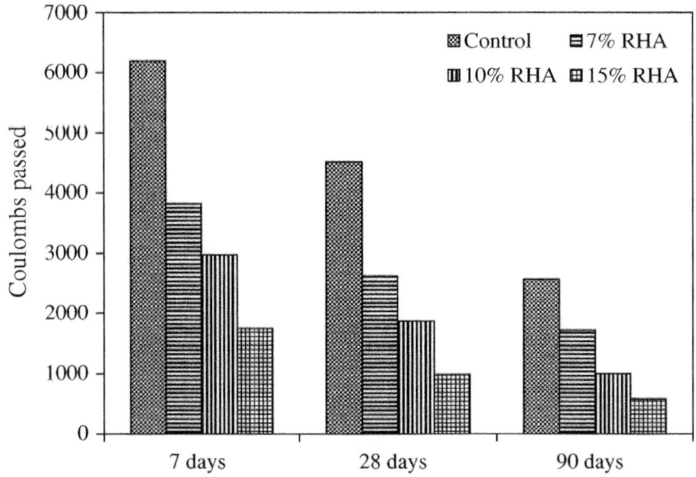

Fig. 5.27 Resistance to chloride ion penetration (Coulomb) at various ages of control and RHA mixtures [43]

The incorporation of 20 and 40% RHA reduced the charge passed to 750 and 200 C at 20 and 40% replacement levels.

Ramezanianpour et al. [43] investigated the influence of RHA on the chloride-ion penetration of concrete up the age of 90 days. Concrete mixtures were made with 0, 7, 10 and 15% RHA as partial replacement of cement. Concrete slices of size 100 × 50 mm were cut from 100 × 200 cylinders for RCPT test. Results of rapid chloride permeability of concrete are shown in Fig. 5.27. It was observed that RHA drastically enhanced resistance to chloride penetration compared to control concrete on average, around 4–5 times higher for the 15% RHA. At 7 days, the control concrete showed the highest value of 6,189 C while the charge passed through the 15% RHA concrete was 1,749 C.

5.6.2 Corrosion Resistance

Saraswathy and Song [46] investigated the influence of RHA on the corrosion resistance of concrete made with 0, 5, 10, 15, 20, 25, and 30% RHA as partial replacement of cement. Proportion of control (with out RHA) mix was 1:1.5:3 with w/c ratio of 0.53. Corrosion performance was evaluated by impressed voltage test and open circuit potential measurements. After 28 days of curing, the specimens were subjected to impressed voltage test. In this technique, the concrete specimens were immersed in 5% NaCl solution and embedded steel in concrete was made anode with respect to an external stainless steel electrode serving as cathode by applying a constant positive potential of 12 V to the system from a DC source. The variation of current was recorded with time. For each specimen, the time taken for

Table 5.39 Impressed voltage test results of OPC and various percentages of rice husk ash replaced concrete [46]

RHA (%)	Time to cracking (h)
0	42
5	72
10	74
15	No cracking even after 144 h of exposure
20	No cracking even after 144 h of exposure
25	No cracking even after 144 h of exposure
30	No cracking even after 144 h of exposure

initial crack and the corresponding maximum anodic current flow was recorded. For open circuit potential test, specimens of size 100 × 100 × 100 mm were cast. 12-mm diameter rebar of 120-mm length were embedded at a cover of 25 mm from one side of the specimen. After casting, the specimens were subjected to water curing for 28 days. After 28 days of curing, cubes were taken out and dried for 24 h and subjected to alternate wetting and drying in 3% NaCl solution. One cycle consisted of 7 days immersion in 3% NaCl solution and 7 days drying in open atmosphere. The tests were continued over a period of 200 days. Open circuit potential measurements were monitored with reference to saturated calomel electrode (SCE) periodically with time as per ASTM C876 [5]. Impressed voltage results are given in Table 5.39. It can be observed from the results that there was no cracking in concretes made with 15, 20, 25 and 30% rise husk even after 144 h of exposure. Whereas in ordinary Portland cement concrete, the specimen was cracked even after 42 h of exposure in 5% NaCl solution. The concrete specimens containing 5 and 10% rise husk also failed within 72 and 74 h of exposure. This indicated that the replacement of rice husk ash refined the pores and thereby the permeability and corrosion gets reduced.

Open circuit potential measurements values (OCP) as per ASTM C876 were lesser than −275 mV versus saturated calomel electrode (SCE), was considered to be passive in condition. They concluded that all the rice husk ash replaced concretes had shown less negative potential than −275 mV even up to 100 days of exposure indicating the passive condition of the rebars. Beyond 100 days of exposure, all the systems showed a more negative potential than −275 mV versus SCE irrespective of the replacement ratio showing the active condition of rebars.

Chindaprasirt and Rukzon [16] investigated the accelerated corrosion with impressed voltage (ACTIV) of mortars made with blends of ordinary Portland and ground rice husk ash (RHA). Cement was replaced with 0, 10, 20 and 40% RHA. Sand-to-binder ratio of 2.75 by weight, water to binder ratio (w/b) of 0.5 was used. Mortar prisms of dimensions 40 × 40 × 160 mm in length with embedded steel of 10-mm diameter and 160 mm in length were used. The steel was protected such that it protruded from the top surface of the prism by 15 mm; thus, providing sufficient mortar cover of 15 mm. At the age of 28 days, the prisms were subjected to the accelerated corrosion test with impressed voltage (ACTIV) using a 5% NaCl solution and a constant voltage of 12 V dc The condition of prism was continuously monitored and the time of initiation of first crack was recorded. This was used as a measurement of the specimen's relative resistance against chloride attack and reinforcement corrosion. The results of the time of first crack of mortar subjected to ACTIV are shown in Fig. 5.28. The time of first crack of OPC mortar was lowest at 89 h. The time to initial crack of mortars is found to increase with the incorporation RHA. RHA was found to be very effective in increasing the time of first crack. The time of first crack of 10RHA, 20RHA and 40RHA mortars were 167, 168 and 166 h, respectively,

5.6.3 Carbonation

Cizer et al. [17] studied the carbonation of calcium hydrxide and calcium silicate binders with rice husk ash. Several cement mortar compositions blended with RHA and lime were prepared. Reference cement mortar (Cref) was prepared in 1:3 cement/sand ratio by weight. In RHA-cement mortars, cement was replaced with 30, 50, and 70% RHA and were designated as (RHA-C.3-7), (RHA-C.5-5) and (RHA-C.7-3), respectively. Two types of ternary blended mortars were prepared with RHA, cement and lime. The ratio of the cement was kept 10% for both mixtures. While one (RHA-C-L.7-1-2) was composed of 70% RHA and 20% lime, the other (RHA-C-L.5-1-4) contained 50% RHA and 40% lime. Standard mortar beams (40 × 40 × 160 mm) were cast and cast and carbonation was studied up to 120 days. Carbonation depth of the RHA-cement mortars increased as the content of the cement decreased (Fig. 5.29). Hardening of RHA-C.3-7 was mostly governed by the hydration reaction as carbonation was relatively lower than the other blended mortars. RHA-C.7-3 containing the lowest cement content (30%-wt.) reached a full carbonation depth at the age of 60 days while carbonation of the rest still continued.

Gastaldini et al. [24] examined the role of chemical activators on the carbonation of concrete containing 20% of rice husk ash as partial replacement of cement.

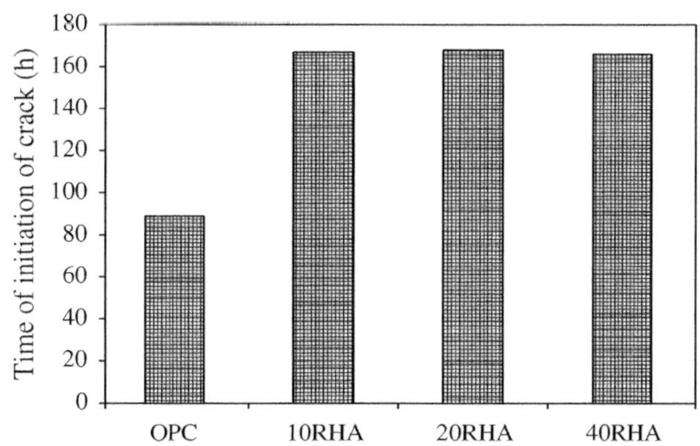

Fig. 5.28 Time to initiation of crack of mortars subjected to ACTIV [16]

Table 5.40 Test results of concrete to freezing and thawing cycles [55]

Mix No.	Type of concrete	Compressive strength (MPa)	Change at the end of 300 freezing/thawing cycles				Durability factor (%)
			Weight	Length	Pulse velocity	Resonant frequency	
CO-D	Control	36.5	0.08	0.006	−0.55	−0.84	98.0
R10-D	10% RHA	45.5	0.02	0.001	0.01	−0.86	98.3
SF10-D	10% SF	42.8	0.12	0.001	0.19	0.47	101.0

Water/binder ratios used were 0.35, 0.50 and 0.65 and binder/aggregate ratios were 1:3.75, 1:5.25 and 1:6.9. Potassium sulfate (K_2SO_4), sodium sulfate (Na_2SO_4) and sodium silicate (Na_2SiO_3) were used as chemical activators in concentrations of 1% by weight of cement. Specimens of size 100 × 100 mm were cured for 28 days. Their top surface was sealed and they underwent a preconditioning cycle as required by RILEM TC 116-PCD [40]. They were then placed in a controlled atmosphere chamber with 5% CO_2, at 23 ± 1°C and RH 65 ± 1%. The carbon dioxide penetration depth was measured at different exposure times, 4, 8 and 12 weeks by means of the phenolphthalein test carried out on the transversely split section of the cylinders specimens using the RILEM CPC-18 [41] method. Figure 5.30 shows the changes in these coefficients for the w/b ratios. For all the mixtures, Kc increased with the increase in w/b ratio because of the increase in concrete porosity and the lower concentration of cement. For the same w/b ratio (0.35, 0.50 and 0.65), the lowest carbonation coefficients were seen in the mixture with RHA and 1% K_2SO_4 showed. The values obtained were lower than those in the reference concrete.

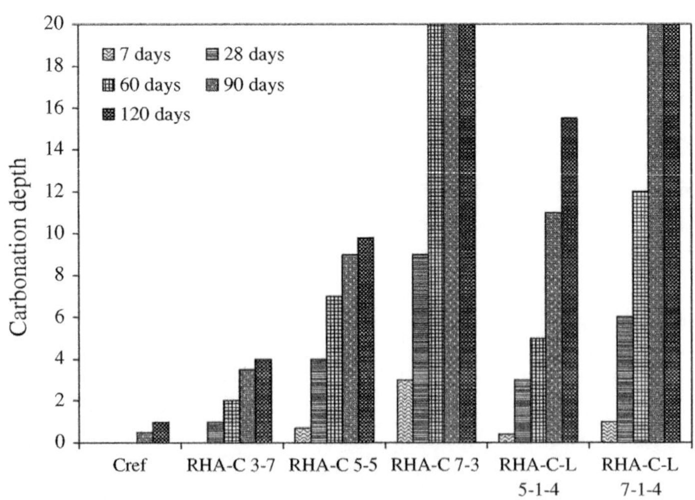

Fig. 5.29 Carbonation depth for the reference and blended cement mortars [17]

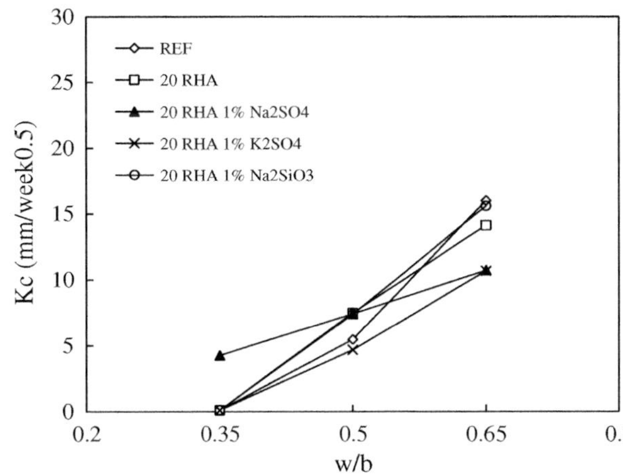

Fig. 5.30 Carbonation coefficients with reference to the w/b ratios [24]

5.6.4 Freezing and Thawing Resistance

Zhang and Malhotra [55] studied the freezing and thawing resistance of concretes made with 10% RHA and 10% silica fume (SF) as per ASTM C666 Procedure A [3]. Freezing/thawing results are given in Table 5.40. The control, RHA, and silica fume concretes showed excellent performance in the freezing and thawing test. The RHA concrete had a durability factor of 98.3 and very small changes in length, mass, pulse velocity, and resonant frequency after 300 cycles of freezing and thawing.

5.6.5 Sulfate Resistance

Sakr [45] studied the effect of RHA on the sulfate resistance of heavyweight concrete. The aggregates used were special natural heavy weight mineral ores, mainly ilmenite and baryte. They were used as the fine and coarse aggregates for the heavy weight concrete. Gravel and sand were used as the coarse and fine aggregate for the gravel concrete. Ilmenite, barite and gravel concrete mixtures were made with 0 and 15% RHA. For the sulfate resistance test, 100 mm cubes were immersed in a 5% $MgSO_4$ solution for different periods (1, 3, and 6 months) and the loss in compressive strength due to sulfate attack was determined. It was observed that (1) reductions in compressive strength of the gravel, baryte, and ilmenite control mixtures was 8.5, 7.0, and 8.0%, respectively, after immersion in 5% $MgSO_4$ for 28 days, while the reductions after 90 days of immersion were 16, 14, and 14%, respectively; (2) reduction percentages of the compressive strength of gravel, baryte and ilmenite concrete incorporating 15% of RHA when immersed in a sulfate solution for 28 days were 5, 6, and 6%, respectively; and (3) decrease rate of the effect of sulfate ions on compressive strength was generally increased as the immersion time in the sulfate solution increased.

Chatveera and Lertwattanaruk [12] investigated the effect of black rice husk ashes (BRHAs), generated from an electricity generating power plant and a rice mill on the sulfate resistance of mortar. The ashes were ground and used as a partial replacement of cement. Replacement levels were 0, 10, 30, and 50% by weight of binder. The water-to-binder ratios were 0.55 and 0.65. For sulfate resistance, mortar specimens were subjected to 5% sodium sulfate (Na_2SO_4) and magnesium sulfate ($MgSO_4$) solutions. The length change due to sulfate attacks was performed in accordance with ASTM C1012 [6] by using 25 × 25 × 285-mm bars. Sand-to-binder materials ratio of 2.75 by weight was used. Expansions of the mortars were measure up to 26 weeks. Expansion of mortars immersed in 5% Na_2SO_4 and $MgSO_4$ solutions are shown in Figs. 5.31, 5.32, 5.33 and 5.34. It was found that the expansion rates are low at the beginning, and increases substantially after 4 weeks of curing.

As shown in Fig. 5.31, the expansion of the OPC mortar was higher than the mortars mixed with BRHA from a rice mill. Increase in BRHHA content tends to reduce the expansion due to the decreased amount of $Ca(OH)_2$ for reacting with sulfate ion to produce gypsum and ettringite. For the aspect of sulfate resistance, the mortars mixed with 10% replacement of BRHA (10MRHA-0.55) yielded the expansion more than that of SPC mortar, whereas the expansions of the mortars with 30 and 50% replacement of BHRA were lower than that of the SPC mortar.

In the case of $MgSO_4$ attack, after the mortars being immersed in $MgSO_4$ solution, the reaction produces brucite ($Mg(OH)_2$) and gypsum depositing on the surface of mortar. This increases the accessibility of sulfate ion to attack the mortar matrix. As a result, the structure of C–S–H is prone to be changed into M–S–H, which does not have binding capacity. Overall, the expansion of mortar under $MgSO_4$ attack is lower than that under Na_2SO_4 attack due to the different mechanisms. In Na_2SO_4 solutions, the deterioration is the result of expansion associated with ettringite formation. In $MgSO_4$ solutions, a main cause of attack is decomposition of C–S–H. When increasing the water-to-binder ratio to 0.65 (as shown in

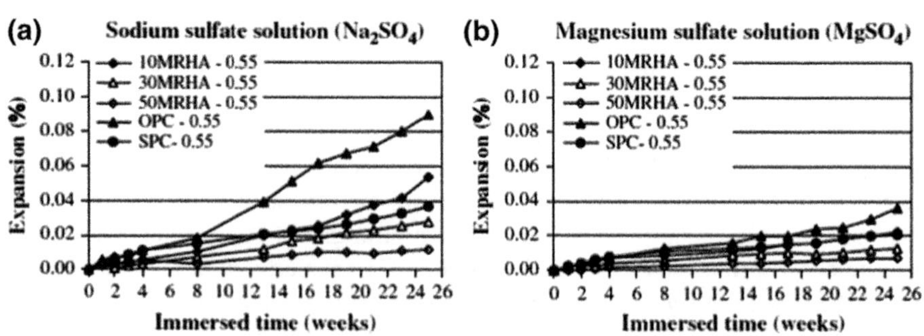

Fig. 5.31 Expansion of mortars mixed with BRHA from a rice mill, water-to-binder ratio of 0.55 [12]

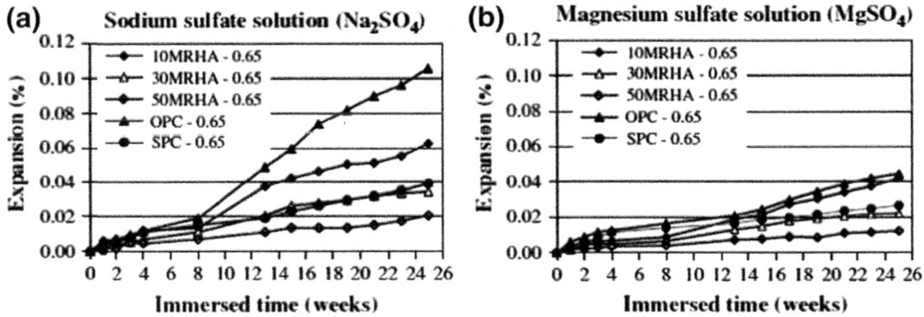

Fig. 5.32 Expansion of mortars mixed with BRHA from a rice mill, water-to-binder ratio of 0.65 [12]

Fig. 5.32), the expansion of mortar increases due to the higher porosity of mortar structure (Neville 1995), allowing more sulfate ion penetration. For the expansions of mortars mixed with BRHA from an electricity generating plant as shown in Figs. 5.33 and 5.34, it was found that the results are similar to those mixed with BRHA from a rice mill.

5.6.6 Deicing Salt Scaling Resistance

Zhang and Malhotra [55] studied the deicing salt-scaling resistance of concretes made with 10% RHA and 10% silica fume (SF). Scaling resistance results are given in Table 5.41. The visual evaluation of test slabs showed that the performance of the RHA concrete was similar to that of the control concrete but marginally better than SF concrete. For both control and RHA concrete, no coarse aggregate was visible after 50 cycles whereas for SF concrete some coarse

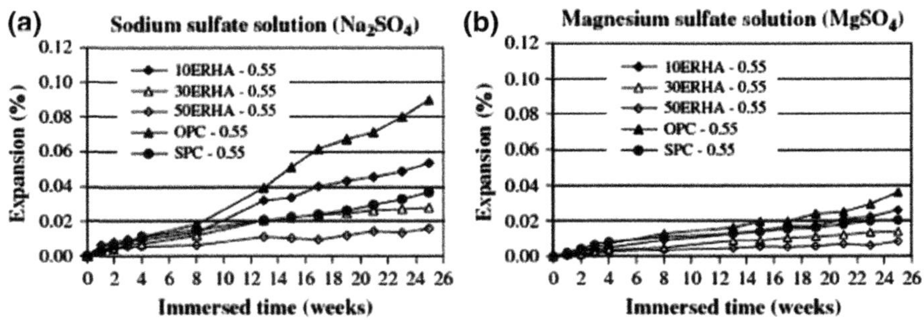

Fig. 5.33 Expansion of mortars mixed with BRHA from an electricity generating power plant, water-to-binder ratio of 0.55 [12]

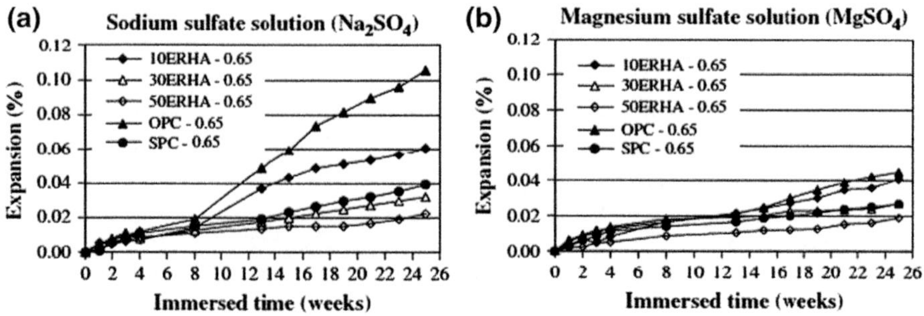

Fig. 5.34 Expansion of mortars mixed with BRHA from an electricity generating power plant, water-to-binder ratio of 0.65 [12]

Table 5.41 Test results of deicing salt scaling [55]

Mix No.	Type of concrete	W/C ratio	Compressive strength (MPa)	Visual rating	Total scaling residue (kg/m^3)
CO-D	Control	0.40	36.5	2	0.3
R10-D	10% RHA	0.40	45.5	2	0.6
SF10-D	10% SF	0.40	42.8	3	0.8

aggregates were visible. All the three concretes had a total mass of scaling residue of equivalent to or less than 0.8 kg/m^3 after 50 cycles in the presence of deicing salts. The control, RHA, and silica fume concretes showed excellent performance in the freezing and thawing test. The RHA concrete had a durability factor of 98.3 and very small changes in length, mass, pulse velocity, and resonant frequency after 300 cycles of freezing and thawing.

Nehdi et al. [38] investigated the effect of RHA and SF on the deicing scaling resistance of concrete. Concrete mixtures were made with 10% RHA and 10% SF as partial replacement of cement. Tests were conducted up to 50 cycles as per ASTM C672 [4]. Results showed that RHA concrete specimens (visual rating 2) performed similar to the control concrete mixture (visual rating 2) but little better than SF concrete mixtures (visual rating 3). These results were in agreement with those reported by Zhang and Malhotra [55].

5.6.7 Alkali–Silica Reaction

Hasparyk et al. [27] investigated the expansion of mortar bars made with different levels of cement replacement with rice husk ash (RHA). Two types of reactive aggregates (quartzite and basalt) were used. Percentages of RHA were 0, 4, 8, 12, and 15%. Tests were conducted as per ASTM C 1260. Expansion at 16 and

Table 5.42 Expansion of mortar bars containing RHA [27]

Replacement (%)	Expansion (%)			
	Quartzite		Basalt	
	16 days	30 days	16 days	30 days
0	0.28	0.53	0.84	1.10
4	0.16	0.36	0.76	0.99
8	0.21	0.45	0.32	0.81
12	0.09	0.25	0.08	0.33
15	0.06	0.15	0.04	0.25

30 days are given in Table 5.42. It can be seen from these results that inclusion of RHA was very effective in controlling the expansion of mortar bars at the age of 16 and 30 days.

References

1. ASTM C 642.: Standard Test Method for Density, Absorption, and Voids in Hardened Concrete. Annual Book of ASTM Standards, Philadelphia (1997)
2. ASTM C 618.: Standard Specification for Coal Fly Ash and Raw or Calcined Natural Pozzolan for Use as a Mineral Admixture in Concrete. Annual Book of ASTM Standards, Philadelphia (1993)
3. ASTM C 666.: Standard Test Method for Resistance of Concrete to Rapid Freezing and Thawing. Annual Book of ASTM Standards, Philadelphia, USA
4. ASTM C 672.: Standard Test Method for Scaling Resistance of Concrete Surfaces Exposed to Deicing Chemicals. Annual Book of ASTM Standards, Philadelphia (1998)
5. ASTM C 876.: Standard Test Method for Half-Cell Potentials of Uncoated Reinforcing Steel in Concrete. Annual Book of ASTM Standards, Philadelphia (1991)
6. ASTM C 1012.: Standard Test Method for Length Change of Hydraulic-cement Mortars Exposed to a Sulfate Solution. American Society for Testing and Materials, Philadelphia (2002)
7. ASTM C 1202.: Electrical Indication of Concrete's Ability to Resist Chloride Ion Penetration. Annual Book of American Society for Testing Materials Standards, C04.02 (2000)
8. ASTM C 1260.: Standard Test Method for Potential Alkali Reactivity of Aggregates (Mortar-Bar Method). Annual Book of American Society for Testing Materials Standards, C04.02 (2001)
9. Agarwal, S.K.: Pozzolanic activity of various siliceous materials. Cem. Concr. Res. **36**(9), 1735–1739 (2006)
10. Ahmadi, M.A., Alidoust, O., Sadrinejad, I., Nayeri, M.: Development of mechanical properties of self-compacting concrete contain rice husk ash. World Acad. Sci. Eng. Technol. **34**, 168–171 (2007)
11. Bui, D.D., Hu, J., Stroeven, P.: Particle size effect on the strength of rice husk ash blended gap-graded Portland cement concrete. Cem. Concr. Compos. **27**(3), 357–366 (2005)
12. Chatveera, B., Lertwattanaruk, P.: Evaluation of sulfate resistance of cement mortars containing black rice husk ash. J. Environ. Manag. **90**(3), 1435–1441 (2009)
13. Chindaprasirt, P., Hornwuttiwong, S., Sirivivatnanon, V.: Influence of fly ash fineness on strength, drying shrinkage and sulfate resistance of blended cement mortar. Cem. Concr. Res. **34**(7), 1087–1092 (2004)

14. Chindaprasirt, P., Kanchanda, P., Sathonsaowaphak, A., Cao, H.T.: Sulfate resistance of blended cements containing fly ash and rice husk ash. Construct. Build. Mater. **21**(6), 1356–1361 (2007)
15. Chindaprasirt, P., Rukzon, S., Sirivivatnanon, V.: Resistance to chloride penetration of blended Portland cement mortar containing palm oil fuel ash, rice husk ash and fly ash. Construct. Build. Mater. **22**(5), 932–938 (2008)
16. Chindaprasirt, P., Rukzon, S.: Strength, porosity and corrosion resistance of ternary blend Portland cement, rice husk ash and fly ash mortar. Construct. Build. Mater. **22**(8), 1601–1606 (2008)
17. Cizer, O., Van Balen, K., Elsen, J., Gemert, D.V.: Carbonation and hydration of calcium hydroxide and calcium silicate binders with rice husk ash. In: 2nd International Symposium on Advances in Concrete through Science and Engineering September 11–13, Quebec City, Canada (2006)
18. Chopra, S.K., Ahluwalia, S.C., Laxmi, S.: Technology and manufacture of rice-husk ash masonry (RHAM) cement. In: Proceedings of ESCAP/RCTT Workshop on Rice-Husk Ash Cement, New Delhi (1981)
19. Cook, D.J.: Development of microstructure and other properties in rice husk ash—OPC systems. In: Proceedings of the 9th Australasian Conference on the Mechanics of Structures and Materials, pp. 355–360. University of Sydney, Sydney (1984)
20. Della, V.P., Kuhn, I., Hotza, D.: Rice husk ash as an alternate source for active silica production. Mater. Lett. **57**(4), 818–821 (2002)
21. Coutinho, J.S.: The combined benefits of CPF and RHA in improving the durability of concrete structures. Cem. Concr. Compos. **25**(1), 51–59 (2003)
22. Feng, Q., Yamamichi, H., Shoya, M., Sugita, S.: Study on the pozzolanic properties of rice husk ash by hydrochlortic acid pretreatment. Cem. Concr. Res. **34**(3), 521–526 (2004)
23. Ganesan, K., Rajagopal, K., Thangavel, K.: Rice husk ash blended cement: assessment of optimal level of replacement for strength and permeability properties of concrete. Construct. Build. Mater. **22**(8), 1675–1683 (2008)
24. Gastaldini, A.L.G., Isaia, G.C., Gomes, N.S., Sperb, J.E.K.: Chloride penetration and carbonation in concrete with rice husk ash and chemical activators. Cem. Concr. Compos. **29**(3), 176–180 (2007)
25. Gastaldini, A.L.G., Isaia, G.C., Hoppe, T.F., Missau, F., Saciloto, A.P.: Influence of the use of rice husk ash on the electrical resistivity of concrete: a technical and economic feasibility study. Construct. Build. Mater. **23**(11), 3411–3419 (2009)
26. Habeeb, G.A., Fayyadh, M.M.: Rice Husk Ash Concrete: the effect of RHA average particle size on mechanical properties and drying shrinkage. Aust. J. Basic Appl. Sci. **3**(3), 1616–1622 (2009)
27. Hasparyk, N.P., Monteiro, P.J.M., Carasek, H.: Effect of silica fume and rice husk ash on alkali–silica reaction. ACI Mater. J. **97**(4), 486–491 (2000)
28. Hewlett, P.C.: Chemistry of Cement and Concrete, pp. 471–601. Wiley, New York (1998)
29. Hwang, C.L., Wu, D.S.: Properties of cement paste containing rice husk ash, pp. 733–765. American Concrete Institute SP-114 (1989)
30. Ikpong, A.A., Okpala, D.C.: Strength characteristics of medium workability ordinary Portland cement-rice husk ash concrete. Build. Environ. **27**(1), 105–111 (1992)
31. Ismail, M.S., Waliuddin, A.M.: Effect of rice husk ash on high strength concrete. Construct. Build. Mater. **10**(7), 521–526 (1996)
32. Jaturapitakkul, C., Roongreung, B.: Cementing material from calcium carbide residue-rice husk ash. J. Mater. Civil Eng. **15**(5), 470–475 (2003)
33. Jauberthie, R., Rendell, F., Tamba, S., Cisse, I.: Origin of the pozzolanic effect of rice husks. Construct. Build. Mater. **14**(8), 419–423 (2000)
34. Luxan, M.P., Mndruga, M., Seavedra, J.: Rapid evaluation of pozzolanic activity of natural products by conductivity measurement. Cem. Concr. Res. **19**(1), 63–68 (1989)
35. Mehta, P.K.: The chemistry and technology of cement made from rice husk ash. In: Proceedings UNIDO/ESCAP/RCTT Workshop on Rice Husk Ash Cements, Peshawar,

Pakistan, January 1979, pp. 113–122 . Regional Centre for Technology Transfer, Bangalore (India) (1979)
36. Mehta, P.K.: Rice husk ash—a unique supplementary cementing material. In: Proceedings of the International Symposium on Advances in Concrete Technology, pp. 407-430. Athens, Greece (1992)
37. Mehta, P.K., Monteiro, P.J.M.: Concrete: Microstructure, Properties, and Materials, 3rd edn. McGraw-Hill, New York (2006)
38. Nehdi, M., Duquette, J., Damatty, A.El: Performance of rice husk ash produced using a new technology as a mineral admixture in concrete. Cem. Concr. Res. **33**(8), 1203–1210 (2003)
39. Rahman, M.A.: Use of rice husk ash in sandcrete blocks for masonry units. Mater. Struct. **20**(5), 361–366 (1987)
40. RILEM Recommendations of TC116-PCD.: Tests for gas permeability of concrete. Mater. Struct. **32**(217), 163–179 (1999)
41. RILEM CPC 18.: Measurement of hardened concrete carbonation depth. Mater. Struct. **21**(126), 453–455 (1988)
42. Real, C., Alcala, M.D., Criado, J.M.: Preparation of silica from rice husks. J. Am. Ceram. Soc. **79**(8), 2012–2016 (1996)
43. Ramezanianpour, A.A., Khani, M.M., Ahmadibeni, Gh: The effect of rice husk ash on mechanical properties and durability of sustainable concretes. Int. J. Civil Eng. **7**(2), 83–91 (2009)
44. Rizwan, S.A.: High performance mortars and concretes using secondary raw materials, p. 132. Technischen Universitat Bergakademie Freiberg, Freiberg (2006)
45. Sakr, K.: Effects of silica fume and rice husk ash on the properties of heavy weight concrete. Int. J. Civil Eng. **18**(3), 367–376 (2006)
46. Saraswathy, V., Song, H.W.: Corrosion performance of rice husk ash blended concrete. Construct. Build. Mater. **21**(8), 1779–1784 (2007)
47. Sensale, G.R.D.: Strength development of concrete with rice-husk ash. Cem. Concr. Compos. **28**(2), 158–160 (2006)
48. Sensale, G.R.D., Ribeiro, A.B., Gonçalves, A.: Effects of RHA on autogenous shrinkage of Portland cement pastes. Cem. Concr. Compos. **30**(10), 892–897 (2008)
49. Singh, N.B., Singh, V.D., Rai, S., Chaturvedi, S.: Effect of lignosulfonate, calcium chloride and their mixture on the hydration of RHA-blended Portland cement. Cem. Concr. Compos. **32**(3), 387–392 (2002)
50. Tashima, M.M., Silva, C.A.R., Akasaki, J.L., Barbosa, M.B.: Influence of rice husk ash in mechanical characteristics of concrete. In: 4th International ACI/CANMET Conference on Quality of Concrete Structures and Recent Advances in concrete Materials and Testing, Brazil, Paper XII.08: 780–790 (2005)
51. Torrent, R., Frenzer, G.A.: A method for rapid determination of the coefficient of permeability of the covercrete. In: Proceedings of the International Symposium on Non-Destructive Testing in Civil Engineering, (NDT-CE), pp. 985–92 (1995)
52. Torrent, R.: Gas permeability of high-performance concretes-site and laboratory test. High-performance concrete and performance and quality of concrete structures, Gramado. In: Proceedings second CANMET/ACI Int., pp. 291–308 (1999)
53. Yeoh, A.K., Bidin, R, Chong, C.N., Tay, C.Y.: The relationship between temperature and duration of burning of rice-husk in the development of amorphous rice-husk ash silica. In: Proceedings of UNIDO/ESCAP/RCTT, Follow-up Meeting on Rice-Husk Ash Cement, Alor Setar, Malaysia (1979)
54. Zhang, M.H., Lastra, R., Malhotra, V.M.: Rice-husk ash paste and concrete: some aspects of hydration and the microstructure of the interfacial zone between the aggregate and paste. Cem. Concr. Res. **26**(6), 963–977 (1996)
55. Zhang, M.H., Malhotra, V.M.: High-performance concrete incorporating rice husk ash as a supplementary cementing material. ACI Mater. J. **93**(6), 629–636 (1996)

Index

Printed by Printforce, the Netherlands